Handbook of Experimental Pharmacology

Volume 122

Springer
Berlin
Heidelberg
New York
Barcelona
Budapest
Hong Kong
London
Milan
Paris
Santa Clara
Singapore
Tokyo

Mechanisms of Drug Interactions

Contributors

G.L. Amidon, H.H. Blume, J.R. Crison, P.F. D'Arcy
P.A.G.M. De Smet, J.D. Ferrero, J.P. Griffin, C.M. Hughes
E. Lipka, J.C. McElnay, M. Schorderet, B.S. Schug, A. Somogyi
P.G. Welling, S. Yosselson-Superstine

Editors

P.F. D'Arcy, J.C. McElnay and P.G. Welling

Springer

Professor emerit. PATRICK F. D'ARCY, OBE, Ph.D., D.Sc.
School of Pharmacy
The Queen's University of Belfast
Medical Biology Centre
97 Lisburn Road
Belfast BT9 7BL
N. Ireland

Professor JAMES C. MCELNAY, Ph.D.
Director, School of Pharmacy
The Queen's University of Belfast
Medical Biology Centre
97 Lisburn Road
Belfast BT9 7BL
N. Ireland

PETER G. WELLING, Ph.D., D.Sc.
Vice President, Pharmacokinetics and Drug Metabolism
Parke-Davis Pharmaceutical Research
2800 Plymouth Road
Ann Arbor, MI 48105
USA

With 58 Figures and 39 Tables

ISBN-13: 978-3-642-64658-4 e-ISBN-13: 978-3-642-61015-8

DOI: 10.107/978-3-642-61015-8

Library of Congress Cataloging-in-Publication Data. Mechanisms of drug interactions/contributors, G.L. Amidon . . . [et al.]: editors, P.F. D'Arcy, J.C. McElnay, and P.G. Welling. p. cm. – (Handbook of experimental pharmacology: v. 122) Includes bibliographical references and index. ISBN 3-540-60204-6 1. Drug interactions. I. Amidon, Gordon L. II. D'Arcy, P.F. (Patrick Francis) III. McElnay, J.C. (James C.), 1954– . IV. Welling, Peter G. V. Series. [DNLM: 1. Drug Interactions. W1 HA51L v. 122 1996/QV 38 M4864 1996] QP905.H3 vol. 122 [RM302] 615'.1 s – dc20 [615'.7045] DNLM/DLC for Library of Congress 96-4873

softcover reprint of the hardcover 1st edition 1996

Cover design: Springer-Verlag, Design & Production

Typesetting: Best-set Typesetter Ltd., Hong Kong

SPIN: 10469387 27/3136/SPS – 5 4 3 2 1 0 –

Preface

Over the years a number of excellent books have classified and detailed drug-drug interactions into their respective categories, e.g. interactions at plasma protein binding sites; those altering intestinal absorption or bioavailability; those involving hepatic metabolising enzymes; those involving competition or antagonism for receptor sites, and drug interactions modifying excretory mechanisms. Such books have presented extensive tables of interactions and their management. Although of considerable value to clinicians, such publications have not, however, been so expressive about the individual mechanisms that underlie these interactions.

It is within this sphere of "mechanisms" that this present volume specialises. It deals with mechanisms of in vitro and in vivo, drug-drug, drug-food and drug-herbals interactions and those that cause drugs to interfere with diagnostic laboratory tests. We believe that an explanation of the mechanisms of such interactions will enable practitioners to understand more fully the nature of the interactions and thus enable them to manage better their clinical outcome.

If mechanisms of interactions are better understood, then it may be possible for the researcher to develop meaningful animal/biochemical/tissue culture or physicochemical models to which new molecules could be exposed during their development stages. The present position, which largely relies on patients experiencing adverse interactions before they can be established or documented, can hardly be regarded as satisfactory.

This present volume is classified into two major parts; firstly, pharmacokinetic drug interactions and, secondly, pharmacodynamic drug interactions. Within these parts, we have been fortunate to enlist the help of acknowledged experts in preparing specific chapters focusing on aspects of the interaction spectrum.

We believe that this volume will add much to that which is currently known about drug interactions and will directly enhance the safer use of medicines.

Belfast, Northern Ireland — P.F. D'Arcy
Belfast, Northern Ireland — J.C. McElnay
Ann Arbor, MI, USA — P.G. Welling

List of Contributors

AMIDON, G.L., The University of Michigan, 2012 Pharmacy Bldg., Ann Arbor, MI 48109-1065, USA

BLUME, H.H., Zentrallaboratorium Deutscher Apotheker, Ginnheimer Straße 20, D-56760 Eschborn, Germany

CRISON, J.R., TSRL Inc., Ann Arbor, MI 48108, USA

D'ARCY, P.F., The Queen's University of Belfast, School of Pharmacy, Medical Biology Centre, 97 Lisburn Road, Belfast BT9 7BL, Northern Ireland

DE SMET, P.A.G.M., Drug Information Center, KLMP, Alexanderstraat 11, NL-2514 JL The Hague, The Netherlands

FERRERO, J.D., Department of Pharmacology, University Medical Center, CH-1211 Geneva 4, Switzerland

GRIFFIN, J.P., "Quartermans", Digswell Lane, Digswell, Welwyn Garden City, Hertfordshire, AL7 1SP, UK

HUGHES, C.M., The Queen's University of Belfast, School of Pharmacy, Medical Biology Centre, 97 Lisburn Road, Belfast BT9 7BL, Northern Ireland

LIPKA, E., The University of Michigan, 2032 Pharmacy Bldg., Ann Arbor, MI 48109-1065, USA, and TSRL Inc., Ann Arbor, MI 48108, USA

MCELNAY, J.C., The Queen's University of Belfast, School of Pharmacy, Medical Biology Centre, 97 Lisburn Road, Belfast BT9 7BL, Northern Ireland

SCHORDERET, M., Department of Pharmacology, University Medical Center, CH-1211 Geneva 4, Switzerland

SCHUG, B.S., Zentrallaboratorium Deutscher Apotheker, Ginnheimer Straße 20, D-56760 Eschborn, Germany

SOMOGYI, A., Department of Clinical and Experimental Pharmacology, University of Adelaide, Adelaide 5005, Australia

WELLING, P.G., Pharmacokinetics/Drug Metabolism, Parke-Davis Pharmaceutical Research, 2800 Plymouth Road, Ann Arbor, MI 48105, USA

YOSSELSON-SUPERSTINE, S., Clinical Pharmacy Studies, School of Continuing Medical Education, Sackler Faculty of Medicine, Tel Aviv University, P.O.B. 39040, Tel Aviv 69978, Israel

Contents

Section I: Pharmacokinetic Drug Interactions

CHAPTER 2

Drug Interactions in the Gastrointestinal Tract and Their Impact on Drug Absorption and Systemic Availability: A Mechanistic Review

CHAPTER 3

Drug-Food Interactions Affecting Drug Absorption

CHAPTER 4

Drug Interactions at Plasma and Tissue Binding Sites

CHAPTER 5

Drug Interactions and Drug-Metabolising Enzymes

CHAPTER 6

Drug Interactions Involving Renal Excretory Mechanisms

CHAPTER 8

Synergistic Drug Interactions

CHAPTER 9

In Vitro Drug Interactions

CHAPTER 1
Introduction

P.F. D'Arcy

A. A Widening Problem

Over 20 years ago, an editorial on drug interactions in the *Lancet* said that the "publication of huge lists and tables will induce in doctors a drug-interaction-anxiety syndrome and lead to therapeutic paralysis". This prediction has not come about although the problem of drug interactions is still with us and the spectrum is widening as new drugs are introduced. Indeed it could be said that the nature of the problem has also widened for in the intervening years drug interactions have come to embrace interactions with food and with herbal medicines as well as the more numerous and better recognised drug-drug interactions.

One of the advantages of the attention that has been focused on adverse drug interactions over the past 20 years or so is that many drug-drug and drug-food interactions are now predictable and many of the unwanted consequences of using drug combinations can be avoided by simply adjusting the dosage of one or more of the interactants. As a result of this, there has been a considerable improvement in the safety and efficacy of therapy with drug combinations.

Nowadays the focus on drug interactions tends to be more on their hazards than their advantages. This is probably the correct orientation since the vast majority of interactions are hazardous. An awareness of the possible hazards of medication and possible interactions between drugs on the part of those who use them, doctors and other health professionals, can only result in better therapy with benefit to the patient in terms of both safety and efficacy. Many excellent publications have explored the nature of drug interactions and their message has filtered down to the patient level, being expressed by the maxim "do not use two drugs when one will do".

Whereas over the years books have classified and detailed drug-drug interactions into their respective categories, e.g., interactions at plasma protein-binding sites; interactions altering intestinal absorption or bioavailability; interactions involving the hepatic-metabolising enzymes; interactions involving competition or antagonism for receptor sites, and finally drug interactions modifying excretory mechanisms, they have not been so expressive about the individual mechanisms that underlie these interactions. Their prime objective was to inform about the interaction and its clinical management.

I. Scope of This Present Volume

In their preliminary thoughts and discussions in the planning of this book, the editors decided to attempt to produce a volume that dealt, specifically, with the mechanisms underlying drug interactions. It was decided not to produce extensive tables of interactions since this has already been competently done by others. It was thought that an explanation of mechanisms would enable practitioners, in some instances, to work out from first principles the nature and management of interactions rather than to rely entirely on consulting long tables of interactions and their management.

In recent years there has been intensive research into the mechanisms of established drug-drug interactions in man. This present volume is classified into two major parts: firstly, pharmacokinetic drug interactions and, secondly, pharmacodynamic drug interactions.

B. Pharmacokinetic Drug Interactions

I. Drug-Drug and Nutrient-Drug Interactions at the Absorption Site

Chapter 2 discusses the many facets of drug-drug interactions at the absorption site. For the majority of drugs this is the stomach and small intestine. The original intent of this book was to produce a single chapter on interactions affecting drug absorption which would include drug-drug as well as drug-food interactions. However, the quantity of material to be covered was too great and, for convenience, drug-drug and drug-food interactions have been presented in two separate chapters. Of necessity a degree of overlap is unavoidable and it is convenient in this present context to deal with Chaps. 2 and 3 together.

The influence of wide-ranging factors such as the effect of fluid volume of gastrointestinal physiology, splanchnic blood flow, passive diffusion, gastric emptying time, pH of the intestinal contents, intestinal motility, ionic content of foods, dietary fat, and gastrointestinal disease can have considerable effect on the absorption of drugs and they are mechanisms common to both drug-drug and drug-food interactions in the gut.

The two topics of drug-food interactions affecting drug absorption and the effect of drugs on food and nutrient absorption are closely related but quite distinct. In this present volume only the former topic is addressed (Chap. 3). No attempt has been made to discuss the latter topic; material on this latter topic has been reviewed elsewhere (Basu 1988; Roe 1989; Trovato et al. 1991).

It is only in relatively recent years that studies on food-drug interactions have been actively pursued. This work was heralded by the major review of Welling (1977). It is surprising that the study of drug-food interactions should have lagged behind that of drug-drug interactions. It is noticeable that food-drug interactions are not as often reported as drug-drug interactions although their potential frequency is far greater since food is by far the most common substance associated with the ingestion of oral doses of medicine. In my experience, medical and pharmacy practitioners are not well informed about food-drug interactions and therefore may fail to pass on cautionary advice to their patients.

Perhaps one of the most troublesome and well reported of food-drug interactions was the interaction between tyramine-containing foods and monoamine oxidase inhibitor (MAOI) antidepressants during the early 1960s (see reviews by D'ARCY 1979; BROWN et al. 1989; LIPPMAN and NASH 1990). Deaths from eating cheese whilst taking MAOIs was given great publicity and this type of interaction did much to create an awareness of the potential of foods to interact with medicines.

Monoamine oxidase type A preferentially oxidatively de-aminates adrenaline, noradrenaline and serotonin, while type B preferentially metabolises benzylamine and phenylethylamine. Dopamine and tyramine are de-aminated by both forms of the enzyme. The traditional MAOIs, such as phenelzine and tranylcypromine, inhibit both types A and B of the enzyme. Selective MAOIs, such as brofaromine, moclobemide and clorgyline, have been introduced with a selective and reversible inhibition of monoamine oxidase (type A), and selegiline, which reversibly inhibits type B, is an adjunct in the treatment of parkinsonism. The dietary restrictions that need to be observed with the traditional MAOIs (type A and B) are much less stringent with the new selective inhibitors.

It was largely the interactions with food that led to a virtual replacement of the use of MAOI antidepressents by the tri- and polycyclic antidepressants. That this may have been a collective error of judgement is now being realised; perhaps the MAOIs are not so bad after all!

The effect of heavy metal ions in milk and other dairy products on the absorption of tetracyclines was another milestone which raised the sensibility of clinicians to the dangers of food-drug interactions (BRAYBROOKS et al. 1975; CHIN and LACH 1975; NEUVONEN 1976).

I well recall in the late 1940s, after tetracycline had been first introduced, advising mothers to open the capsules of tetracycline and dissolve the contents in a little milk before giving the antibiotic to their infants. But it took almost 20 years before the milk-tetracycline interaction was recognised. Once this interaction was realised, it was only a relatively short time before the spectrum of such interactions was widened and it was realised that it was not uncommon that the absorption from the gut of other antimicrobial agents, or reduction of their bioavailability, could be seriously influenced by food or metal ions. Even more recent work has shown that the bisphosphonates, used to treat a broad range of bone disorders, including osteolytic bone metastases, hyperparathyroidism, Paget's disease of bone and established vertebral osteoporosis in women, can be completely antagonised by food and vitamin-mineral supplements with a high calcium content (FOGELMAN et al. 1984, 1985, 1986; FELS et al. 1989; COMPSTON 1994).

II. Drug Interactions at Plasma- and Tissue-Binding Sites

Drug interactions at plasma- and tissue-binding sites and their mechanisms are presented in Chap. 4. The importance of plasma protein-binding displacement as a clinically important drug interaction mechanism is controversial.

When a displacing agent interacts with a primary drug the result is an increase in the free concentration of the displaced drug in the plasma. In the past, it has been wrongly assumed that these free drug concentrations will remain raised. This is not so. What in fact occurs is that the increased free drug in the plasma quickly distributes throughout the body and may localise in the tissues. When equilibrium is again reached (assuming the volume into which the drug distributes is large) the free concentration in the plasma will be the same, or near to its pre-interaction level. The "total" drug concentration (i.e. bound plus free drug) in the plasma will decrease after the equilibrium for free drug is reached. The ultimate consequence, therefore, of a plasma protein-binding displacement interaction is to lower total drug concentration in the plasma and to leave free drug concentration unchanged. Since pharmacological actions, including toxicities, correlate with free drug in the plasma, any increased effect seen after displacement is usually of a transient nature and, in most cases, is clinically unimportant.

Generally, therefore, although current opinion accepts that the binding of drugs to protein is an important pharmacokinetic parameter, it is of the view that its importance as a mechanism of drug-drug interaction has been overestimated and overstated.

III. Drug Interactions and Drug-Metabolising Enzymes

It is now realised that many interactions can be explained by alterations in the metabolic enzymes that are present in the liver and other extra-hepatic tissues. Induction or inhibition of the collection of isoenzymes, collectively known as cytochrome P450 enzymes, are mechanisms that have been shown to underlie some of the more serious drug-drug interactions.

Chapter 5 presents a chronological account of research in this area, and discusses extrahepatic microsomal forms of cytochrome P450, genetic polymorphism, age and disease and cytochrome P450, and the clinical importance of enzyme induction or inhibition. Although the chapter concentrates on the P450 enzyme systems, it emphasises that certain oxidative reactions of xenobiotics are also catalysed by enzymes other than cytochrome P450s. A brief account of these other enzymes is also presented.

IV. Interactions Involving Renal Excretory Mechanisms

The kidney represents the final elimination organ for virtually all foreign substances irrespective of whether they are cleared unchanged or as metabolites formed in the body predominantly by the liver. Chapter 6 presents an account of the mechanisms of renal excretory clearance, drug interactions involving tubular secretion and tubular reabsorption.

Such interactions can have important clinical implications in terms of patient mortality and morbidity. It is fortunate that an understanding of the molecular mechanisms of renal drug transport has allowed predictions to be made of potential drug interactions in early phase drug development. This understanding together with the availability of in vitro models of renal drug transport has also been used as an early guide to potential in vivo drug interactions. This chapter has also highlighted inconsistencies in preconceived knowledge on the renal tubular secretion of drugs.

C. Pharmacodynamic Drug Interactions

I. Drug-Drug Interactions at the Receptor and Other Active Sites

Pharmacodynamic interactions at receptor sites have been extensively reviewed in Chap. 7. Although the majority of drug-drug interactions are potentially, if not actually, hazardous, some are synergistic and may be to the patient's therapeutic advantage. The chapter concentrates on adverse drug interactions resulting in either a diminished therapeutic effect or an increased toxicity. The chapter describes the mechanisms by which pharmacodynamic interactions occur and illustrates them with clinically relevant interactions. Theoretically, at least, since interactions are based on supposedly established mechanisms, they should be more easily prevented and should have less clinical impact than pharmacokinetic interactions. However, the relevance of such a view may be questioned in the light of recent advances in the characterisation and detection of kinetic interactions.

II. Synergistic Interactions

Not all drug-drug interactions are hazardous, and some synergistic interactions when they occur can be of clinical value in therapy. Indeed some food-drug interactions can be valuable especially when food improves the bioavailability of the active drug substance. In other examples, the inhibition of the oxidative phase of hepatic drug metabolism by, for example, cimetidine may potentiate the effect and/or duration of a variety of drugs. Unfortunately, increasing the plasma concentration of a primary drug may well also increase the probability of enhanced toxicity.

In current medical practice there are a number of examples of combined formulations and/or co-prescribing of active ingredients on the basis of their believed synergistic action. For many of these the evidence for efficacy may not stand up to critical analysis. A brief account of the more important treatments with believed synergistic drug combinations is presented in Chap. 8.

III. Drug Interactions In Vitro

Currently much information is appearing in the literature usually from pharmaceutical research sources on drug interactions in vitro. Chapter 9 deals with this and will update the reader on the mechanisms involved in the latest developments. Much of this information in the past has been interactions between drug and drug, and drug and fluid in intravenous infusions. Newer work has concentrated on interactions between specific drugs, notably chloroquine, cyclosporin insulin and vasodilator nitrates, and pharmaceutical packaging materials (glass and plastics), and the mechanisms by which these in vitro interactions occur.

Also included in the chapter are drug interactions with contact lens materials and drug-excipient interactions. In the past, it has not always been appreciated that the so-called "inactive" excipients used in product formulations are physicochemically active substances which are often quite capable of interacting or producing complexes with drug substances in the formulation. They can cause changed absorption and altered bioavailability. The often quoted outbreak of phenytoin intoxication in Australia during 1968–1969 was caused by increased bioavailability due to a simple change in capsule filler from calcium sulphate to lactose (Tyrer et al. 1970).

For some time now, a great deal of effort has been given to the complexing of drug substances to a variety of polymers with the object of producing sustained-release products. Some of this research has been done with proteins, the ultimate goal, I believe, being the production of an orally effective insulin.

IV. Age and Genetic Factors

Particular attention in this volume has been given to the influences of age and genetic factors in drug interaction (Chap. 10). It has long been established that elderly patients use more medicines than younger age groups. It is, however, clear from this chapter that age and genetic factors are not directly responsible for causing drug-drug interactions and that, qualitatively, the elderly share similar drug interactions with younger age groups. It is the severity of the interaction that may be more serious in the elderly patient or in the patient with a genetic abnormality. Also the increased use of medication by the elderly allows a greater possibility of drug-drug interactions occurring.

It may be noted that most of the drugs involved in clinically relevant interactions are those upon which patients are carefully stabilised for relatively long periods of time, for example, anticoagulants, anticonvulsants, beta-blockers and other hypotensives, cardiac glycosides, H_2-receptor blockers, hypoglycaemics, lithium salts, psychotropics and immunosuppressants. It may be deduced that many of these patients will be in the older age bracket. Examination of clinical reports of drug interactions often shows that it is these drug-stabilised patients who are at special risk of any changes in therapy or environment that will influence the potency or availability of their normal medication. It must also be appreciated that removal of a drug from an otherwise stabilised regimen of treatment may also initiate a serious interaction sequel.

V. Interference with Laboratory Testing

Laboratory tests, physical examination and the patient's medical history often provide the main key to accurate diagnosis and to subsequent rational therapy. Chapter 11 relates how one of these key features – laboratory tests – can be influenced in sensitivity or specificity (or both) by the drugs consumed by the patient. Frequently the influence of medication on the reliability of subsequent laboratory tests is not realised or merely overlooked. As the chapter details, the mechanisms of such interactions can be generally classified into two areas: pharmacological and methodological interferences. The author of this chapter has emphasised that whenever a drug-laboratory test interaction is established it should be communicated to the medical community so that they may avoid possible errors in diagnosis. Chapter 11 summarises much information about a field of interaction, the importance of which is little realised or appreciated.

VI. Herbal and Other Non-orthodox Medicines

There are many types of herbal remedies. At one end of the scale there are the self-made teas prepared from self-collected herbs. At the other end there are officially registered drug products which have passed through a rigid registration procedure. In between there are a bewildering range of nostrums that have obscure origins and even more obscure constituents. The latter classes may be a mixture of herbal products with other ingredients of non-herbal origin (e.g. arsenic or lead) or undeclared Western drugs [e.g., corticosteroids, non-steroidal anti-inflammatory/antirheumatic agents (NSAIDS) or paracetamol]. Although herbal medicines are by far the largest component of non-orthodox remedies, they do not have exclusive claims. Other non-orthodox remedies range from preparations of animal origin, minerals, vitamins and amino acids; many of these are capable of interfering with orthodox medicines.

A survey by Wharton and Lewith (1986) showed that only 5% of British doctors claimed more than a poor knowledge of herbal medicines. Medical practitioners generally believed that herbal preparations were residues from the remote past, that they were harmless and largely ineffective. Many patients believed that they were safe, simply because they were natural products, and that they were often miraculously effective. Unfortunately both patients and doctors are misinformed since, as Chap. 12 relates, many herbal products can be exceedingly toxic and may indeed present a peculiar hazard if taken in combination with orthodox medicines.

D. Comment

I. Drug Interactions: Hazardous and Expensive Use of Resources?

There is little doubt that drug-drug interactions can often be serious even life-threatening. They can also be very expensive and evidence from the Medical Defence Union's Reports of over 10 years ago reveals that one case which settled for £44000 was due to phenylbutazone-induced potentiation of warfarin which was followed by an intraspinal haemorrhage, resulting in an incomplete tetraplegia at the level of C7. It is interesting in this case that the solicitor's letter stated that "Butazolidin is a well known potentiator of counmarin anticoagulants of which warfarin is one" . . . "the prescribing of Butazolidin for a patient known to be taking warfarin routinely was a breach of your professional duty to him".

Knowledge of the "state of the art" has much advanced since the 1980s and what were then novel and scientifically interesting interactions are now established and well-recorded interactions in standard books of reference. Knowledge has advanced but their has been little progress in the way in which it is generated.

II. Sources of Information

Most reports of drug-drug or drug-food interaction arise primarily from experience in the clinic. Some of these are anecdotal and may not be reported again, whilst others represent the first warning signs of a wider cohort of cases that is yet to be realised. Their purpose is to report the interaction and explain its hazard so that others can avoid it occurring. However, few of these reports are able to indicate the precise mechanism of the interaction.

Generally, animal experiments have not contributed much of substance to knowledge about interactions. Admittedly, the early work of CONNEY et al. (1956, 1957) in rats did much to explore the nature and organisation of enzyme systems in liver microsomes, but it was the human studies of LEVY (1970) and NEUVONEN et al. (1970) that provided the now classical evidence that the salts of divalent or trivalent metals formed non-absorbable complexes with tetracyclines and reduced their absorption, resulting in subtherapeutic levels in the plasma. It was observation in patients that gave the first indication that concomitant medication could antagonise the efficacy of oral contraceptives (DOSSETOR 1975). It was in tuberculosis patients that the apparent interaction between *para*-aminosalicylic acid (PAS) and rifampicin was first recognised and later to be explained that the interaction was not due to PAS but to the bentonite content of its granules (BOMAN et al. 1971). I was reminded of this latter interaction in recent months when I met a lady in Belfast who had been a TB patient at that time and had been receiving rifampicin and PAS and without warning or explanation was changed to rifampicin plus isoniazid.

III. The Literature on Drug Interactions

The literature on drug interactions tends to be very non-specific and in the last 20 years or so it has become voluminous and has been cluttered up with a sticky mass of largely irrelevant studies of interactions in animal models and in single-dose pharmacokinetic studies in animals and in young adult age groups of volunteers. Such studies can be of predictive value, but only if they mimic the clinical situation and if they relate to drug combinations and dosage regimens that are normally used in sick patients.

It is clear that the literature requires some judgmental element to be exercised as to its clinical significance. It is the single reports in the correspondence columns of medical and pharmaceutical journals that, although often anecdotal and uncorroborated, have, in my view, made a major contribution to elucidating the nature and extent of interactions. Such reports are also useful since they often stimulate other clinicians to report similar experiences in their own patients.

There is still need to focus attention on those drug-drug or drug-nutrient interactions that really do influence the safety or efficacy of human drug therapy in all age groups. If mechanisms of interaction are clearly understood then it may be possible to develop animal/biochemical/tissue culture/physicochemical or other types of models of interactions to which new molcules could be exposed during their development stages. It is hardly satisfactory or safe to have to rely on patients experiencing adverse interactions before they can be established and documented. It is hoped that, by discussing mechanisms of interactions, this volume will stimulate interest in investigating the possibility and potential of developing meaningful models.

References

Basu TV (1988) Drug nutrient interactions. Croom Helm, London

Boman G, Hanngren A, Malmborg Ä, Borgå O, Sjöqvist F (1971) Drug interaction: decreased serum concentrations of rifampicin when given with PAS. Lancet I:800

Braybrooks MP, Barry BW, Abbs ET (1975) The effect of mucin on the bioavailability of tetracycline from the gastrointestinal tract: in vivo, in vitro correlation. J Pharm Pharmacol 27:508–515

Brown C, Taniguchi G, Yip K (1989) The monoamine oxidase inhibitor-tyramine interaction. J Clin Pharmacol 29:529–532

Chin TF, Lach JL (1975) Drug diffusion and availability: tetracycline metallic chelation. Am J Hosp Pharm 32:625–629

Compston JE (1994) The therapeutic use of bisphosphonates. Br Med J 309:711–715

Conney AH, Miller EC, Miller JA (1956) The metabolism of methylated aminoazo dyes. V. Evidence for induction of enzyme synthesis in the rat by 3-methylcholanthrene. Cancer Res 16:450–459

Conney AH, Miller EC, Miller JA (1957) Substrate-induced synthesis and other properties of benzpyrenehydroxylase in rat liver. J Biol Chem 228:753–766

D'Arcy PF (1979) Drug interactions. In: D'Arcy PF, Griffin JP (eds) Iatrogenic diseases, 2nd edn. Oxford University Press, Oxford, pp 45–76

Dossetor J (1975) Drug interactions with oral contraceptives. Br Med J 4:467–468

Fels JP, Necciari J, Toussain P, Debry G, Luckx A, Scheen A (1989) Effect of food intake on kinetics and bioavailability of (4-chlorophenyl) thiomethylene bisphosphonic acid (Abstr). Calcif Tissue Int 44 [Suppl]:S-104

Fogelman I, Smith ML, Mazess R, Bevan JA (1984) Absorption of diphosphonate in normal subjects (Abstr). Calcif Tissue Int 36 [Suppl 2]:574

Fogelman I, Smith ML, Mazess R, Bevan JA (1985) Absorption of oral diphosphonate in normal subjects (Abstr). Bone 6:54

Fogelman I, Smith ML, Mazess R, Wilson MA, Bevan JA (1986) Absorption of oral diphosphonate in normal subjects. Clin Endocrinol (Oxf) 24:57–62

Levy G (1970) Biopharmaceutical considerations in dosage form and design. In: Sprowls JB (ed) Prescription pharmacy, 2nd edn. Lippincott, Philadelphia, pp 70, 75, 80

Lippman SB, Nash K (1990) Monoamine oxidase inhibitor update: potential adverse food and drug interactions. Drug Saf 5:195–204

Neuvonen PJ (1976) Interactions with the absorption of tetracyclines. Drugs 11:45–54

Neuvonen PJ, Gothoni G, Hackman R, Björksten K (1970) Interference of iron with the absorption of tetracyclines in man. Br Med J 4:532–534

Roe AR (1989) Diet and drug interactions. Van Nostrand Reinhold, New York

Trovato A, Nuhlicek DN, Midtling JE (1991) Drug-nutrient interactions. Am Fam Physician 44:1651–1658

Tyrer JH, Eadie MJ, Sutherland JM, Hooper WD (1970) Outbreak of anticonvulsant intoxication in an Australian city. Br Med J 2:271–273

Welling PG (1977) Influence of food and diet on gastrointestinal drug absorption. A review. J Pharmacokin et Biopharm 5:291–334

Wharton R, Lewith G (1986) Complementary medicine and the general practitioner. Br Med J 292:1498–1500

Section I
Pharmacokinetic Drug Interactions

CHAPTER 2

Drug Interactions in the Gastrointestinal Tract and Their Impact on Drug Absorption and Systemic Availability: A Mechanistic Review

E. LIPKA, J.R. CRISON, B.S. SCHUG, H.H. BLUME, and G.L. AMIDON

A. Introduction

The effect of drug interactions with other drugs, excipients and gastrointestinal contents on their oral absorption and systemic availability have been extensively investigated over the past 50 years. These interactions can be classified as direct interactions between the drug and other drugs or components in the formulation, e.g., excipients in the gastrointestinal tract altering the drugs' thermodynamic activity for absorption or, alternatively, the effects may be indirect through alteration of gastrointestinal transit, gastrointestinal secretions, or gastrointestinal/hepatic metabolism or elimination. Recently attention has been directed to metabolism interactions in the liver and even more recently in the gastrointestinal mucosal tissue. These interactions can be caused by direct inhibition of enzyme metabolism or enzyme induction, resulting in changes in the absorption rate and leading to altered oral bioavailability. These metabolic effects also can be indirect, through the position dependence of metabolizing enzyme levels in the gastrointestinal tract; consequently transit changes can alter metabolic profiles and systemic availability. Recent interest has been directed to carrier-mediated absorption, where drug interactions can occur through competition for carrier(s) responsible for absorption and, finally, it has been proposed that the P-glycoprotein exporter may be responsible for some of the observed drug interactions and nonlinear absorption effects. This review will focus on examples of the various types of mechanistic interactions noted above. The interaction of drugs with foods, a very important component in drug regulation today, is reviewed elsewhere in this volume. A more comprehensive survey of relevant drug-drug interactions than can be covered in this review can be found in Table 1.

B. Physicochemical Interactions

The two physicochemical interactions that affect oral absorption arise due to complexation of the drug with opposite-charged ion species or to nonspecific adsorption of the drug. In this section, examples of complexation of drugs with

Table 1. Summary of drug-drug interactions influencing absorption. (Modified from WELLING 1984)

Drug affected	Interfering agent	Observed change in absorption	Possible cause	Reference
p-Aminosalicylate (PAS)	Diphenhydramine	Delayed	Delayed stomach emptying	LAVIGNE and MARCHAND (1973)
Aspirin	Charcoal	Reduced	Adsorption	NEUVONEN et al. (1978); LEVY and TSUCHIYA (1972)
Aspirin (enteric-coated)	Antacids	Increased rate	Faster drug release	FELDMAN and CARLSTEDT (1974)
Carbamazepine	Charcoal	Reduced	Adsorption	NEUVONEN and ELONEN (1980)
Cefdinir	Iron ion	Reduced rate and extent	Chelation	UENO et al. (1993)
Cephalexin	Cholestyramine	Reduced	Adsorption or steatorrhea	PARSONS and PADDOCK (1975)
Chlorothiazide	Colestipol	Reduced	Binding	KAUFFMAN and AZARNOFF (1973)
	Metoclopramide	Decreased	Absorption window or dissolution	OSMAN and WELLING (1983)
	Propantheline	Increased	Absorption window or dissolution	OSMAN and WELLING (1983)
Chlorpromazine	Antacids	Reduced	Adsorption	FORREST et al. (1970); FANN et al. (1973) RIVERA-CALIMLIM et al. (1973)
	Benzhexol (trihexphenidyl)	Reduced	Delayed stomach emptying	
Chlortetracycline	Antacids	Reduced	Adsorption	SEED and WILSON (1950); GREENSPAN et al. (1951); WAISBREN and HUECKEL (1950)
Ciprofloxacin	Antacids	Reduced	Chelation	LODE (1988) and NIX et al. (1989)
Clindamycin	Kaolin-pectin	Delayed	Adsorption	ALBERT et al. (1978b)
Diazepam	Antacids	Delayed	Adsorption	GREENBLATT et al. (1978)
Dicoumarol (bishydroxycoumarin)	Magnesium hydroxide	Increased	Chelation	AMBRE and FISCHER (1973)
Digitoxin	Cholestyramine	Increased elimination rate	Interrupted enterophepatic circulation	CALDWELL et al. (1971)

Digoxin	Neomycin	Reduced	Sprue-like syndrome, malabsorption	Lindenbaum et al. (1972)
	Antacids	Reduced	Adsorption and faster stomach emptying (partly)	Brown and Juhl (1976); Albert et al. (1978a)
	Charcoal	Reduced	Adsorption	Neuyonen et al. (1978)
	Metoclopramide	Reduced	Limited dissolution	Manninen et al. (1973)
	Propantheline	Increased	Dissolution	Manninen et al. (1973)
Doxycycline	Ferrous sulfate	Reduced	Chelation, influencing absorption and elimination	Neuyonen and Penttilä (1974); Neuvonen et al. (1970)
Enoxacin	Ranitidine	Reduced	Increased pH	Lebsack et al. (1992)
Ethanol	Metoclopramide	Increased rate	Faster stomach emptying	Gibbons and Lant (1975)
	Propantheline	Reduced rate	Delayed stomach emptying	Gibbons and Lant (1975)
	Atropine	Reduced rate	Delayed stomach emptying	Gibbons and Lant (1975)
Hydrochlorothiazide	Propantheline	Delayed, but increased	Delayed stomach emptying	Beerman and Groschinsky-Grind (1978)
Ferrous ion	Tetracycline	Reduced	Chelation	Neuvonen et al. (1975); Heinrich et al. (1974)
Isoniazid	Antacids	Delayed and reduced	Adsorption and first-pass metabolism	Hurwitz and Schlozman (1974)
Levodopa	Homatropine	Reduced patient response	Delayed stomach emptying, increased metabolism in stomach	Fermaglich and O'Doherty (1972)
	Antacids	Increased	Faster stomach emptying	Revera-Calimlim et al. (1971)
	Metoclopramide	Increased rate	Faster stomach emptying	Morris et al. (1976)
Lignocaine (lidocaine)	General anesthetics	Delayed	Delayed stomach emptying	Adjepon-Yamoah et al. (1973)
Lincomycin	Sodium cyclamate	Reduced		Wagner (1969)
Lithium	Metoclopramide	Increased rate	Faster stomach emptying	Crammer et al. (1974)
	Propantheline	Delayed	Delayed stomach emptying	Crammer et al. (1974)
Methacycline	Ferrous sulfate	Reduced	Chelation	Neuvonen et al. (1970)
Nadolol	Trihydroxy bile salts, sodium cholate	Reduced	Poorly soluble micelle formation	Yamaguchi (1986)

Table 1. *Continued*

Drug affected	Interfering agent	Observed change in absorption	Possible cause	Reference
Nitrofurantoin	Magnesium trisilicate	Reduced	Adsorption	Naggar and Khalil (1979)
Norfloxacin	Antacids	Reduced	Chelation	Nix et al. (1990)
Ofloxacin	Antacids	Reduced	Chelation	Lode (1988)
Oxytetracycline	Ferrous sulfate	Reduced	Chelation	Neuvonen et al. (1970)
Paracetamol (acetaminophen)	Metoclopramide	Increased rate	Faster stomach emptying	Nimmo et al. (1973)
	Propantheline	Delayed	Delayed stomach emptying	Nimmo et al. (1973)
	Pethidine	Delayed	Delayed stomach emptyin	Nimmo et al. (1973)
	Morphine	Delayed	Delayed stomach emptying	Nimmo et al. (1973)
Penicillamine	Ferrous sulfate	Reduced	Chelation	Schuna et al. (1983)
	Antacids	Reduced	Adsorption, chelation	Schuna et al. (1983)
Penicillin V	Neomycin	Reduced	Sprue-like syndrome, malabsorption	Cheng and White (1962)
Phenobarbitone	Charcoal	Reduced absorption and increased elimination rate	Adsorption (reduces availability and prevents intestinal reabsorption)	Neuvonen and Elonen (1980)
Phenylbutazone	Charcoal	Reduced absorption and increased elimination rate	Adsorption (reduces availability and prevents intestinal reabsorption)	Neuvonen and Elonen (1980)
Phenytoin	Charcoal	Reduced		
Pseudoephedrine	Aluminum hydroxide	Increased rate	Raised GI pH	Lucarotti et al. (1972)
	Kaolin	Delayed	Adsorption	
Promazine	Attapulgite and pectin	Reduced	Adsorption	Sorby and Liu (1966)
Proquazone	Antacids	Delayed		Ohnhaus (1980)
Quinine	Antacids	Reduced (in rats)	Raised pH, delayed stomach emptying	Hurwitz (1971)
Riboflavine-5′-phosphate	Sodium alginate	Increased	Slow GI transit because of high viscosity	Levy and Rao (1972)

Rifampicin	*p*-Aminosalicylate (PAS)	Delayed and reduced	Adsorption to bentonite in PAS granules	BOMAN et al. (1971, 1975)
Sulfadiazine	Sodium bicarbonate	Increased rate	Faster dissolution rate	PETERSON and FINLAND (1942)
	Magnesium hydroxide	Increase (in rats)	Faster dissolution rate	HURWITZ (1971)
Sulfadiazine sodium	Antacids	Decreased (in rats)	Delayed stomach emptying	HURWITZ (1971)
Sulfamethoxazole	Cholestyramine	Reduced	Adsorption or steatorrhea	PARSONS and PADDOCK (1975)
Sulfathiazole	Antacids	Increased	Faster dissolution rate	BARLOW and CLIMENKO (1941)
Tetracycline	Ferrous sulfate	Reduced	Chelation	NEUVONEN et al. (1970); GOTHONI et al. (1972)
	Ferrous salts	Variable reduction (depending on salt)	Chelation	NEUVONEN and TURAKKA (1974)
	Zinc sulfate	Reduced	Chelation	PENTTILÄ et al. (1975)
	Magnesium sulfate	Reduced		HARCOURT and HAMBURGER (1957)
	Proteolytic enzymes	Unaffected or increased		BRADBROOK et al. (1978); SENECA and PEER (1965)
	Fe^{3+}, Fe^{2+}, Al^{3+}, Mg^{2+}, Ca^{2+}	Reduced	Chelation	NEUVONEN (1976)
	Sodium bicarbonate	Reduced	Poor dissolution	BARR et al. (1971)
Thyroxine	Cholestyramine	Reduced	Adsorption	NORTHCUTT et al. (1969)
Trimethoprim	Cholestyramine	Delayed	Adsorption	PARSONS and PADDOCK (1975)
Vitamin A	Neomycin	Reduced	Sprue-like syndrome, malabsorption	BARROWMAN et al. (1973)
	Aluminum hydroxide	Reduced	Absorption oxidation	HOFFMAN and DYNIEWICZ (1945)
Vitamin B_{12}	Colchicine	Reduced	Ileal blockade	FALOON and CHODOS (1969)
	Neomycin	Reduced	Malabsorption	FALOON and CHODOS (1969)
	p-Aminosalicylate	Reduced		PALVA et al. (1972)
Warfarin	Magnesium hydroxide	No effect		AMBRE and FISCHER (1973)
	Cholestyramine	Reduced	Adsorption	ROBINSON et al. (1971)

metal ions, resins and bile salts as well as adsorption of drugs on charcoal will be discussed.

I. Complexation with Metal Ions

Most tetracycline derivatives are rapidly, but incompletely, absorbed and the extent of absorption is further reduced when the drug is coadministered with other medications or nutrients that contain bi- or trivalent cations, such as calcium, magnesium and iron (DITCHBURN and PRITCHARD 1956). These cations form complexes with tetracyclines that are either poorly absorbed and/or poorly soluble, leading in some cases to clinically relevant reduction of plasma levels up to 90% (WELLING 1984). The extent of absorption inhibition depends upon the solubility and stability of the drug-metal complexes formed, which in turn are determined by the physicochemical characteristics of both the tetracyclines and the metal ions. For tetracycline, the order of decreasing complex stability has been reported to be $Fe^{3+} > Al^{3+} > Cu^{2+} > Ni^{2+} > Fe^{2+} > Co^{2+} > Zn^{2+} > Mn^{2+}$ (NEUVONEN 1976). Additionally, the stability of the metal salt administered, which defines the free fraction of the cation, also plays a role with regard to drug-metal complex formation. In a series of iron salts, ferrous sulfonate has been shown to reduce tetracycline plasma levels by up to 90%, whereas ferric sodium edetate induced a reduction of only 30% (NEUVONEN and TURAKKA 1974). Tetracycline derivatives contain multiple potential binding sites for chelate formation, the most important one being the 1,3-keto-enol

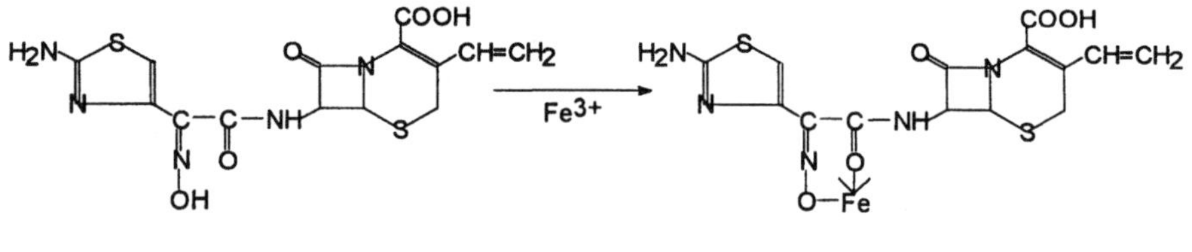

Parameters	Study 1 ($n = 6$)	Study 2 ($n = 6$)	Study 3 ($n = 6$)
C_{max} (μg/ml)	1.71 ± 0.42	0.16 ± 0.14	1.28 ± 0.23
t_{max} (h)	4.2 ± 1.0	1.8 ± 2.1	3.3 ± 0.5
AUC_{0-12} (μg h/ml)	10.3 ± 1.35*,**	0.78 ± 0.25	6.55 ± 1.61
AUC_{0-3} (μg h/ml)	2.13 ± 0.83***	0.38 ± 0.3	1.95 ± 0.24
AUC_{3-12} (μg h/ml)	8.03 ± 1.72*,**	0.4 ± 0.31	4.60 ± 1.54

Data are mean values ± SD. C_{max}, maximum plasma concentration; t_{max}, time to reach C_{max}; AUC, area under the plasma concentration-time curve. Study 1, cefdinir alone; study 2, cefdinir and iron ion preparation (simultaneous administration); study 3, iron preparation 3 h after cefdinir administration.
* $P < 0.01$, study 1 versus study 2; ** $P < 0.05$, study 1 versus study 3; *** $P < 0.05$, study 1 versus study 2.

Fig. 1. Structures of cefdinir and the formation of cefdinir-iron ion complex. The table summarizes the pharmacokinetic parameters for cefdinir. (Adapted from UENO et al. 1993)

structure (WEINBERG 1957). Drug-metal complexes of 1:1 and at higher pH of 1:2 might be formed. With trivalent cations tetracycline forms complexes of 1:3. Some of these complexes are poorly soluble in water, while others, although moderately soluble, exhibit a reduced diffusion rate as a consequence of complexation (CHIN and LACH 1975).

Cefdinir, a new oral cephalosporin, has been shown to undergo a 93% reduction of bioavailability when coadministered with iron ions (UENO et al. 1993). Since the antibiotic is administered as a sustained-release dosage form, it reaches maximum plasma levels after 4h, and even administration of iron preparations 3h after cefdinir intake reduces bioavailability by about 40%. The authors proposed a tetracycline-like chelate formation as the underlying mechanism for this drug-drug interaction, where cefdinir chelates with iron via its 7-hydroxyimino radical and its carbonyl oxygen, forming a poorly soluble complex. A schematic of the complexation and the bioavailability data with and without iron treatment are shown in Fig. 1.

The coadministration of metal ions present in antacids, iron formulations and nutrients such as dairy products with drug formulations might therefore affect the therapeutic efficacy of the drugs. An 80%–90% reduction in plasma levels of chlortetracycline was observed when the drug was coadministered with aluminum-containing antacid gels, resulting in concentrations below the therapeutic level (WAISBREN and HUECKEL 1950). Dairy products have been shown to reduce dimeclocycline bioavailability by up to 80% (SCHREINER and ALTEMEIER 1962).

However, in addition to solubility reduction as a consequence of chelate formation, tetracycline dissolution might become impaired in the presence of other compounds, depending on the formulation. Barr et al. have reported that sodium bicarbonate, which lacks polyvalent cations, reduces tetracycline absorption by 50% when the antibiotic was given as a capsule, but had no effect when tetracycline was administered as a solution (BARR et al. 1971). This was attributed to a reduced dissolution.

Another group of drugs reported to be affected by complexation with metal ions are the anticoagulants. In contrast to the examples discussed above, bishydroxycoumarin (BHC), for example, exhibits an increase in absorption after oral coadministration together with milk of magnesia (AMBRE and FISCHER 1972). Surprisingly, no effect was noted after aluminum coadministration. A Mg-BHC chelate has been described previously, and administration of the chelate to two subjects resulted in significantly increased plasma levels, suggesting the preferential absorption of the chelate compared to the parent drug. However, the mechanism by which the chelate increases BHC plasma levels is unknown and whether the chelate is absorbed intactly or releases BHC before absorption remains to be clarified. This interaction potentially results in severe undesired pharmacological effects, considering the low therapeutic index of this class of compounds.

As a final example, an up to tenfold reduction in absorption of quinolones after antacid administration has been reported and has been primarily attributed to formation of chelate complexes (LODE 1988).

II. Binding to Resins

Significant reduction in warfarin plasma levels of up to 30% after oral coadministration with cholestyramine has been noted in a clinical trial with six subjects (ROBINSON et al. 1970). Since patients with hyperlipidemia are quite frequently treated with a cholesterol-lowering combination therapy that includes cholestyramine and might also receive anticoagulants, such interactions are again clinically important due to the narrow therapeutic interval of anticoagulant agents. In vitro binding studies indicate a significant binding of warfarin to cholestyramine at a pH above its pK_a of 5.5. This suggests that the bioavailability of warfarin is reduced as a result of a smaller fraction being available for absorption. Furthermore, due to reduced lipid absorption, vitamin K uptake is impaired, which makes the overall pharmacological response highly unpredictable.

Highly stable binding of thyroxine, another drug with a low therapeutic index, to cholestyramine has also been suggested as the underlying mechanism for the reduced oral bioavailability when the two drugs are combined in the treatment. A time interval of at least 4h between administration of the two drugs is recommended (NORTHCUTT et al. 1969).

III. Complexation with Bile Salts

The absorption of nadolol has been shown to be strongly inhibited by trihydroxy, but not by dihydroxy, bile salts (Fig. 2) (YAMAGUCHI et al. 1986a,b).

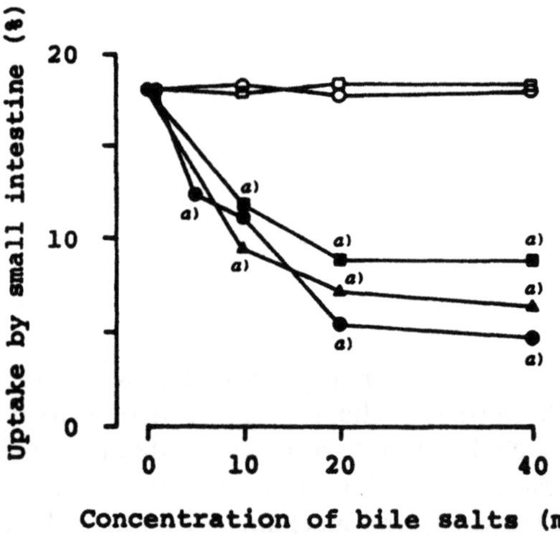

Fig. 2. Effect of bile salts on nadolol absorption 4h after injection into the rat jejunum loop. (Adapted from YAMAGUCHI et al. 1986b). *closed circles*, sodium cholate; *closed squares*, sodium glycocholate; *closed triangles* sodium taurocholate; *open circles*, sodium chenodeoxycholate; *open squares*, sodium deoxycholate; *open triangles*, sodium lithocholate. Dose of nadolol was 0.01 mmol/rat, and that of bile salts was 0.005–0.1 mmol/rat. *Each point* represents the mean value for at least four rats. a) $P < 0.01$ compared to the control (without bile salts)

Furthermore, no absorption inhibition was noted for other β-blocking agents, implying that specific structural features are responsible for this interaction. Subsequent reports demonstrated that nadolol forms a tenfold more stable micellar complex with trihydroxy bile salts than other β-blockers (Yamaguchi et al. 1986b). This complex was poorly soluble in water, resulting in a significant decrease in nadolol uptake from the intestine. After performing magnetic resonance imaging (MRI) spectroscopic studies with a series of nadolol derivatives, the authors concluded that the *cis*-2,3-diol moiety of the molecule is crucial for stabilizing the micellar complex (Yamaguchi et al. 1986c).

IV. Nonspecific Adsorption

Adsorption of therapeutic drugs to so-called inert excipients in formulations as well as adsorption to charcoal administered for diarrhea therapy might significantly interfere with the chronic treatment of patients. Rifampicin, for example, is prescribed in a capsule formulation that contains bentonite, which adsorbs the drug to a considerable extent (Boman et al. 1975). Consequently, rifampicin bioavailability is reduced by 50% compared to administration of the bentonite-free tablet.

An outbreak of anticonvulsant intoxication was observed in epileptic patients in Australia during 1968–1969 (Tyrer et al. 1970). The clinical picture of all 51 patients was consistent with the diagnosis of anticonvulsant intoxication and it could be shown that all patients took 100-mg sodium phenytoin (diphenylhydantoin) capsules produced by one particular manufacturer. The excipients in this capsule formulation had been changed from calcium sulfate dihydrate to lactulose. This results of subsequent investigations showed that the patient's blood phenytoin concentration fell rapidly to one-fourth of its former value when the phenytoin capsules containing calcium sulfate as excipients were administered instead of phenytoin capsules with lactulose. The blood concentration rose again when the capsules containing lactulose as excipient were administered. This phenomenon was of relevant therapeutic importance, but it is not completely understood whether it is the result of unspecific adsorption, complexation with metal ions, or other physicochemical factors.

The inhibitory effect of charcoal on absorption was demonstrated in two studies where the oral bioavailability of digoxin, phenytoin, phenobarbitone, carbamazepine and phenylbutazone was reduced by more than 95% (Neuvonen et al. 1978; Neuvonen and Elonen 1980). In addition, a reduced plasma half-life was reported for these compounds, which is most likely due to an interruption of the enterohepatic circulation.

C. Interactions with Drugs That Influence GI Motility and pH

I. Alteration of Gastrointestinal Motility

Drugs that influence gastrointestinal motility and therefore transit time can have a significant effect on drug uptake. Since very few drugs are absorbed from the stomach, gastric emptying rate as well as small intestinal transit time may become the rate-limiting steps to absorption. The effect of drugs that alter intestinal pH and motility upon other drugs depends on the physicochemical as well as the biological properties of these coadministered compounds. A decrease in motility might result in an increased extent of absorption as a result of a prolonged dissolution time and increased contact time with the intestinal surface. Alternatively, a decrease in absorption rate could result as a consequence of delayed gastric emptying and low pH instability. The following examples will illustrate some of the mechanisms described.

Propantheline, an anticholinergic drug, decreases motility and thus the rate of gastric emptying, whereas the dopamine antagonist metoclopramide increases gastric emptying rate. NIMMO et al. (1973) demonstrated the effect of both compounds on acetaminophen absorption in healthy volunteers. Propantheline decreased whereas metoclopramide increased the rate of acetaminophen absorption, correlating well with the simultaneously observed changes (decrease and increase, respectively) in gastric emptying. However, the overall extent of absorption as reflected by the urinary recovery did not change, suggesting that small and large intestinal transit did not change.

A marked influence of propantheline and metoclopramide on the absorption kinetics of ethanol has also been reported (GIBBONS and LANT 1975). Rapid gastric emptying induced by intravenous administration of metoclopramide led to a close to 100% increase in peak plasma levels of ethanol. The increased rate of absorption in combination with the saturable first-pass metabolism of ethanol might result in severe undesirable side effects if these compounds are administered simultaneously, e.g., if a formulation contains significant amounts of ethanol. Intravenous administration of propantheline significantly decreased the absorption rate of ethanol and resulted in a 50% reduction of peak plasma levels.

The effects of these motility-modulating compounds on the absorption of digoxin are based on a different mechanism. Here increased plasma levels of digoxin after propantheline and decreased levels after metoclopramide administration have been demonstrated (MANNINEN et al. 1973). Digoxin uptake is rate limited by the drugs' solubility as well as the disintegration and dissolution of the dosage form. Prolonged gastrointestinal residence time induced by propantheline appears to result in a greater fraction of the drug dissolved and consequently in an increased extent of absorption. The opposite effect can be observed after metoclopramide treatment, where impaired extent of drug absorption occurs as a result of the increased rate of gastrointestinal transit.

These observations can be theoretically described employing a simple mass balance model that represents the intestine as a tube to predict extent of absorption as a function of dissolution and permeability parameters (Ho et al. 1983; Ni et al. 1980). In order to develop a more quantitative and predictive model for drug absorption rates, it is necessary to include in the model flow, dissolution, absorption and reaction processes occurring in the intestine. In general this is quite complex. However, a simple model that considers a segment of intestine over which the permeability may be considered constant, a plug flow fluid with the suspended particles moving with the fluid, no significant particle-particle interactions (i.e., aggregation) and dissolution in the small particle limit, leads to the following pair of differential equations in dimensionless form (Amidon et al. 1995):

$$dr^*/dz^* = -(Dn/3)(1 - C^*)/r^* \tag{1}$$

and

$$dC^*/dz^* = DoDnr^*(1 - C^*) - 2AnC^* \tag{2}$$

where

$$z^* = z/L = (v_z/L)t = t^*$$

$$t^* = t/(L/v_z) = t/(AL/Q) = t/(V/Q)$$

where L = tube length, v_z = axial fluid velocity in the tube, A = tube surface area = $2\pi RL$, R = tube radius, Q = axial fluid flow rate = Av_z. The three important dimensionless groups are:

$$D_o = \text{dose number} = \frac{M_o/V_o}{C_s}$$

$$D_n = \text{dissolution number} = \frac{DC_s}{r_o} \cdot \frac{4\pi r_o^2}{\frac{4}{3}\pi r_o^3 \rho} \cdot t_{res} = t_{res} \cdot 3DC_s/\rho r_0^2$$

$$= t_{res}/t_{Diss}$$

$$A_n = \text{absorption number} = \frac{P_{eff}}{R} \cdot t_{res} = t_{abs}^{-1} \cdot t_{res}$$

$$t_{res} = \pi R^2 L/Q = \text{mean residence time}$$

$$t_{Diss} = \frac{r_0^2 \rho}{3DC_s} = \text{time required for a particle to dissolve}$$

$$t_{abs}^{-1} = k_{abs} = (S/V)P_{eff} = \frac{2 \times P_{eff}}{R} = \text{the effective absorption rate constant}$$

It is clear from the definition of the dissolution number that if the dissolution rate and absorption rate in the intestine are constant (constant permeability,

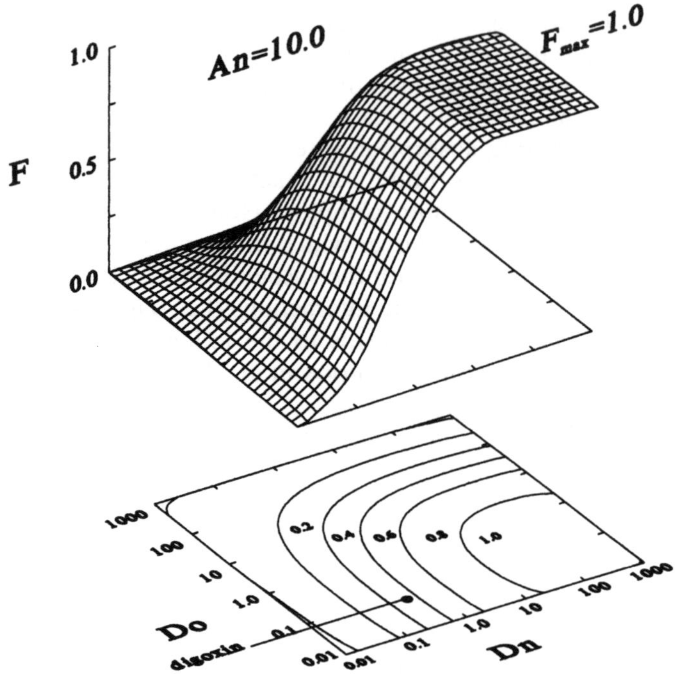

Fig. 3. Fraction dose absorbed versus dose and dissolution number

particle size and solubility, e.g., a nonpolar non-ionizable drug), gastrointestinal residence time will be the primary source of variation. Figure 3 is a plot of the fraction dose absorbed versus the dose and dissolution number for drugs with low solubility and high permeability estimated by Eqs. 1 and 2. The most important compounds affected by transit time are those that fall on the steep area of the surface, i.e., dose number ≈ 1–10 and a dissolution number ≈ 0.1–10. For these drugs, small changes in transit time can cause large changes in the fraction absorbed. In general, the average small intestinal transit time (SITT) is 3.2 ± 1.3h, whereas the large intestinal transit time (LITT) is 32 ± 18h and represents a variability inherent in the population (DAVIS et al. 1986). Referring back to digoxin as an example, which has an aqueous solubility of 0.024mg/ml and a dose of 0.5mg, one can expect about 50% absorption, assuming an average particle diameter of 50μm. Changing the dissolution number by small increments, either positively or negatively, will result in large changes in the fraction dose absorbed. JOHNSON et al. (1978) studied the effect of transit time and particle size on the absorption of digoxin. Their study showed that the micronized product (d_{ave}< 10μm) was not affected by changes in the intestinal transit time whereas the unmicronized product (d_{ave} ≈ 100μm) was significantly affected. For the 100-μm particles, the cumulative 4-day urinary excretion of digoxin was increased by 14% when coadministered with propantheline (reduces GI motility and transit time) and reduced by 27%

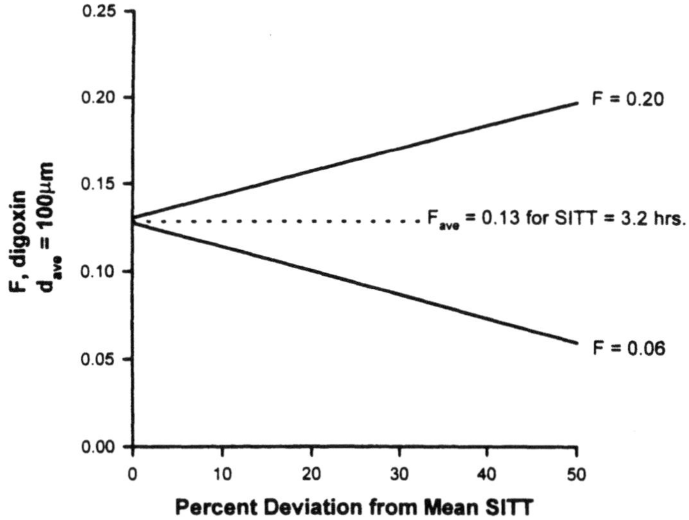

Fig. 4. Predicted fraction dose absorbed for 100-μm particles of digoxin at different small intestinal transit times (SITT)

when coadministered with metoclopramide (increases GI motility and transit time). The predicted fractions of the dose absorbed for different transit times are shown in Fig. 4. The model predicts 100% absorption for the micronized digoxin.

The existence of an "absorption window" in the upper part of the small intestine represents the third hypothesis discussed as the underlying mechanism of drug interactions with transit time altering compounds. RESETARITS and BATES (1979) reported a possible absorption-enhancing effect of propantheline bromide on orally administered chlorothiazide in dogs, which they attributed to a longer residence time at the absorption site. Based on the relatively low solubility of chlorothiazide at pH values comparable to small intestinal fluids (pH 6.2) of 0.65 mg/ml, however, the authors also considered an increase in dissolution time as a possible explanation. Furthermore, chlorothiazide exhibits a dose-dependent decrease in absorption at higher doses, which could be in part reversed by simultaneous administration of propantheline bromide, again suggesting that dissolution might be the rate-limiting step to chlorothiazide absorption. Increased urinary excretion of cholorothiazide after propantheline administration and decreased excretion after metoclopramide administration has also been reported in humans, with saturable or site-specific absorption as well as solubility limitations being considered as possible mechanisms (OSMAN and WELLING 1983). Simulation studies with chlorothiazide that included drug parameters such as pK_a, solubility and intestinal intrinsic permeability as well as varying physiological parameters (pH profile, volume of intestinal contents and intestinal flow rates) also indicated that the dose dependency of chlorothiazide may be attributed to solubility-ionization effects rather than a saturable absorption mechanism (DRESSMAN et al. 1984).

One of the few examples where prolonged gastric residence time as a result of anticholinergic drug administration leads to reduced oral availability is L-dopa. Both trihexylphenidyl and atropine significantly lower plasma levels of L-dopa after oral administration (ALGERI et al. 1976). Metabolism by the gastric mucosa was reported for L-dopa; therefore a delay in gastric emptying might result in decreased oral bioavailability of the parent drug in this case (RIVERA-CALIMLIM et al. 1970a–c). Alternatively very slow gastric emptying could lead to higher small intestinal metabolism if the metabolism is nonlinear and/or higher in the upper small intestine.

It is widely known that ingestion of ethanol influences gastrointestinal motility. Significantly delayed gastric emptying has been shown, e.g., as a result of administration of highly concentrated alcoholic beverages (BARBORIAK and MEADE 1970). However, other possible mechanisms have also been discussed for the ethanol-influencing bioavailability of certain drugs. Thus, ethanol may modify/affect solubility of the drug substance itself. Furthermore, permeability of absorptive membranes might be increased (WHITEHOUSE et al. 1975). In a two-period crossover study, HAYES et al. (1977) demonstrated that bioavailability of diazepam was significantly increased when the drug was coadministered with 30ml of 50% ethanol. In this study maximum plasma concentration as well as extent of bioavailability were significantly higher. Comparable effects were reported for sulfapyridine, chloral hydrate and chlordiazepoxide (RIETBROCK et al. 1983).

II. pH Alteration in the Gastrointestinal Tract

Aluminum hydroxide, widely used in a variety of antacid formulations, represents an intermediate compound in terms of its effects on the gastrointestinal system. As an antacid, aluminum hydroxide increases gastric pH; in addition it also delays gastric emptying. Thus, it could affect drug absorption via two possible mechanisms. HURWITZ (1971) investigated the effects of aluminum hydroxide and magnesium hydroxide on the oral absorption of sulfadiazine, a weak acid, and quinine, a weak base, in rats. After pretreatment with aluminum hydroxide, administration of sodium sulfadiazine resulted in significantly lower plasma levels, which was attributed to a delay in gastric emptying, whereas magnesium hydroxide did not alter the plasma profile. In contrast, an increase in plasma levels was reported when sulfadiazine was administered as the free acid after pretreatment with magnesium hydroxide. It was proposed that magnesium hydroxide raises gastric pH sufficiently to increase the solubility of the free acid. Both magnesium hydroxide and aluminum hydroxide decreased quinine absorption, the former predominantly by raising gastric pH and precipitating the quinine, the latter additionally by delaying gastric emptying.

Decreased oral bioavailability of quinolone antibiotics after simultaneous administration of antacids has been widely reported. Aluminum-magnesium

Table 2. Mean ± SD values of enoxacin pharmacokinetic parameters alone and after coadministration with ranitidine, pentagastrin and ranitidine + pentagastrin. (Adapted from LEBSACK et al. 1992)

Treatment	C_{max} (μg/ml)	t_{max} (h)	AUC (μg h/ml)	$t_{1/2}$ (h)	Cl_r (ml/min)
Enoxacin alone	2.74 ± 0.68	1.3 ± 0.5	16.1 ± 3.6	5.2 ± 1.0	182 ± 73
Enoxacin + ranitidine	1.70 ± 0.66*	2.0 ± 1.0	11.9 ± 5.2*	5.7 ± 1.0*	211 ± 46
Enoxacin + pentagastrin	2.49 ± 0.55	2.1 ± 0.9	16.2 ± 4.9	5.2 ± 0.8	198 ± 81
Enoxacin + ranitidine + pentagastrin	2.93 ± 0.66	1.5 ± 0.5	17.7 ± 5.6	5.5 ± 0.9*	212 ± 86

C_{max}, peak plasma concentration; t_{max}, time to reach peak concentration; AUC, area under the plasma concentration-time curve; $t_{1/2}$, half-life; Cl_r, renal clearance.
* Significantly different from enoxacin alone ($P < 0.05$).

hydroxide, for example, reduced the mean oral bioavailability of norfloxacin and ciprofloxacin by 91% and 85%, respectively, which was previously attributed to chelate complexation (NIX et al. 1989, 1990). However, earlier studies demonstrated a 40% decrease in enoxacin bioavailability also after ranitidine pretreatment, therefore indicating that increased gastric pH could also be a factor when antacids are coadministered with quinolones (GRASELA et al. 1989). Recently, LEBSACK et al. (1992) reported a 26% decrease in the oral bioavailability of enoxacin after intravenous administration of ranitidine (Table 2). This effect could be reversed by the simultaneous administration of pentagastrin, suggesting that a decrease in gastric acidity rather than a direct interaction of enoxacin with ranitidine is responsible for this interaction. This was further supported by the pH-dependent aqueous solubility of enoxacin. At pH values below 4.5, enoxacin is highly soluble (up to 16 mg/ml at pH 2.0), whereas its solubility drops below 1 mg/ml at pH values above 5. The pH-dependent solubility might therefore be the rate-limiting step to enoxacin absorption. However, to what extent antacids as well as H_2-antagonists alter the pH in the jejunum and ileum, the major absorption sites, and possibly influence reprecipitation of the drug remains to be evaluated.

Bioavailability of drugs may also be influenced by coadministration of H_2-antagonists when dissolution of the active ingredient from the formulation is pH-dependent. This was shown for a controlled/modified-release verapamil formulation where in vitro dissolution experiments showed markedly reduced dissolution in media buffered at pH 8.0, 6.8 or 4.5 compared to pH 1.2 (BLUME 1990; BLUME et al. 1988). A comparative bioavailability investigation of this product showed that intravenous coadministration of famotidine resulted in a reduction of bioavailability of verapamil during the first 2 h after administration, indicating an effect on the dosage-form-dependent release rate due to an increase of gastric pH.

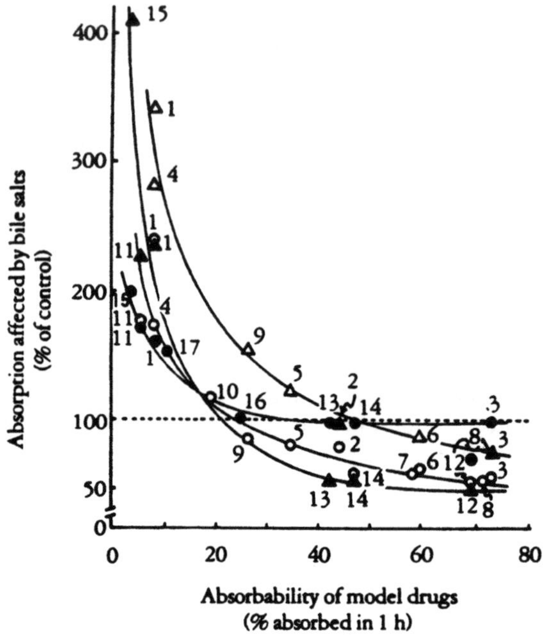

Fig. 5. Plot of the effect of bile salts on drug absorption against the absorbability of the model drugs. (Adapted from KIMURA et al. 1985). Intestinal absorption of drugs was examined in the presence of 20m*M* bile salts or sodium lauryl sulfate. *Each point* represents the mean of at least three experiments. Drugs: *1*, sulfaguanidine; *2*, sulfanilamide; *3*, sulfadimthoxine; *4*, sulfanilacetamide; *5*, sulfisoxazole; *6*, sulfapyridine; *7*, sulfamethoxypyridazine; *8*, sulfaphenazole; *9*, sulfisomidine; *10*, sulfamethizole; *11*, phenol red; *12*, imipramine; *13*, quinine; *14*, 2-allyloxy-4-chloro-*N*-(2-diethylaminoethyl)-benzamide; *15*, N^1,N^1-anhydro-bis-(β-hydroxy-ethyl)biguanide; *16*, metoclopramide; *17*, procainamide. *Filled circles*, with sodium taurocholate; *open circles*, with sodium cholate; *filled triangles*, with sodium taurodeoxycholate; *open triangles*, with sodium lauryl sulfate

D. Interactions Between Drugs That Share the Same Absorption Mechanism

I. Passive Absorption

Various compounds with proposed enhancing effects on passive transcellular as well as paracellular absorption have been investigated over recent years, including oleic acid, sodium lauryl sulfate and bile salts (MURANASHI 1990). KIMURA et al. (1985) summarized in a detailed report the effects of di- and trihydroxy bile salts on the in situ absorption of a broad range of compounds as depicted in Fig. 5. They proposed different mechanisms with regard to both the observed enhancing as well as inhibitory effects of bile salts on drug absorption. Sodium taurodeoxycholic acid (STDC) interacts most likely directly with the intestinal membrane by forming mixed aggregates with the phospholipids in the lipid bilayer, therefore altering the membrane structure.

Compounds that are poorly absorbed by the unperturbed membrane usually exhibit an increased rate of absorption in the presence of bile salts. Furthermore, both STDC and STC (sodium taurocholic acid) appear to induce calcium depletion of the tight junctions, resulting in an opening of the cell junctions, therefore facilitating paracellular diffusion of small polar drugs. Well-absorbed compounds, in contrast, may exhibit a decrease of their absorption rate when administered with bile salts. This observation might be in part due to a reduction in thermodynamic activity as a consequence of micelle formation as well as to depletion of some mucosal components, lipids and/or proteins to which the drug may have a high affinity.

Interactions of drugs that are absorbed via the paracellular pathway have been rarely described. As mentioned above, an opening of the tight junctions due to calcium depletion induced by bile salts might lead to an enhanced uptake of small hydrophilic compounds. Another mechanism where opening of the intercellular space facilitates drug uptake is what is commonly known as solvent drag effect. Electrolytes and glucose induce convective transport of other molecules by increasing the osmolarity in the aqueous channels between the cells and therefore increasing net water uptake. The intestinal permeability of acetaminophen, a small water-soluble drug, in rats increased significantly as a function of glucose concentration (Fig. 6) (FLEISHER 1995). The solvent drag approach was also utilized as an attempt for increasing the oral bioavailability of small peptides and peptide analogues. Permeability increase in the presence of glucose was demonstrated for the dipeptide D-kytorphin and the tripeptide analogue cephadrine (FLEISHER 1995). However, no effect of glucose was observed for the uptake of a growth hormone releasing hexapeptide and a cyclic octapeptide somatostatin analogue, which probably reflects the size cut-off of the paracellular route. Studies in humans of the effect of water transport on drug absorption are relatively few, but recent permeability studies suggest that there is only a limited effect. In fact, there appears to be evidence that D-glucose as well as L-leucine fail to stimulate water absorption in the human jejunum, leading to a reevaluation of the "solvent drag" concept in humans (LENNERNAS 1995).

II. Carrier-Mediated Absorption

The amino acid carrier as well as the small peptide carrier, which transports di- and tripeptides, are two well-characterized active transport systems in the small intestine (MAILLARD et al. 1995; WILLIAMSON and OXENDER 1995). While the former exhibits a high substrate specificity, the latter is known to facilitate the transport of a diverse group of peptides and peptide analogues, which might result in possible interactions during absorption. Compounds that are supposed to be absorbed via the small peptide carrier include amino β-lactam antibiotics and angiotensin-converting enzyme (ACE) inhibitors (TSUJI 1995; YEE and AMIDON 1995).

Representative drugs suggested to be absorbed by the amino acid pathway are L-dopa and α-methyldopa. Studies using L-leucine as an inhibitor

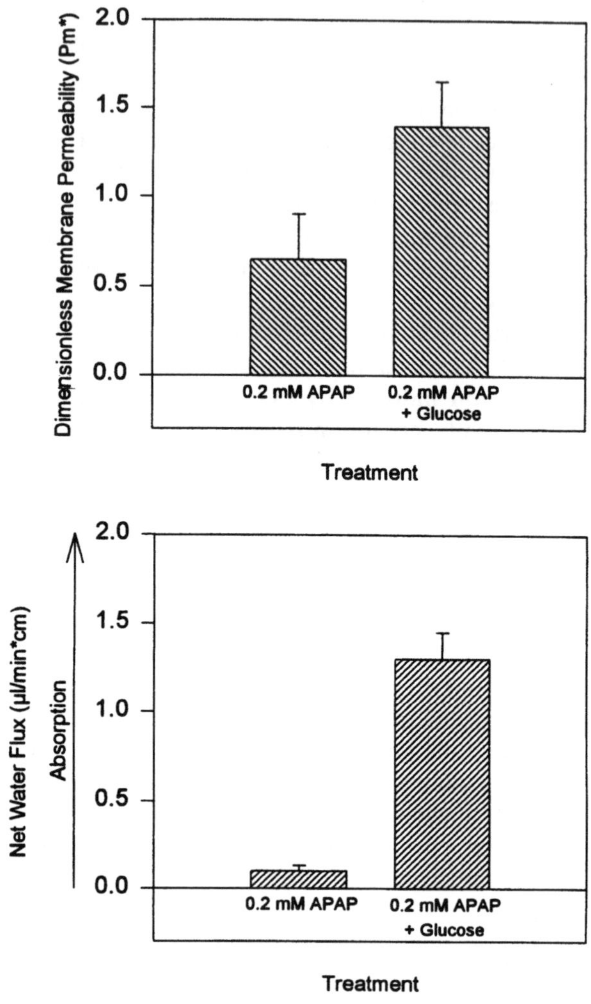

Fig. 6. Acetaminophen (APAP) permeability correlates with jejunal water absorption induced by glucose in rat perfusion, suggesting solvent drag of drug through paracellular pathways. (Adapted from FLEISHER 1995)

of L-dopa uptake in single-pass perfusion studies in rats demonstrated a tenfold decrease in intestinal wall permeability, indicating transport of L-dopa by the amino acid carrier (Fig. 7) (SINKO et al. 1987). LENNERNAS et al. (1993) provided evidence in humans for a competitive uptake via this carrier by demonstrating a decrease in L-dopa intestinal absorption from 40% to 21% in the presence of L-leucine.

Cefatrizine absorption was significantly inhibited by the dipeptide Phe-Phe, whereas L-Phe had no effect on its absorption, suggesting that cefatrizine uptake occurs via the small peptide transporter (SINKO et al. 1987).

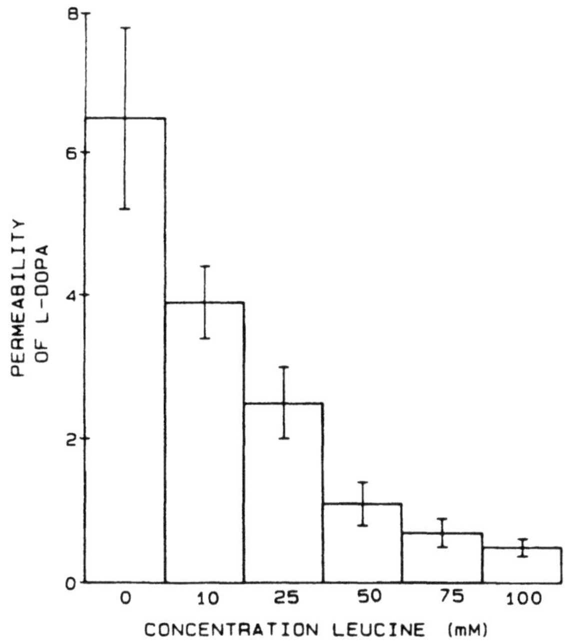

Fig. 7. Plot of the effect of the inhibitor L-leucine on the wall permeability of L-dopa. The concentration of L-dopa remained constant at 0.1 m*M* whereas the concentration of L-leucine varied from 0 to 100 m*M*. (Adapted from SINKO et al. 1987)

Mutual absorption inhibition among amino β-lactam antibiotics was also hypothesized as a result of competition for the small peptide transporter (ISEKI et al. 1984), but conflicting results were obtained from rat perfusion studies. The absorption of amoxicillin was significantly inhibited by cyclacillin, cephadrine and cephalexin, but no effect of these compounds on ampicillin absorption was observed. Furthermore, cephalexin absorption was reduced by L-carnosine, but not by glycylglycine, and cephadrine uptake was not affected by any of the dipeptides investigated. While there appears to be considerable evidence that β-lactam antibiotics are transported by the small peptide carrier, there is still no general agreement about the absolute number of transporters and their substrate selectivity (WALTER et al., accepted). Therapeutically significant interactions of drugs that are primarily absorbed by this carrier system have not been demonstrated to date, but further elucidation of the substrate specificity and selectivity of this transport pathway(s) is required.

A clear inhibition of the absorption of the ACE inhibitor captopril by various compounds including dipeptides and cephadrine was evident from in situ perfusion studies in rats (Table 3) (HU and AMIDON 1988). However, a clinical study in humans failed to show a significant effect of cephradine coadministration on captopril bioavailability (Yee and Amidon, unpublished

Table 3. Effect of various compounds on captopril permeability. (After HU and AMIDON 1988)

Compound coperfused with captopril (1 m*M*)	Pw* (SEM)	*P* value (to A)[a]	*P* value (to B)	*P* value (to C)
A = none	2.63 (0.1)	–	–[b]	–
B = 5 m*M* Pro-Phe	2.22 (0.34)	–	–	–
C = NAP and NAC (50 m*M*)	2.19 (0.43)	–	–	–
D = 85 m*M* Gly-Gly	1.10 (0.27)	0.005	0.1	0.1
E = 5 m*M* 2,4-DNP	0.64 (0.14)	0.001	0.025	0.025
F = 10 m*M* cephadrine	0.45 (0.15)	0.001	0.025	0.05
G = no sodium	0.98 (0.18)	0.005	0.05	0.01
H = 15 m*M* Gly-Pro	0.65 (0.22)	0.001	0.025	0.05
I = peptide mixture[c]	0.58 (0.09)	0.001	0.025	0.025

Pw*, dimensionless intestinal wall permeability.
[a] Two-sample *t*-test performed comparing condition A with other conditions (B, C, etc.): $P < 0.05$, significant; $P < 0.01$, very significant; $P > 0.05$, not significant.
[b] Not statistically different.
[c] A mixture of 2 m*M* each of Gly-Pro, Pro-Phe, Asp-Phe and Pro-Tyr.

results). Since there is a significant passive component in the absorption of captopril, conclusions with regard to clinically important interactions between drugs transported by the dipeptide carrier must be tentative.

P-glycoprotein (Gp 170) is a glycoprotein coded by the multidrug resistance gene (*MDR*), which was first described in cancer tissue, where it served as an efflux transporter for exogenous compounds including chemotherapeutics, leading to drug resistance and therapy failures. Recently the existence of P-glycoprotein was also demonstrated in normal tissue such as the blood-brain barrier, the kidneys, the liver and the jejunum (CORNWELL 1991). Furthermore, it was suggested that the transporter might protect against exogenous compounds and contribute to the rarity of small intestinal cancer in humans, a hypothesis that needs further investigation (HSING et al. 1992).

A variety of therapeutic drugs are substrates for this secretory carrier, with the chemotherapeutics vinblastine and vincristine and the calcium channel blocker verapamil being the most intensely studied drugs (INCE et al. 1988). Early work suggested a competitive interaction at the transport protein, leading to an increased antitumor efficacy of vincristine in the presence of verapamil. More recently, HUNTER et al. (1993) demonstrated a marked increase of vinblastine absorption after simultaneous administration of verapamil in Caco-2 cells. The significant efflux of vinblastine from the basolateral to the apical side of the cell model could be inhibited by verapamil, whereas the almost nondetectable flux of vinblastine from the apical to the basolateral side was significantly increased after addition of the calcium channel blocker.

It has been hypothesized that drugs other than chemotherapeutics might be transported by P-glycoprotein, consequently offering an alternative

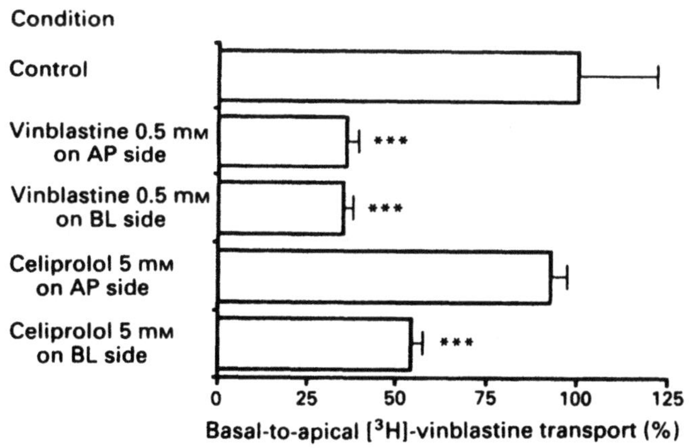

Fig. 8. Effect of unlabeled celiprolol and vinblastine on basal-to-apical transport of [^{3}H]-vinblastine in Caco-2 cell monolayers .The basal-to-apical transport of 10 n*M* [^{3}H]-vinblastine was determined after 30 min in the presence of excess unlabeled celiprolol and vinblastine on the basal (*BL*) or apical (*AP*) side. Data are presented as a percentage of control transport. *Each bar* represents mean ± SD; $n = 4$. Significant differences: ***$P < 0.001$. (Adapted from KARLSON et al. 1993)

explanation for the frequently observed nonlinear oral absorption kinetics of drugs. Examples of such drugs include celiprolol, pafenolol and talinolol, three hydrophilic β-blockers that exhibit a dose-dependent increase in their oral pharmacokinetics. Celiprolol was shown to be secreted into the intestinal lumen in a time-, temperature- and pH-dependent manner in rats (KUO et al. 1994). Studies in Caco-2 cells indicated that celiprolol secretion could be inhibited by vinblastine and verapamil, further supporting the involvement of P-glycoprotein in the secretory mechanism (Fig. 8) (KARLSON et al. 1993). For pafenolol, significant intestinal secretion in rats has been reported as a possible explanation for the nonlinear oral bioavailability (LENNERNAS and REGARDH 1993). A very recent study with talinolol in humans as well as inhibition studies in Caco-2 cells also suggests that the variable and dose-dependent oral absorption of this drug may be due to a potential interaction with P-glycoprotein (WETTERICH et al. 1995). However, since the Caco-2 cell line is a transformed colon cell line, the extent to which this transport protein might be involved in clinically significant drug interactions in vivo remains to be further investigated.

E. Interactions as a Function of Intestinal Metabolism

Numerous in vitro and in vivo studies have demonstrated that many drugs and xenobiotics are metabolized by the gastrointestinal tract, either in the gut lumen or in the intestinal wall, resulting in high inter- and also intraindividual variations in oral bioavailability (ILETT et al. 1990). Metabolic enzymes in the lumen originate from exocrine glands and are predominantly active in the upper parts of the intestine (RENWICK 1982; KRISHNAMOORTHY and MITRA

1995). Enzymatic reactions as a function of bacterial organisms include degradation, reduction and hydrolysis and take place in the terminal portions of the gut. A variety of enzymatic reactions also occur in the intestinal mucosal cell. These include phase I biotransformations such as oxidation, reduction and hydrolysis and several phase II conjugation reactions, with glucuronidation and sulfation representing the most prominent ones. While the concentrations of phase I enzymes were generally assumed to be relatively low, conjugative phase II enzyme activity is comparable to that in the liver (TAM 1993). Some examples with regard to drug interactions can be found in the early literature, such as the undesired increase in ethinylestradiol plasma levels after concomitant administration of either acetaminophen or ascorbic acid, two compounds that compete with ethinylestradiol for the limited sulfate pool (BACK and ROGERS 1987). However, the extent of interactions in the gut wall might have been overestimated in these studies, since liver contribution was not taken into account. These considerations might also be an explanation for interactions reported for morphine and fenoterol, as well as morphine and orciprenaline (KOSTER et al. 1985a,b).

The contribution of intestinal phase I metabolism was regarded as unlikely to be important until recently, since the ratio of hepatic to intestinal cytochrome P450 concentration has been reported to be approximately 20 (TAM 1993). However, recent reports focus on phase I biotransformations, predominantly via cytochrome P450, in the intestinal wall and provide a more quantitative and mechanistic understanding of these metabolic reactions. WATKINS et al. (1987) were the first to report that a major cytochrome P450, CYP3A, which appears to account for approximately 70% of the total P450 present in human enterocytes, might be present in a higher concentration in the intestine than in the liver. This high level of a cytochrome P450 in the intestine may be even more important considering that more than 50% of drugs used in humans might be substrates for this enzyme (WACHER et al. 1995). Consequently, metabolic interactions could be very likely. In subsequent studies the same authors reported that cyclosporin, which was assumed to be metabolized exclusively in the liver, undergoes at least 50% metabolism in the intestinal wall as measured from portal and femoral blood in anhepatic liver transplant patients (Table 4) (KOLARS et al. 1991). Considering that not all metabolites of cyclosporin were measured in this study, the contribution of intestinal metabolism might be even higher. Furthermore, intestinal CYP3A has been shown to be inducible by rifampicin, leading to an abolished oral bioavailability of cyclosporin in another transplant patient studied (KOLARS et al. 1992). GOMEZ et al. (1995) confirmed these findings in a study where ketoconazole, a CYP3A inhibitor, induced an increase of cyclosporine bioavailability from 22.4% to 56.4%. Grapefruit juice and its possibly active component, naringin, have been used for investigating interactions of drugs metabolized by CYP3A. Felodipine area under the plasma concentration-time curve (AUC) and peak plasma concentrations increased about 200% and 170%, respectively, when coadministered with grapefruit juice, whereas

Table 4. Cyclosporin and metabolite (M1 and M21) concentrations in portal vein (PV) and femoral artery (FA) after duodenal instillation during anhepatic phase. (Adapted from KOLARS et al. 1991)

Time after instillation (min)	Cyclosporin (ng/ml)		M1 + M21 (ng/ml)	
	PV	FA	PV	FA
Patient A[a]				
0	0	0	0	0
30	56	37	12	16
60	124	92	42	32
Pateint B				
0	0	0	0	0
40	14	29	0	0
120	22	18	23	10

[a] In patient A the cyclosporin solution was mixed with 12 ml bile before instillation to facilitate absorption.

naringin produced much less of an effect in this study (Fig. 9) (BAILEY et al. 1993). AUC ratios between dehydrofelodipine (the main metabolite of felodipine) and felodipine were reduced, suggesting inhibition of presystemic metabolism. The relative contribution of intestinal metabolism in contrast to hepatic metabolism remains to be determined for felodipine.

For cyclosporine (DUCHARME et al. 1995) and midazolam (KUPFERSCHMIDT et al. 1995) on the other hand, it was demonstrated that grapefruit juice had no effect on pharmacokinetic parameters such as AUC, systemic clearance and elimination half-life when the drugs were administered intravenously. However, after oral administration of cyclosporine as well as midazolam together with grapefruit juice an increase in bioavailability of 50% and 30%, respectively, has been observed. The lack of effects on systemic clearance after intravenous dosing suggests that oral bioavailability is improved by either increased absorption or inhibition of gut-wall first-pass metabolism, with the latter being more likely considering previous reports (KOLARS et al. 1991, 1992). However, one cannot completely rule out the effect on gastrointestinal transit and increased solubilization due to bilary secretions as also being partial explanations of these observations.

Clinically significant and severe interactions between azole antifungal drugs and terfenadine, both substrates for CYP3A, have been reported (HONIG et al. 1993; POHJOLA-SINTONEN et al. 1993). The decreased metabolism of terfenadine to its active acid metabolite in the presence of either itraconazole or ketoconazole results in increased plasma levels of unmetabolized terfenadine. Terfenadine itself prolongs the QT interval and increases the risk of torsade de pointes ventricular tachycardia. This metabolic interaction might already occur on the intestinal level, but again the percent-

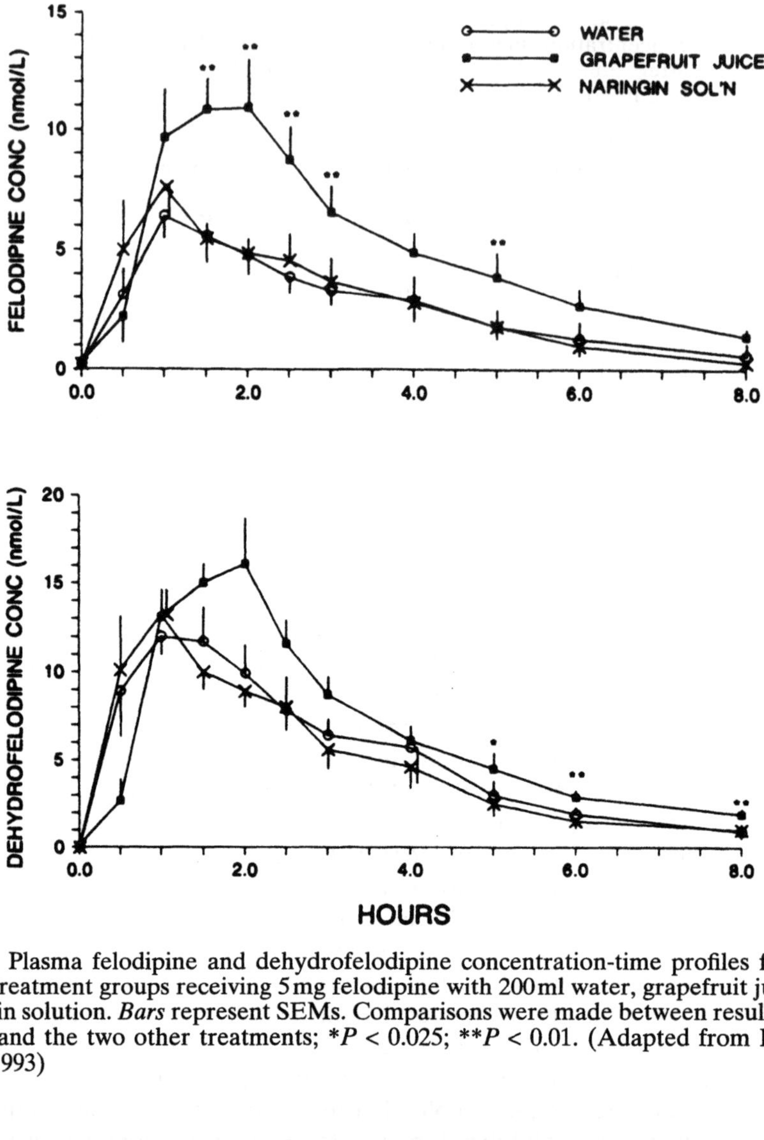

Fig. 9. Plasma felodipine and dehydrofelodipine concentration-time profiles for the three treatment groups receiving 5 mg felodipine with 200 ml water, grapefruit juice or naringin solution. *Bars* represent SEMs. Comparisons were made between results with water and the two other treatments; $*P < 0.025$; $**P < 0.01$. (Adapted from BAILEY et al. 1993)

age of intestinal metabolism in contrast to hepatic metabolism remains to be determined. These combined findings suggest that this intestinal metabolic pathway might be relevant for the low and highly variable oral bioavailability of a large number of substrates of cytochrome CYP3A, including erythromycin, lignocaine and estrogen.

There are few examples in the literature that illustrate presystemic removal of drugs after oral administration as a function of bacterial metabolism in the distal intestine. The clinically most relevant interaction resulting from decreased bacterial metabolism is the coadministration of estrogen and antibiotics. β-Glucuronidase, β-glycosidase and sulfatase in intestinal microorganisms are responsible for hydrolyzing biliary-excreted estrogen conjugates and allow enterohepatic circulation. Failure of oral contraceptives has been associated with a significantly lower quantity of microorganisms in the lower gastrointestinal tract after antibiotic treatment (BACK et al. 1978). In contrast, digoxin has been demonstrated to be inactivated via enzymatic reduction by the intestinal flora. This process is inhibited by antibiotic therapy, resulting in an increased oral absorption (DOBKIN et al. 1983). It could be demonstrated that a 5-day coadministration of erythromycin or tetracycline markedly reduced the inactivation of digoxin by gut bacteria (LINDENBAUM et al. 1981).

F. Altered Absorption as a Result of Drug-Induced Mucosal Changes

Since administration of cytostatic drugs often results in pronounced changes in the intestinal mucosa, the bioavailability of drugs may be affected by such concomitant therapy. Bioavailability may be reduced in particular in the case of lipophilic drugs, which are mainly passively absorbed. Such an effect is of primary relevance for drugs with a small therapeutic range, e.g., digoxin. It has been shown that coadministration of cytostatic drugs (combination of cyclophosphamide, oncovin and procarbazine or cyclophosphamide, oncovin and prednisone) diminished steady-state plasma concentration of digoxin in patients with a daily oral dosage of 0.3 mg β-acetyldigoxin (KUHLMANN et al. 1981). In addition after single-dose administration of digoxin to patients under cytostatic therapy the maximum plasma concentration was significantly delayed. These results could be explained by changes in membrane permeability and/or an effect on overall gastrointestinal transit.

References

Adjepon-Yamoah KK, Scott DB, Prescott LF (1973) Impaired absorption and metabolism of oral lignocaine in patients undergoing laparoscopy. Br J Anaesth 45:143–147

Albert KS, Ayres JW, Disanto AR, Weidler DJ, Sakmar E, Halmark MR, Stll RG, DeSante KA, Wagner JG (1978a) Influence of kaolin-pectin suspension of digoxin bioavailability. J Pharm Sci 67:1582–1585

Albert KS, DeSante KA, Welch RD, DiSanto AR (1978b) Pharmacokinetic evaluation of a drug interaction between kaolin, pectin and clindamycin. J Pharm Sci 67:1579–1582

Algeri S, Cerletti C, Curcio M, Morselli PL, Bonollo L, Guniva G, Minazzi M, Minoli G (1976) Effect of anticholinergic drugs on gastro-intestinal absorption of L-dopa in rats and in man. Eur J Pharmacol 35:293–299

Ambre JJ, Fischer LJ (1972) Effect of coadministration of aluminum and magnesium hydroxides on absorption of anticoagulants in man. Clin Pharmacol Ther 14(2):231–237

Amidon GL, Lennernas H, Shah VP, Crison JR (1995) A theoretical basis for a biopharmaceutics drug classification: the correlation of in-vitro drug product dissolution and in-vivo bioavailability. Pharmacol Res 12(3):413–420

Back DJ, Rogers SM (1987) First-pass metabolism by gastrointestinal mucosa. Aliment Pharmacol Ther 1:339–357
Back DJ, Breckenridge MC, Crawford FE, Challiner M, Orme ML (1978) The effect of antibiotics on the enterohepatic circulation of ethinyl-estradiol and norethisterone in rats. J Steroid Biochem 9:527–531
Bailey DG, Arnold JM, Munoz C, Spence JD (1993) Grapefruit juice-felodipine interaction: mechanism, predictability and effect of naringin. Clin Pharmacol Ther 53(6):637–642
Barboriak JJ, Meade RC (1970) Effect of alcohol on gastric emptying in man. Am J Clin Nutr 23:1151–1153
Barlow OW, Climenko DR (1941) Studies on the pharmacology of sulfapyridine and sulfathiazole. JAMA 116:282–286
Barr WH, Adir J, Garrettson L (1971) Decrease of tetracycline absorption in man by sodium bicarbonate. Clin Pharmacol Ther 12:779–784
Barrowman JA, D'Mello A, Herxheimer A (1973) A single dose of neomycin impairs absorption of vitamin A (retinoil) in man. Eur J Clin Pharmacol 5:199–202
Beerman B, Groschinsky-Grind M (1978a) Enhancement of the gastrointestinal absorption of hydrochlorothiazide by propantheline. Eur J Clin Pharmacol 13:385–387
Blume H (1990) Einfluss von Mahlzeiten auf die Bioverfuegbarkeit von Retardarzneimitteln: sind "food-studies" erforderlich? In: Blume H (ed) Bioverfuegbarkeit und Bioaequivalenz von Retardarzneimitteln; proceedings of the 8th ZL-expert meeting, 20 March 1990, Govi, Frankfurt
Blume H, Siewert M, Eichelbaum M (1988) Vortrag anlaesslich des workshops "In-vitro and in-vivo testing and correlation of controlled/modified release dosage forms". 14–16 Dec 1988, Washington DC, USA
Boman G, Hanngren A, Malmborg A-S, Borgå O, Sjöqyist F (1971) Drug interaction: decreased serum concentrations of rifampicin when given with P.A.S. Lancet II:800
Boman G, Lundgren P, Stjernström G (1975) Mechanism of the inhibitory effect of PAS granules on the absorption of rifampicin: adsorption of rifampicin by an excipient, bentonite. Eur J Clin Pharmacol 8:293–299
Bradbrook ID, Morrison PJ, Rogers HJ (1978) The effect of bromelain on the absorption of orally administered tetracycline. Br J Clin Pharmacol 6:552–554
Brown DD, Juhl RP (1976) Decreased bioavailability of digoxin due to antacids and kaolin-pectin. N Engl J Med 295:1034–1037
Caldwell JH, Bush CA, Greenberger NJ (1991) Interruption of the enterohepatic circulation of digitoxin by cholestyramine. II. Effect on metabolic disposition of tritium-labeled digitoxin and cardia systolic intervals in man. J Clin Invest 50:2638–2644
Cheng SH, White A (1962) Effect of orally administered neomycin on the absorption of penicillin V. N Engl J Med 267:1296–1297
Chin T-F, Lach JL (1975) Drug diffusion and bioavailability: tertracycline-metallic chelation. Am J Hosp Pharm 32:625–629
Cornwell MM (1991) Molecular biology of P-glycoprotein. In: Ozols R (ed) Molecular and clinical advances in anticancer drug. Kluwer Academic, Boston, pp 37–56
Crammer JL, Rosser RM, Crane G (1974) Blood levels and management of lithium treatment. Br Med J 3:650–654
Davis SS, Hardy JG, Fara JW (1986) Transit of pharmaceutical dosage forms through the small intestine. Gut 27:886–892
Ditchburn RW, Pritchard RM (1956) Avidity of the tetracyclines for the cations of metals. Nature 177:433–434
Dobkin JF, Saha JR, Butler VPJR, Neu HC, Lindenbaum J (1983) Digoxin-inactivating bacteria: identification in human gut flora. Science 220:325–327
Dressman JB, Fleisher D, Amidon GL (1984) Physicochemical model for dose-dependent drug absorption. J Pharm Sci 73(9):1274–1279

Ducharme MP, Warbasse LH, Edwards DJ (1995) Disposition of intravenous and oral cyclosporine after administration with grapefruit juice. Clin Pharmacol Ther 57(5):485–491
Faloon WW, Chodos RB (1969) Vitamin B_{12} absorption studies using colchicine, neomycin and continuous $^{57}CoB_{12}$ administration. Gastoenterology 56:1251
Fann WE, Davis JM, Janowsky DS, Sekerke HJ, Schmidt DM (1973) Chlorpromazine: effects of antacids on its gastrointestinal absorption. J Clin Pharmacol 13:388–390
Feldman S, Carlstedt BC (1974) Effect of antacid on absorption of enteric-coated aspirin. JAMA 227:660–661
Fermaglich J, O'Doherty SS (1972) Effect of gastric motility on levodopa. Dis Nervous Syst 33:624–625
Fleisher D (1995) Gastrointestinal transport of peptides: experimental systems. In: Taylor MD, Amidon GL (eds) Peptide-based drug design. American Chemical Society, Washington DC, pp 501–525
Forrest FM, Forrest IS, Serra MT (1970) Modification of chlorpromazine metabolism by some other drugs frequently administered to psychiatric patients. Biol Psychiatr 2:53–58
Gibbons DO, Lant AF (1975) Effects of intravenous and oral propantheline and metoclopramide on ethanol absorption. Clin Pharmacol Ther 17(5):578–584
Gomez DY, Wacher VJ, Tomlanovich SJ, Hebert MF, Benet LZ (1995) Clin Pharmacol Ther (in press)
Gothoni G, Neuvonen PJ, Mattila M, Hackman R (1972) Iron tetracycline interactions: effect of time interval between the drugs. Acta Med Scand 191:409–411
Grasela TH, Schentag JJ, Sedman AJ, Wilton JH, Thomas DJ, Schulz RW, Lebsack ME, Kinkel AW (1989) Inhibition of enoxacin absorption by antacids or ranitidine. Antimicrob Agents Chemother 33(5):615–617
Greenblatt DJ, Allen MD, MacLaughlin DS, Harmatz JS, Shader RI (1978) Diazepam absorption: effect of antacids and food. Clin Pharmacol Ther 24:600–609
Greenspan R, MacLean H, Milzer A, Necheles H (1951) Antacids and aureomycin. Am J Dig Dis 18:35037
Harcourt RS, Hamburger M (1975) The effect of magnesium sulfate in lowering tetracycline blood levels. J Lab Clin Med 50:464–468
Hayes SL, Pablo G, Radomski T, Palmer RF (1977) Ethanol and oral diazepam absorption. N Engl J Med 296:186–189
Heinrich HC, Oppitz KH, Gabbe EE (1974) Hemmung der Eisenabsorption beim Menschen durch Tetracyclin. Klin Wochenschr 52:493–498
Ho NFH, Merkle HP, Higuchi WI (1983) Quantitative, mechanistic and physiologically realistic approach to the biopharmaceutical design of oral drug delivery systems. Drug Dev Ind Pharm 9(7):1111–1184
Hoffman WS, Dyniewicz HA (1945) The effect of alumina gel upon the absorption of vitamin A from the intestinal tract. Gastoenterology 5:512–522
Honig PK, Wortham DC, Zamani K, Conner DP, Mullin JC, Cantilena LR (1993) Terfenadine-ketoconazole interaction. Pharmacokinetic and electrocardiographic consequences. JAMA 269(12):1513–1518
Hsing S, Gatmaitan Z, Arias IM (1992) The function of Gp 170, the multidrug-resistance gene product, in the brush border of rat intestinal mucosa. Gastroenterology 102:879–885
Hu M, Amidon GL (1988) Passive and carrier-mediated intestinal absorption components of captopril. J Pharm Sci 77(12):1007–1011
Hunter J, Hirst BH, Simmons NL (1993) Drug absorption limited by P-glycoprotein-mediated secretory drug transport in human intestinal epithelial Caco-2 cell layers. Pharmacol Res 10(5):743–749
Hurwitz A (1971) The effects of antacids on gastrointestinal drug absorption. II. Effect on sulfadiazine and quinine. J Pharmacol Exp Ther 179(3):485–489
Hurwitz A, Schlozman DL (1974) Effects of antacids on gastrointestinal absorption of isoniazid in rat and man. Am Rev Resp Dis 109:41–47

Ilett KF, Tee LBG, Reeves PT, Minchin RF (1990) Metabolism of drugs and other xenobiotics in the gut lumen and wall. Pharmacol Ther 46:67–93

Ince P, Appleton DR, Finney KJ, Moorghen M, Sunter JP, Watson AJ (1988) Verapamil sensitises normal and neoplastic rodent intestinal tissue to the stathmokinetic effect of vincristine in vivo. Br J Cancer 57:348–352

Iseki K, Iemura A, Sato H, Sunada K, Miyazaki K, Arita T (1984) Intestinal absorption of several β-lactam antibiotics. V. Effect of amino β-lactam analogues and dipeptides on the absorption of amino β-lactam antibiotics. J Pharm Dyn 7:768–775

Johnson BF, O'Grady J, Bye C (1978) The influence of digoxin particle size on absorption of digoxin and the effect of propantheline and metoclopramide. Br J Clin Pharmacol 5:465–467

Karlson J, Kuo SM, Ziemniak J, Artursson P (1993) Transport of celiprolol across human intestinal epithelial (Caco-2) cells: mediation of secretion by multiple transporters including P-glycoprotein. Br J Pharmacol 110(3):1009–1016

Kauffman RE, Azarnoff DL (1973) Effect of colestiopol on gastrointestinal absorption of chlorothiazide in man. Clin Pharmacol Ther 14:886–889

Kimura H, Inui K-I, Sezaki H (1985) Differences in effects on drug absorption between dihydroxy and trihydroxy bile salts. J Pharmacobio Dyn 8:578–585

Kolars JC, Awni WM, Merion RM, Watkins PB (1991) First-pass metabolism of cyclosporin by the gut. Lancet 338:1488–1490

Kolars JC, Schmiedlin-Ren P, Schuetz JD, Fang C, Watkins PB (1992) Identification of rifampicin-inducible P450IIIA4 (CYP3A4) in human small bowel enterocytes. J Clin Invest 90:1871–1878

Koster AS, Frankhiujzen-Sierevogel AC, Noordhoek J (1985a) Glucuronidation of morphine and six beta 2-sympathomimetics in isolated rat intestinal epithelial cells. Drug Metabol. Dispos. 13:232–238

Koster AS, Hofman GA, Frankhiujzen-Sierevogel AC, Noordhoek J (1985b) Presystemic and systemic intestinal metabolism of fenoterol in the conscious rat. Drug Metabol Dispos 13:464–470

Krishnamoorthy R, Mitra A (1995) Peptide metabolism by gastric, pancreatic, and lysosomal proteinases. In: Taylor MD, Amidon GL (eds) Peptide-based drug design. American Chemical Society, Washington DC, pp 501–525

Kuhlmann J, Zilly W, Wilke J (1981) Effects of cytostatic drugs on plasma level and renal excretion of beta-acetyldigoxin. Clin Pharmacol Ther 30:518–527

Kuo SM, Whitby BR, Artursson P, Ziemniak JA (1994) The contribution of intestinal secretion to the dose-dependent absorption of celiprolol. Pharmcol Res 11(5):648–653

Kupferschmidt HHT, Ha HR, Ziegler WH, Meier PJ, Kraehenbuehl S (1995) Interaction between grapefruit juice and midazolam in humans. Clin Pharmacol Ther 58(1):20–28

Lavigne J-G, Marchand C (1973) Inhibition of gastointestinal absorption of p-aminosalicylate (PAS) in rats and humans by diphenhydramine. Clin Pharmacol Ther 14:404–411

Lebsack ME, Nix D, Ryeerson B, Toothaker RD, Welage L, Norman AM, Schentag JJ, Sedman AJ (1992) Effects of gastric acidity on enoxacin absorption. Clin Pharmacol Ther 52(3):252–256

Lennernas H (1995) Does fluid flow across the intestinal mucosa affect quantitative oral drug absorption? Is there time for a reevaluation? Pharmacol Res 12

Lennernas H, Regardh CG (1993) Dose-dependent intestinal absorption and significant intestinal secretion (exsorption) of the beta-blocker pafenolol in the rat. Pharmacol Res 10(5):727–731

Lennernas H, Nilsson D, Aquilonius SM, Ahrenstedt O, Knutson L, Paalzow LK (1993) The effects of L-leucine on the absorption of L-dopa, studied by regional jejunal perfusion in man. Br J Clin Pharmacol 35(3):243–250

Levy G, Rao BK (1972) Enhanced intestinal absorption of riboflavin from sodium alginate solution in man. J Pharm Sci 61:279–280

Levy G, Tsuchiya T (1972) Effect of activated charcoal on aspirin absorption in man, part I. Clin Pharmacol Ther 13:317–322
Lindenbaum J, Maulitz RM, Saha JR, Shea N, Butler VP (1972) Impairment of digoxin absorption by neomycin. Clin Res 20:410
Lindenbaum J, Rund DG, Butler VP, Tse-Eng D, Saha JR (1981) Inactivation of digoxin by the gut flora: reversal by antibiotic therapy. N Engl J Med 305:789–794
Lode H (1988) Drug interactions with quinolones. Rev Infect Dis 10 [Suppl 1]:S132–S136
Lucarotti RL, Colaizzi JL, Barry H, Poust RI (1972) Enhanced pseudoephedrine absorption by concurrent administration of aluminium hydroxide gel in humans. J Pharm Sci 61:903–905
Mailliard ME, Stevens BR, Mann GE (1995) Amino acid transport by small intestinal, hepatic and pancreatic epithelia. Gastroenterology 108:888–910
Manninen V, Apajalahti A, Melin J, Karesoja M (1973) Altered absorption of digoxin in patients given propantheline and metoclopramide. Lancet I(800):398–400
Muranishi S (1990) Absorption enhancers. Crit Rev Ther Drug Carrier Syst 7(1):1–33
Naggar VF, Khalil SA (1979) Effect of magnesium trisilicate on nitrofurantoin absorption. Clin Pharmacol Ther 25:857–863
Neuvonen PJ (1976) Interactions with the absorption of tetracyclines. Drugs 11:45–54
Neuvonen PJ, Elonen E (1980) Effect of activated charcoal on absorption and elimination of phenobarbitone, carbamazepine and phenylbutazone in man. Eur J Clin Pharmacol 17:51–57
Neuvonen PJ, Turakka H (1974) Inhibitory effect of various iron salts on the absorption of tetracycline in man. Eur J Clin Pharmacol. 7:357–360
Neuvonen PJ, Gothini G, Hackma R, af Björksten K (1970) Interference of iron with the absorption of tetracyclines in man. Br Med J 4:532–534
Neuvonen PJ, Pentikäinen PJ, Gothoni G (1975) Inhibition of iron absorption by tetracycline. Br J Clin Pharmacol 2:94–96
Neuvonen PJ, Elfing SM, Elonen E (1978) Reduction of absorption of digoxin, phenytoin and aspirin by activated charcoal in man. Eur J Clin Pharmacol 13:213–218
Ni F, Ho NFH, Fox JL, Leuenberger H, Higuchi WI (1980) Theoretical model studies of intestinal drug absorption V. Non-steady-state fluid flow and absorption. Int J Pharmacol 5:33–47
Nimmo J, Heading RC, Tothill P, Prescott LF (1973) Pharmacological modification of gastric emptying: effects of propantheline and metoclopramide on paracetamol absorption. Br Med J 1:587–589
Nix DE, Watson WA, Lener ME, Frost RW, Krol G, Goldstein H, Lettieri J, Schentag J (1989) Effects of aluminum and magnesium antacids and ranitidine on the absorption of ciprofloxancin. Clin Pharmacol Ther 46(6):700–705
Nix DE, Wilton J, Ronald B, Distlerrath L, Williams VC, Norman A (1990) Inhibition of norfloxacin absorption by antacids. Antimicrob Agents Chemother 34:432–435
Northcutt RG, Stiel JN, Hollifield JW, Stant EG Jr (1969) The influence of cholestyramine on thyroxine absorption. JAMA 208(10):1857–1861
Ohnhaus EE (1980) The effect of antacid and food on the absorption of proquazone (Biarison) in man. Int J Clin Pharmacol 18:136–139
Osman MA, Welling PG (1983) Influence of propantheline and metoclopramide on the bioavailability of chlorothiazide. Curr Ther Res 34(2):404–408
Palva IP, Rytkönen U, Alatulkkila M, Palva HLA (1972) Drug-induced malabsorption of vitamin B_{12}. Scand J Haematol 9:5–7
Parsons RL, Paddock GM (1975) Absorption of two antibacterial drugs, cephalexin and co-trimoxazole, in malabsorption syndromes. J Antimicrob Chemother I [Suppl]:59–67
Penttilà O, Hurme H, Neuvonen PJ (1975) Effect of zinc sulphate on the absorption of tetracycline and doxycycline in man. Eur J Clin Pharmacol 9:131–134

Peterson OL, Finland M, Ballou AN (1942) The effect of food and alkali on the absorption and excretion of sulfonamide drugs after oral and duodenal administration. Am J Med Sci 204:581–588

Pohjola-Sintonen S, Viitasalo M, Toivonen L, Neuvonen P (1993) Itraconazole prevents terfenadine metabolism and increases risk of torsade de pointes ventricular tachycardia. Eur J Clin Pharmacol 45(2):191–193

Renwick AG (1982) First-pass metabolism within the lumen of the gastrointestinal tract. In: George et al (eds) Clinical pharmacology and therapeutics: presystemic drug elimination. Butterworth, London, pp 3–28

Resetarits D, Bates TR (1979) Apparent dose-dependent absorption of chlorothiazide in dogs. J Pharmacol Biopharmacol 7(5):463–470

Rietbrock N, Kuhlmann J, Leopold G (1983) Factors influencing the enteral absorption of drugs. Inn Med 10:112–120

Rivera-Calimlim L, Dujovne CA, Morgan JP, Lasagna L, Bianchine JR (1970a) L-dopa absorption and metabolism by the human stomach. Pharmakologist 12:269

Rivera-Calimlim L, Dujovne CA, Morgan JP, Lasagna L, Bianchine JR (1970b) L-dopa treatment failure: explanation and correction. Br Med J 4:93

Rivera-Calimlim L, Morgan JP, Dujovne CA, Bianchine JR, Lasagna L (1970c) L-dopa absorption and metabolism by the human stomach. J Clin Invest 49:79a

Rivera-Calimlim L, Dujovne CA, Morgan JP, Lasagna L, Bianchine JR (1971) Absorption and metabolism of L-dopa by the human stomach. Eur J Clin Invest 1:313–320

Rivera-Calimlim L, Castañeda L, Lasagna L (1973) Effects of mode of management on plasma chlorpromazine in psychiatric patients. Clin Pharmacol Ther 14:978–985

Robinson DS, Benjamin DM, McCormack JJ (1970) Interaction of warfarin and nonsystemic gastrointestinal drugs. Clin Pharmacol Ther 12(3):491–495

Schreiner J, Altemeier WA (1962) Experimental study of factors inhibiting absorption and effective therapeutic levels of declomycin. Surg Gynecol Obstet 114:9–14

Schuna A, Osman MA, Patel RB, Welling PG, Sundstrom WP (1983) Influence of food on the bioavailability of penicillamine. J Rheumatol 10:95–97

Seed JC, Wilson CE (1950) The effect of aluminum hydroxide on serum aureomycin concentrations after simultaneous oral administration. Bull Johns Hopkins Hosp 86:415–418

Seneca H, Peer P (1965) Enhancement of blood and urine tetracycline levels with a chymotrypsin-tetracycline preparation. J Am Geriatr Soc 13:708–717

Sinko PJ, Hu M, Amidon GL (1987) Carrier mediated transport of amino acids, small peptides, and their drug analogs. J Contr Rel 6:115–121

Sorby DL, Liu G (1966) Effects of adsorbents on drug absorption. II. Effect of an antidiarrhea mixture on promazine absorption. J Pharm Sci 55:504–510

Tam YK (1993) Individual variation in first-pass metabolism. Clin Pharmacokinet 25(4):300–328

Tsuji A (1995) Intestinal absorption of β-lactam antibiotics. In: Taylor MD, Amidon GL (eds) Peptide-based drug design. American Chemical Society, Washington DC, pp 501–525

Tyrer JH, Eadie MJ, Sutherland JM, Hooper WD (1970) Outbreak of anticonvulsant intoxication in an Australian city. Br Med J 4:271–273

Ueno K, Tanaka K, Tsujimura K, Morishima Y, Iwashige H, Yamazaki K, Nakata I (1993) Impairment of cefdinir absorption by iron ion. Clin Pharmacol Ther 54(5):473–475

Wacher VJ, Wu CY, Benet LZ (1995) Mol Carcinog (submitted)

Wagner JG (1969) Aspects of pharmacokinetics and biopharmaceutics in relation to drug activity. Am J Pharm 141:5–20

Waisbren BA, Hueckel JS (1950) Reduced absorption of aureomycin caused by aluminium hydroxide gel (Amphojel). Proc Soc Exp Biol Med 73:73–74

Walter E, Kissel T, Amidon GL The intestinal peptide carrier: a potential transport system for small peptide derivated drugs. Adv Drug Deliv Rev (accepted)

Watkins PB, Wrighton SA, Schuetz EG, Molowa DT, Guzelian PS (1987) Identification of glucocorticoid-inducible cytochromes P-450 in the intestinal mucosa of rats and man. J Clin Invest 80:1029–1036
Weinberg ED (1957) The mutual effects of antimicrobial compounds and metallic cations. Bacteriol Rev 21:46–68
Welling PG (1984) Interactions affecting drug absorption. Clin Pharmacol 9:404–434
Wetterich U, Mutschler E, Spahn-Langguth H, Langguth P (1995) Evidence for intestinal secretion of the β-adrenoceptor antagonist talinolol: data from humans and studies with Caco-2 cells. Naunyn Schmiedenbergs Arch Pharmacol 351 [Suppl]:R1
Whitehouse LW, Paul CJ, Coldwell BB (1975) Effects of ethanol on diazepam distribution in rats. Res Commun Chem Pathol Pharmacol 12:221–242
Williamson RM, Oxender DL (1995) Molecular biology of amino acid, peptide and oligopeptide transport. In: Taylor MD, Amidon GL (eds) Peptide-based drug design. American Chemical Society, Washington DC, pp 501–525
Yamaguchi T, Ikeda C, Sekine Y (1986a) Intestinal absorption of a β-adrenergic blocking agent nadolol. I. Comparison of absorption behavior of nadolol with those of other β-blocking agents in rats. Chem Pharm Bull 34(8):3362–3369
Yamaguchi T, Ikeda C, Sekine Y (1986b) Intestinal absorption of a β-adrenergic blocking agent nadolol. II. Mechanism of the inhibitory effect on the intestinal absorption of nadolol by sodium cholate in rats. Chem Pharm Bull 34(9):3836–3843
Yamaguchi T, Ikeda C, Sekine Y (1986c) Intestinal absorption of a β-adrenergic blocking agent nadolol. III. Nuclear magnetic resonance spectroscopic study on nadolol-sodium cholate micellar complex and intestinal absorption of nadolol derivatives in rats. Chem Pharm Bull 34(10):4259–4264
Yee S, Amidon GL (1995) Oral absorption of angiotensin-converting enzyme inhibitors and peptide prodrugs. In: Taylor MD, Amidon GL (eds) Peptide-based drug design. American Chemical Society, Washington DC, pp 501–525

CHAPTER 3

Drug-Food Interactions Affecting Drug Absorption

P.G. WELLING

A. Introduction

The intent of this book is to present current information on drug interactions influencing drug absorption, distribution, elimination, and activity. Consistent with this, the original plan of the authors concerned was to present a single chapter on interactions affecting drug absorption which would include drug-drug interactions and also drug-food interactions. After examining the quantity of material that has been published on these two closely related but distinct topics, it was agreed that drug-drug interactions and drug-food interactions affecting drug absorption would be better presented in two separate chapters.

This chapter then, is devoted exclusively to drug-food interactions affecting drug absorption. No attempt has been made to address the related topic of the effect of drugs on food and nutrient absorption. Material on this topic has been reviewed elsewhere (BASU 1988; ROE 1989; TROVATO et al. 1991).

While a number of studies on drug-food interactions were reported earlier, the subject did not achieve full recognition in scientific and regulatory circles until the first major review on this topic (WELLING 1977). Since that time the number of studies examining various aspects of drug-food interactions affecting drug absorption has increased dramatically and many of the results obtained from these studies have been quite spectacular, not only in the extent of some interactions but also in their unexpected and totally unpredictable nature. The topic has been reviewed extensively, not only in terms of interactions of particular drugs or families of drugs (WELLING 1989, 1993; PFEIFER 1993; SÖRGEL and KINZIG 1993; WELLING and TSE 1982; WILLIAMS et al. 1993), but also relating to clinical consequences and the management of drug-food interactions in the clinical environment (LASSWELL and LORECK 1992; NEUVONEN and KIVISTÖ 1989). The number of reviews on this topic reflects increased awareness in scientific, regulatory, and clinical circles of its importance.

The original goal of many investigators in this area, including the writer, was to establish mechanisms involved in drug-food interactions and also to use data obtained from a variety of situations to establish rules or guidelines to predict the nature and extent of interactions in terms of circulating drug or metabolites. As with so many scientific goals this has proven to be elusive.

After almost 20 years of research it is still difficult, and many times impossible, to predict the nature of drug-food interactions and the literature is replete with surprises, and unpredictable results.

Based on current knowledge, the unpredictable nature of various interactions is probably related, not only to the complex environment in which interactions occur, but also to the many different ways in which they have been examined. Some of them are summarized here.

I. Food

The type and size of meal may have a marked effect on the nature of a drug-food interaction. One might not expect that a meal prepared in England, in India, or in Japan would act similarly, nor might a breakfast compared to an evening meal. Similarly liquid meals, which are so often used in an attempt to obtain mechanistic information, might have a totally different effect on drug absorption than solid meals, which are usually more clinically relevant. Also the time interval between eating and medication will influence the nature and extent of a drug-food interaction.

II. Dosage Form

In addition to the interaction between food components or their pharmacological sequelae on the gastrointestinal (GI) tract and drug molecules, must be considered the effect of the formulation. Much of the data summarized in previous reviews, and also to be reviewed in this chapter, have demonstrated the dramatic effect that formulation may have on drug-food interactions. This has been particularly important and controversial with controlled release products. Studies on these formulations have given rise to the unfortunate term "dose dumping," often sadly misinterpreted. The preponderance of evidence shows that the more disperse a formulation, the less the effect of drug-food interactions. Thus, drugs administered in solution are likely to be much less affected by food than drugs formulated in a compressed tablet. Such is the extent of the literature on this topic that this review is restricted for the most part to material published since 1990. Material published before 1990 can be readily obtained from previous reviews or from literature cited here.

B. Influence of Food on the Gastrointestinal Tract

The possible effects that ingested food may have on GI physiology are summarized in Table 1. This subject has recently been reviewed by OOSTERHUIS and JONKMAN (1993). In the fasting state, gastric motility passes through cycles of migrating motor complexes (MMC). Each cycle lasts 2–3 h and comprises four phases, of which phase 3, the "housekeeper wave," is the strongest. Non-nutrient liquids pass through the stomach throughout the MMC but solids are

Table 1. Possible effects of ingested food on gastrointestinal tract physiology. (From WELLING 1993)

Physiological function	Effect of food	Possible effect on drug absorption
Stomach emptying rate	Decreased rate with solid meals, fats, high temperature, acids, solutions of high osmolarity. Increased rate with large fluid volumes	Absorption generally delayed, may be reduced with unstable compounds, may be increased due to drug dissolution in stomach. Absorption increased with large fluid volumes
Intestinal motility	Increased	Faster dissolution and decreased diffusional path promotes absorption. Shorter transit time may inhibit absorption
Splanchnic blood flow	Generally increased, but may be decreased by ingestion of glucose	Absorption increased with faster blood flow. Variable effects on first-pass metabolism, depending on drug characteristics
Bile secretion	Increased	Absorption may increase due to faster dissolution or may decrease due to complexation
Acid secretion	Increased	Increased absorption of basic drugs provided they are acid stable. Decreased absorption of acid-labile compounds
Enzyme secretion	Increased	Increased or decreased absorption depending on drug characteristics
Active absorption progress	Decreased	Active drug absorption reduced by competitive inhibition

moved from the stomach into the intestine mainly during phase 3. Depending on when a solid meal is ingested relative to the MMC, the gastric residence time may vary from a few minutes to 2–3h (EWE et al. 1989). Ingested food changes gastric motility to a postprandial pattern during which time gastric residence time is increased. Gastric residence time is increased by large, particularly solid meals and by chyme of low pH, high osmolality, and high fat content. Residence time is also influenced by hot meals.

While solid foods tend to delay gastric emptying, non-nutrient liquid meals, for example, a fluid volume ingested with medication, may have the opposite effect. As described previously in the review by WELLING (1977), the stomach appears to empty liquid meals into the duodenum in apparent first-order fashion. Distension of the stomach is the only natural stimulus known to increase stomach emptying and the observed faster emptying rate with increasing fluid volume can be rationalized in terms of varying tension at receptors in the stomach wall (HOPKINS 1966). The presence of ingested food also promotes gastric secretion of hydrochloric acid. Once food passes from the stomach into the small intestine it has a stimulating effect on intestinal motility, on digestive enzyme secretion, and also, particularly in the case of fatty meals, on bile secretion.

The influences of altered gastric and intestinal motility, the one decreased and the other increased, and increasing GI secretions with ingested food, might be expected to have a number of possible effects on drug absorption. All of these have been proposed, in one way or another, to explain observed results of drug-food interactions. Delayed gastric emptying will likely delay absorption of drugs that are absorbed predominantly from the small intestine, but not of drugs that are efficiently absorbed from the stomach. It may delay the absorption of acidic compounds or drugs in enteric-coated formulations by delaying drug transit from the acidic gastric contents to the relatively alkaline region of the small intestine. On the other hand, delayed gastric emptying might increase systemic availability of compounds that are slowly soluble at acidic gastric pH by permitting more material to dissolve in the stomach before passing into the small intestine. Compounds that are unstable in acidic pH are likely to be degraded as a result of prolonged residence in the stomach regardless of the extent to which pH may be lowered, or raised, by the intake of food. The latter point is important because, while solid foods stimulate acidic secretion, the buffering effect of different types of meals may minimize any change in pH and could even give rise to a transient increase in gastric pH.

Increased intestinal motility in the fed state may have the effect of increasing dissolution of solid particles and could also decrease the mean diffusional path of dissolved molecules to the intestinal mucosa. On the other hand, it could increase the transit rate of molecules from the efficient absorption region of the small intestine to the less efficient region of the large intestine, thus negatively impacting absorption efficiency.

The effect of fluid volume on GI physiology adds another dimension to the complexity of drug-food interactions. It has already been noted that large volumes of non-nutrient fluids empty from the stomach at a faster rate than small volumes. The large volume itself tends to accelerate dissolution and also to accelerate transfer of both dissolved and undissolved drug into the small intestine. The presence of large fluid volumes in the intestine may continue to increase drug absorption by providing a more liquid environment and also by a solvent drag effect from the serosal to the mucosal side of the intestinal membrane. On the other hand, and from a physicochemical perspective, a large fluid volume might be expected to delay or reduce absorption due to a reduced serosal-mucosal drug concentration gradient. In the original review on this subject (WELLING 1977) it was noted that the positive effects of large fluid volumes appear often to outweigh the negative effects, and more examples of increased drug absorption when they are administered with large fluid volumes have appeared in the literature to support this argument. The role of gastric motility in drug absorption, with particular reference to the definition of "fasting" and "non-fasting" drug administration, has recently been reviewed by WALTER-SACK (1992).

In addition to its effects on GI motility and secretion, ingested food may also affect splanchnic blood flow. However, the degree and direction of change may vary with the types of food. For example, a high-protein meal has been shown to cause a 35% increase in splanchnic blood flow, while a liquid glucose meal caused an 8% decrease. In most cases after solid meals, one would expect splanchnic blood flow to increase (McLean et al. 1978) and this should promote drug absorption via the splanchnic circulation. How much drug subsequently enters the systemic circulation is a function also of presystemic clearance and how this is affected by food-related changes in splanchnic circulation. The most frequently documented food effect has been that of increased drug systemic availability associated with reduced presystemic clearance. This has been reported for propranolol, hydralazine, metoprolol, and propafenone (Melander and McLean 1983; Axelson et al. 1987). A number of studies have investigated this phenomenon and have shown that a transient postprandial increase in splanchnic blood flow is associated with a decrease in presystemic hepatic extraction; the magnitude of this effect correlates with protein content of a meal (Liedholm et al. 1990).

Apart from its indirect and variable inference on presystemic hepatic clearance associated with altered splanchnic blood flow, food may also exert a direct influence on intrinsic drug metabolism, both presystemic and systemic. For example, it has been known for some time that oxidative metabolism may be increased by high-protein diets or diets high in cruciferous vegetables. A more recent example of a direct interaction is acute inhibition by grapefruit juice on the metabolism of dihydropyridine calcium channel blockers. Systemic availability of orally administered felodipine was increased more than twofold by coadministered grapefruit juice (Bailey et al. 1991). A similar but less pronounced effect was observed with nifedipine. A postulated mechanism for this effect is linked to possible inhibition of cytochrome P450 IIIA4 by flavonoids present in grapefruit juice, but this has yet to be confirmed.

C. Direct Effect of Food on Drug Absorption

In addition to the indirect effects that food may exert by altering GI physiology, food may also affect drug absorption by direct interaction. Some of the direct effects of food affecting drug absorption are summarized in Table 2. Food may act as a physical barrier, inhibiting drug dissolution and preventing access of dissolved drug to the mucosal surface of the GI tract. Fluid volume may present a direct physical interaction by its effect on drug or formulation dissolution rate and solubility together with the opposing effects of solvent gradient and transmucosal solvent drag. Specific ions or other substances in food may interact with drug, resulting in reduced drug absorption. Examples are chelation of tetracycline and penicillamine by metal ions and also complex formation by proteins. For the small number of drugs that are actively absorbed, although this number is likely to increase with the introduction of

Table 2. Direct drug-food interactions that may influence drug absorption. (From WELLING 1993)

Interaction	Possible effect on drug absorption
Food and food components	Absorption decreased due to chelation, absorption, adsorption, physical blockade, but may be increased due to increased solubility with some diets. Variable effects due to pH changes
Fluid volume	Absorption rate may decrease with large fluid volumes due to reduced concentration gradient, but absorption efficiency may be increased due to faster dissolution, osmotic effect, and exposure of drug molecules to greater GI surface area

more protein-derived drugs, direct competition for active carriers may occur between food protein and protein fragments and drug molecules, inevitably giving rise to decreased drug systemic availability.

Given the multifactorial nature of drug-food interactions, and the large number of variables that need to be addressed in one way or another in studies of this type, it is not surprising that variable and often conflicting results are obtained from different laboratories. In order to better understand and interpret these types of studies it is essential that study conditions be carefully controlled and completely described. The writer makes no plea for uniformity in studies, on the contrary. Only by studying interactions under widely varying conditions will it be possible to completely understand them, together with their mechanistic and clinical implications. Nonetheless, complete study details must be reported just as in the recording of a chemical or physical experiment. Different studies have different goals. For example administering a drug together with the well-known English breakfast addresses a clear clinical question, relevant at least to those who indulge in extravagances, but does not attempt to address the mechanistic question. Administering a drug together with a high-protein fixed-volume liquid meal, on the other hand, attempts to address the mechanistic question but tends to be clinically irrelevant. Which study is the most important depends on one's perspective. However, both study types are necessary in order to completely understand the phenomenon. It is interesting to note that the extent to which drug absorption may be altered by the presence of food, while in some cases may be slight or even negligible, may in other cases be far greater than would be permitted by any regulatory agency if it were considering interchangeability of generic products. The problem of drug-food interactions has been exacerbated by the introduction of controlled release oral products. These devices may serve a useful clinical role in generating desirable plasma drug profiles, possibly leading to better control of pathological conditions, and also increasing patient compliance. However, they also present a greater quantity of drug to the patient per single dosing unit than conventional dosage forms, and are designed to deliver drug at a controlled rate over a prolonged period. As these dosage forms are also given less frequently than conventional formulations, a marked effect by administered food on systemic availability may have a serious and prolonged consequence on circulating drug levels.

The writer finds it convenient in reviews of this type to divide drugs or drug formulations into those whose absorption is reduced, delayed, increased, or not affected by food. The great majority of cases fall into one or more of these four categories. However, a small number of studies have reported accelerated but not increased absorption, and this category is new. The tables summarizing here the results obtained in these various categories provide as much information as can be practically included, describing the formulation and also study details where they are available. Lack of information in some cases may be due to the brevity of information in published abstracts, rather than full papers. The format of the tables is similar to that used in the original review on this topic (WELLING 1977). Although most of the reported drug studies during the review period are, to the writer's knowledge, included, the treatment is intended to be representative rather than exhaustive.

D. Interactions Causing Reduced Drug Absorption

Studies that have demonstrated reduced drug levels and/or urinary excretion of drugs are summarized in Table 3. In this and the following tables, compounds are listed alphabetically.

Alendronate sodium is a potent biophosphonate under investigation for treatment of osteoporosis. The systemic bioavailability of alendronate averages approximately 0.75% of the administered dose, with considerable intra-, and intersubject variation (GERTZ et al. 1993). Absorption is further reduced by approximately 40% when alendronate is taken 1–2h before food, and by as much as 90% when taken up to 2h postprandially. Although the moderate decrease in absorption when alendronate is taken before meals may have marginal clinical significance, the more dramatic reduction with postprandial dosage may well be significant. This has led to a recommendation that alendronate be taken after overnight fast, at least 30min before food. No explanation was offered for reduced absorption of the already poorly absorbed alendronate, but it was noted that alendronate forms insoluble complexes with multivalent cations such as Ca^{2+} and Mg^{2+}, which may contribute.

Dramatic reduction has been reported in serum levels of the cholinesterase inhibitor ambenonium chloride by food in male and female patients with myasthenia gravis (OHTSUBO et al. 1992). Following ingestion of one or one-half a Mytelase (ambenonium chloride) 10-mg tablet, peak serum concentration (C_{max}) and the area under the serum concentration profile from zero to 3h after dosing [AUC(0–3)] were both reduced by 70% after a conventional breakfast compared to the fasting state, while the time of peak concentration (t_{max}) was moderately increased. Mean serum-time curves following single 10- and 5-mg doses of ambenonium chloride are shown in Figs. 1 and 2.

Table 3. Drug-food interactions resulting in reduced drug absorption

Drug	Dosage form	Dose	Dosage regimen	Subjects	Food details	Fluid volume	Time interval	Sampling duration	Result of food administration	Reference
Alendronate sodium	–[a]	–	SD[b]	–	–	–	0.5–1 h before, with, or up to 2 h after meal	–	Absorption reduced 40% when dosed 0.5–1 h before food, and by 80%–90% after food	Gertz et al. (1993)
Ambenonium chloride	Tablet	5 or 10 mg	SD	13 patients with myasthenia gravis 6M, 7F	Conventional breakfast	150 ml water	30 min after meal	3 h, nerum	Serum C_{max} and AUC values reduced by over 70%, T_{max} moderately increased	Ohtsubo et al. (1992)
Atenolol	Tablet, capsule with bile acids	50 mg	–	8 healthy subjects (6M, 2F)	No food	250 ml water	NR[c]	24 h plasma	C_{max} reduced 28%, AUC reduced 30% by bile acids	Barnwell et al. (1993)
Azithromycin	–	–	–	–	–	–	–	–	52% decrease in C_{max}, 43% decrease in AUC due to large meal	Drew and Gallis (1992)
Cefprozil	–	7.5 mg/kg	–	6 10- to 14-year-old children	–	–	Before/after meal	8 h serum and urine	C_{max} reduced 22%, urine excretion by 16%	Nakamura and Iwai (1992)
Ceftibuten	Capsule	250 mg	SD	–[a]	Enterued[d] Hepan ED	200 ml water	–	10 h plasma 11 h urine	Urinary ceftibuten excretion markedly decreased by both elemental liquid diets	Iseki et al. (1994)
2-Chloro-2′-deoxyadenosine	Solution	0.24 mg/kg	SD	7 male patients with leukemia	Standard breakfast[d]	–	–	24 h plasma	C_{max} reduced 40%, AUC reduced 9%	Albertioni et al. (1993)
2-Choro-2′-deoxyadenosine	–	–	SD	4 patients with chronic lymphatic leukemia	–	–	–	Plasma	C_{max} reduced 40%, systemic availability 11%. T_{max} increased by 0.8 h	Albertioni et al. (1993)
Cicaptrost	Tablet	5, 7.5, 10 mg	MD[e]	8 healthy male subjects	–	–	1 h after meals 9 a.m., 2 p.m., 7 p.m., 9 a.m.	–	Heavier meals (lunch, dinner) may attenuate antiplatelet effect	Belch et al. (1993)
Ciprofloxacin	Tablet	500 mg	SD	7 healthy volunteers 3M, 4F	Milk, yoghurt[d]	300 ml water	Immediately before milk, water, or yoghurt	24 h plasma and urine	Bioavailability reduced 30% by milk, 36% by yoghurt	Neuvonen et al. (1991)
Didanosine	Chewable tablet	300 mg	SD	10 males seropositive for HIV but free of AIDS symptoms	Standard breakfast[d]	–	1 h or 30 min before, 1 h or 2 h after meal	12 h plasma and urine	No effect when drug taken before meal. C_{max} reduced 60% and AUC 55% when dose taken 1 or 2 h after meal	Knupp et al. (1993)
Didanosine	Chewable tablet	375 mg	SD	8 healthy male subjects	Standard breakfast[d]	120 ml water	20 min after starting meal	12 h plasma and urine	C_{max} reduced 54%, AUC 47%, and urinary excretion by 50%	Shyu et al. (1991)
Dideoxycytidine	Tablet	1.5 mg	SD	20 HIV positive patients	Standard breakfast	–	With meal	–	C_{max} reduced 35%, AUC reduced 14%	Nazareno et al. (1992)
Doxazosin	Tablet	1 mg	SD	12 hypertensive subjects	Standard breakfast[d]	200 ml water	With meal	48 h plasma	Moderate reduction in C_{max} and AUC, also moderate reduction in effect on blood pressure	Conway et al. (1993)

Flecainide	–	25 mg q 6 h	–	1- to 6-day-old male infant	Milk[d]	–	–	Single samples	Serum concentrations reduced by milk feeds	Russel and Martin (1989)
Hydralazine	Solution	50 mg	SD	8 healthy subjects 4M, 4F	Standard breakfast[d], liquid nutrient bolus or infusion	100 ml water	With meal	6 h plasma	C_{max} substantially reduced, AUC moderately reduced by breakfast and bolus nutrient	Semple et al. (1991)
Levodopa, carbidopa	Controlled release	Variable	MD	12 patients (7M,5F) with idiopathic Parkinson's disease	Standard luncheon[d]	–	Together with meal	4 h plasma	Lower plasma levodopa and carbidopa levels with meal, but 3-*O*-methyldopa levels increased	Roos et al. (1993)
Metformin	Tablet	850 mg	SD	24 healthy male subjects	–	–	After meal	–	C_{max} reduced 35%, AUC reduced 20%	Brookes et al. (1991)
Methotrexate	–	2.7–18 mg/m²	SD	14 juvenile rheumatoid arthritis patients 4M, 10F	Breakfast of choice	–	Immediately after meal	6 h plasma	C_{max} reduced 38%, AUC reduced 15%	Dupuis et al. (1992)
MK-679	Tablet	500 mg	SD	Healthy male subjects	–	–	–	12 h plasma	Plasma AUC reduced 40%	Yogendran et al. (1992)
Naproxen	Controlled release tablet	750 mg	SD	12 healthy subjects, 7M, 5F	Standard breakfast[d]	200 ml orange juice or water	Immediately after meal	48 h plasma	C_{max} reduced 14%, AUC reduced 10%	Palazzini et al. (1992)
Navelbine	Capsule	80 mg/m²	SD	13 patients, 4M, 9F, with refractory	Standard meal	–	–	–	C_{max} and AUC reduced 10%	Lucas et al. (1993)
Nitrendipine	Tablet, whole or fragmented	Variable	MD	13 juvenile patents 0.5–17 years	–	≤10 ml water	–	6 or 8 h plasma	C_{max} and AUC significantly reduced and T_{max} significantly increased by food	Wells and Sinaiko (1991)
Norfloxacin	Tablet	200 mg	SD	7 healthy subjects 4M,3F	Milk, yoghurt[d]	300 ml water	Dosed immediately before milk, water, or yoghurt	24 h plasma and urine	Serum C_{max} reduced 50% by milk and yoghurt. AUC(0–24) reduced 48% by milk and 58% by yoghurt	Kivistö et al. (1992)
Paracetamol	Solution in coca cola	20 mg/kg	SD	10 healthy Thai subjects on vegetarian diet, 10 on normal diet	Daily diets[d]	400 ml water	After overnight fast	24 h plasma and urine	C_{max} reduced 25%, AUC 12%, and urinary recovery by 16% in vegetarians	Prescott et al. (1993)
Phenytoin	Suspension	360 mg/day	MD	1 67-year-old male patient	Liquid food concentrate	–	–	Single samples	Serum phenytoin levels reduced by nasogastric inquid food	Taylor et al. (1993)
Pravastatin	Tablet	20 mg	MD	74 hyperchosterolemic male patients	Standard 600- to 650-cal. meal	–	with or 1 h before meal, 4 weeks	12 h serum	C_{max} reduced 49%, AUC reduced 31%	Pan et al. (1993)
RO 42-5892	Capsule	600 mg	SD	22 subjects	Standard breakfast	–	Immediately after or 1 h after meal	24 h plasma	C_{max} reduced 59%–63% when dosed immediately after meal, and 10%–61% when dosed 1 h after meal	Weber et al. (1993)
Rufloxacin	–	800 mg	SD	10 healthy male subjects	–	–	–	96 h plasma and urine	C_{max} reduced 17%, AUC (0–96) reduced 33%. Converse increase in Ae(0–96)	Segre et al. (1992)

Table 3. *Continued*

Drug	Dosage form	Dose	Dosage regimen	Subjects	Food details	Fluid volume	Time interval	Sampling duration	Result of food administration	Reference
SK & F 106203	Tablet	150 mg	SD	12 healthy male subjects	–	–	–	–	Plasma levels reduced and delayed	NICHOLS et al. (1992)
Sotalol	–	160 mg	SD	18 healthy subjects	Standard meal	–	–	36 h plasma	Bioavailability reduced approximately 20%	HANYOK (1993)
Sulpiride	Film-coated tablet, solution	100 mg	SD	3 healthy male subjects	Standard light, medium and heavy meals[d]	100 ml water	With meal	48 h, urine	Cumulative urinary excretion reduced by 30% following both dosage forms	SHINKUMA et al. (1990)
Tacrine	Capsule	40 mg	SD	24 healthy subjects	Standard breakfast[d]	8 fluid ounces water	15 min and 2 h after starting meal	14 h plasma	C_{max} reduced 45% and AUC 23% when taken 15 min after starting meal. C_{max} reduced 35% and AUC 21% when taken 2 h after starting meal	WELTY et al. (1994)
Tetracycline	Capsule	250 mg	SD	9 healthy subjects	Standard meal and high-calcium Mexican meal[d]	200 ml water	Immediately after meal	72 h urine	Urine excretion reduced 40% by standard meal and 55% by Mexican meal	COOK et al. (1993)
Verapamil	SR tablet	240 mg	SD	9 halthy male subjects	–	–	Immediately after meal	30 h serum	48% decrease in C_{max} and 30% decrease in AUC, also altered electro cardiographic effects	HOON et al. (1992)
Zidovudine	–	100 mg, 250 mg	SD	8 male AIDS patients	2500 kJ breakfast, fat content 40 g	200 ml water	Mid-meal	6 h plasma	C_{max} and AUC reduced 64% and 29%, respectively	LOTTERER et al. (1991)
Zidovudine	Capsule	250 mg	SD	13 AIDS patients, 10M, 3F	High-fat liquid meal given as milk shake[d]	600 ml water	Immediately after meal	4 h plasma and urine	C_{max} reduced 50%, T_{max} increased approximately threefold	UNADKAT et al. (1990)
Zidovudine	Capsule	100 mg, 250 mg	SD	25 patients, 24M 3F, with HIV infection	Standard breakfast[d]	100 ml water	Immediately after meal	6 h plasma, 24 h urine	C_{max} and AUC reduced 35% at 100-mg dose, C_{max} reduced 73% and AUC 14% at 250-mg dose	RUHNKE et al. (1993)
Zidovudine	–	200 mg	SD	HIV-1-positive patients	–	–	After meal	4 h serum	Serum C_{max} reduced 50% and AUC 38%. Urinary recovery reduced by 19%. Serum levels of glucuronide also reduced but urinary recovery unchanged	UEDA et al. (1993)
882C	–	200 mg	SD	15 healthy male subjects	Standard breakfast	–	–	–	C_{max} reduced 16%, AUC reduced 10%. Food reduced variability in AUC	PECK et al. (1993)

[a] Data not available.
[b] Single dose.
[c] Not relevant.
[d] Meal described.
[e] Multiple doses.

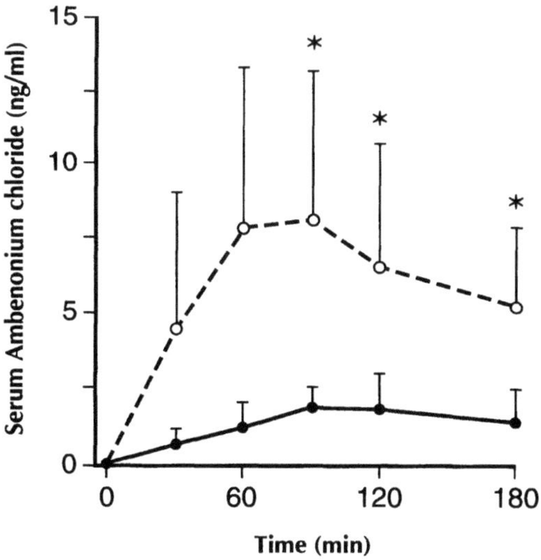

Fig. 1. Mean serum versus time concentrations (±SE) of ambenonium chloride after oral administration of 10 mg ambenonium chloride to six myasthenia gravis patients under fasting (○) and nonfasting (●) conditions. *P, <0.05; **P, <0.01. (From OHTSUBO et al. 1992)

Significant differences were observed between fed and fasted serum concentrations at 1.5, 2, and 3 h ($P < 0.05$) following the 10-mg dose. Following the 5-mg dose, significant differences were observed at 1 ($P < 0.005$), 1.5, 2, and 3 h ($P < 0.05$).

Ambenonium chloride is a quaternary ammonium compound and is unlikely to be well absorbed, even in the fasting state. Further reduction of absorption due to food could be due to a number of factors, but these were not examined in this study. Despite the rather dramatic food effect, relationships between serum drug levels and relative clinical effect (changes in muscle strength) were variable and inconclusive. It is nonetheless recommended that ambenonium chloride dosage be adjusted during non-fasting treatments when the adjusting the optimum regiment for patents with myasthenia gravis.

Previous studies with atenolol have shown that absorption is reduced by food (MELANDER et al. 1979). In this respect the very hydrophilic atenolol molecule differs from the lipophilic compounds of this class, metroprolol and propranolol, both of which undergo extensive presystemic metabolism and whose absorption is increased by food. The mechanism by which food decreases atenolol absorption was addressed in a study conducted in healthy volunteers who received single 50-mg doses of atenolol as a commercial tablet, or as capsules containing bile acids (BARNWELL et al. 1993). In this study the bile acid formula reduced mean atenolol plasma C_{max} by 28% and AUC by

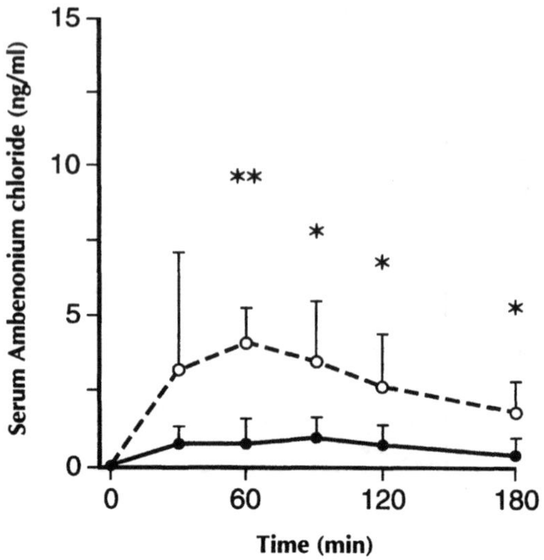

Fig. 2. Mean serum versus time concentrations (±SE) of ambenonium chloride after oral administration of 5 mg ambenonium chloride to seven myasthenia gravis patients under fasting (○) and nonfasting (●) conditions. *P, <0.05; **P, <0.01. (From OHTSUBO et al. 1992)

30%. As both formulations were administered in the fasting state it is claimed that bile acids reduce systemic availability of atenolol to a similar or greater extent to that previously reported by ingested food, although the mechanism of this effect was not postulated. Previous studies have shown that bile acids can have the effect of increasing (S.K. COLE et al. 1992) or decreasing (YAMAGUCHI et al. 1986) drug absorption. The use of different formulations in the present study, i.e., commercial tablets and contemporaneously prepared capsules containing bile acids with atenolol, may also have contributed to the observed results.

An extensive food effect reducing drug absorption is implied by summary data contained in a review article on azithromycin (DREW and GALLIS 1992). Following a 500-mg oral dose, systemic availability of this azalide antibacterial agent is approximately 30%. However, coadministration with a large meal may further reduce absorption by up to 50%. The mechanism of this interaction is unknown. This basic, highly lipid soluble and poorly water soluble compound is degraded by acid-catalyzed hydrolysis of the ether bond to the neutral cladinose sugar (FIESE and STEFFEN 1990). Regardless of the possible mechanisms involved, the manufacturers recommend that azithromycin be taken either 1 h before or 2 h after meals. Two studies have demonstrated moderately reduced absorption of new cephalosporins, one by food in children (NAKAMURA and IWAI 1992a), the other by an elemental diet (ISEKI et al. 1994). In the first study, conducted in six children, administration after a meal caused a 22% reduction in mean C_{max} values and a 16% reduction in urinary excretion of cefprozil. In the second study, simultaneous administration of an elemental diet Enterued, composed of oligopeptide (egg white hydrolysate), an elemental diet Hepan ED composed of amino acids, and also a mineral solution containing neither peptides nor vitamins had a similar effect on ceftibuten. Plasma level data in rats, and also in situ rat jejunal loop studies, confirmed the effect of these test meals on ceftibuten absorption. These results contrasted with those from cephalexin, which was unaffected by Enterued formula, showing that the inhibitory effects of Enterued on absorption were not due to inhibition of a peptide transport system.

Administration of a standard breakfast moderately reduced the systemic availability of oral 2-chloro-2′-deoxyadenosine (CdA) in male patients with leukemia (Albertioni et al. 1993). The C_{max} of CdA in plasma was reduced by 40% while the AUC was reduced by 9% by food. In this study, pretreatment with omeprazole did not significantly influence CdA bioavailability, or interindividual variability in the fasting state. The absolute bioavailability of CdA is only 50% when administered to fasting patients with and without omeprazole. CdA has a low pK_a such that most of the drug will be ionized in the gut, hindering absorption. The drug may also undergo first-pass clearance in the gut wall and/or the liver. Reduced systemic availability in the presence of food may be related to slower absorption, resulting in a greater proportion of absorbed drug undergoing presystemic metabolism. Reduced absorption due to food may be a possible explanation for reduced antiplatelet activity of cicaprost (Belch et al. 1993). Healthy volunteers received varying doses of cicaprost at 9 a.m., 2 p.m., 7 p.m., and again at 9 a.m. the next day. Antiplatelet activity was dose related but the effect was attenuated following the 2 p.m. and 7 p.m. doses, which were taken after heavier meals than the 9 a.m. doses. An alternative explanation is that tachyphylaxis may occur during the second (2 p.m.) and a third (7 p.m.) dose, causing reduced efficacy, while the 14-h gap between the 7 p.m. dose and the next 9 a.m. dose may have been sufficient for platelet sensitivity to return to normal. More studies are necessary to differentiate these possible mechanisms. Although neither a standard breakfast nor a high-fat, high-calcium breakfast altered ciprofloxacin absorption (Frost et al. 1989), absorption was markedly impaired by coadministered milk and yoghurt (Neuvonen et al. 1991). Systemic availability was reduced 35% by coadministered 30 ml milk and 36% by 300 ml yoghurt compared to 300 ml water. Ciprofloxacin together with other fluoroquinolones is known to bind to heavy metals including calcium to form an insoluble chelate. The reason for the different results from the two studies is claimed to be due to calcium in milk and yoghurt being in liquid form compared to the previous study in which most of the high calcium present was in solid form, for example, in cheese. This suggests, that, for at least some dairy products, ciprofloxacin is affected to a similar extent as tetracycline derivatives (Mattila et al. 1972). Plasma levels of ciprofloxacin taken with milk, yoghurt, and water are shown in Fig. 3.

The timing of food ingestion relative to dosing had a marked effect on the absorption of the purine nucleoside analogue didanosine (KNUPP et al. 1993). This study, conducted in ten men seropositive for human immunodeficiency virus (HIV) but free of AIDS symptoms, examined the absorption of didanosine following overnight fast, 13 min before, 1 h before, 1 h after, and 2 h after a standard high-fat, high-calorie breakfast. There were no significant differences among the fasting, 30 min, and 1 h before meal doses. However, C_{max}, AUC(0–∞), and urinary excretion were significantly reduced following the two postprandial doses. The mean profiles following the five treatments are summarized on a logarithmic scale in Fig. 4.

Decreased bioavailability after postprandial administration may have been due to prolonged gastric retention leading to increased degradation of the acid-labile didanosine, possible interference with active transport, or direct interaction of didanosine with one or more meal components. In a separate study using a chewable tablet form of didanosine in male subjects seropositive for HIV, the bioavailability of didanosine was reduced approximately twofold when it was administered 5 min after a substantial, standardized breakfast

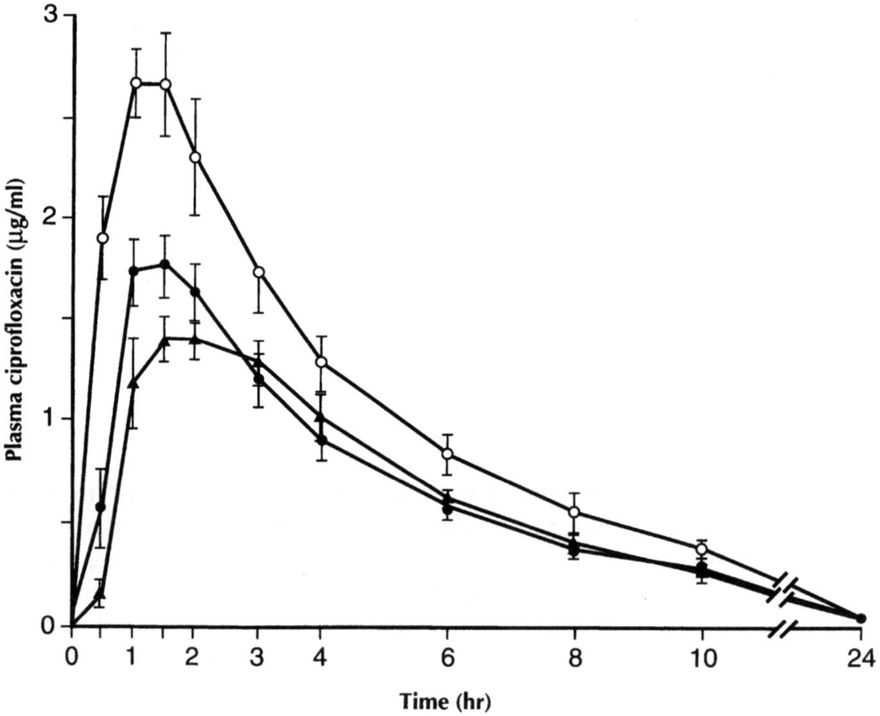

Fig. 3. Mean plasma versus time concentrations (±SE) of ciprofloxacin in seven subjects following a single 500-mg oral dose of ciprofloxacin with 300 ml milk (●), yoghurt (▲), or water (○). (From NEUVONEN et al. 1991)

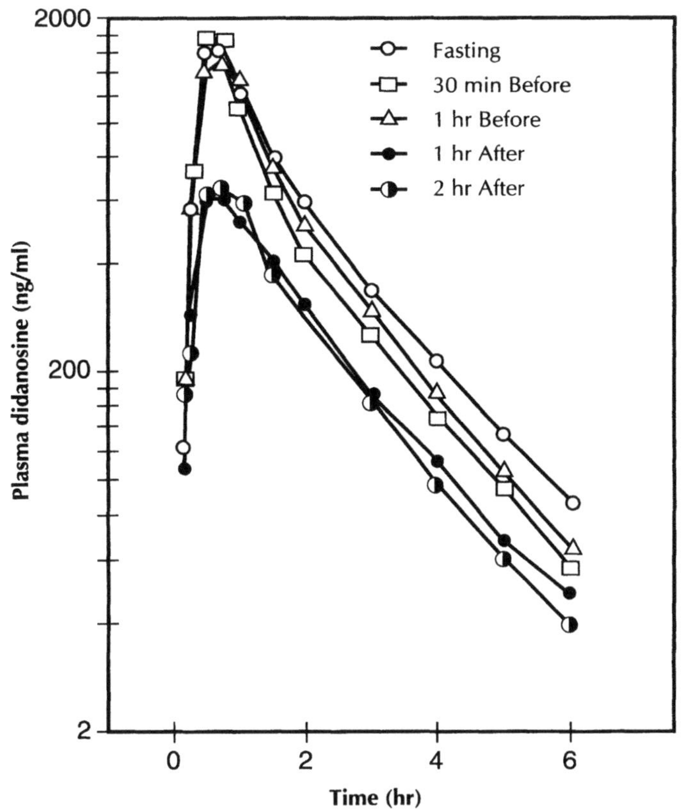

Fig. 4. Mean plasma versus time concentrations of didanosine in ten patients following a single 300-mg oral dose of a chewable tablet under fasting conditions or at various times relative to a high-fat, high-calorie meal. (From KNUPP et al. 1993)

(SHYU et al. 1991). The results obtained in these two studies have led to the recommendation that didanosine be administered under fasting conditions.

A milder food effect was reported for the novel quinazoline antihypertensive agent doxazosin. Following single 1-mg doses with a standardized light breakfast, mean C_{max} in 12 hypertensive subjects was reduced 18% and the AUC by 12% compared to fasting. The acute fall in blood pressure following doxazosine doses was not significantly affected by food, but there was a trend toward lower instances of complaints after this treatment. Apart from that, the modest food effects observed in this study are unlikely to be clinically significant. A more serious clinical effect with flecainide was possibly attributed to a milk drug interaction (RUSSEL and MARTIN 1989). Toxicity in the form of ventricular tachycardia occurred in a baby boy when dextrose was substituted for milk foods during flecainide therapy. Serum flecainide levels, based on single point determinations, were doubled from 990 μg/ml to 1824 μg/

ml 24h following dextrose substitution. The seriousness of the effect would justify close flecainide monitoring in these types of patients who may be on intermittent milk diets. This has obvious implications for other dairy-based diets in this population.

In an attempt to shed some light on variable results achieved in previous studies that have reported increased, unchanged, and decreased hydralazine absorption with foods, a study was conducted to compare the effect of fasting and isocaloric standard breakfast, bolus enteral nutrients, and infused enteral nutrients (SEMPLE et al. 1991). The study was conducted in four healthy men and four healthy women. Both the standard breakfast and bolus enteral nutrient treatments gave rise to much lower blood levels of hydralazine than fasting and infusion of nutrients, although there was considerable interindividual variation. Typically the fasted treatment yielded C_{max} and AUC(0–6) values of 87 ng/ml and 2641 ng·min/ml compared to 11 ng/ml and 999 ng·min/ml and 15 ng/ml and 1189 ng·min/ml for the standard breakfast and enteral bolus treatments, respectively. While not addressing the underlying cause of the conflicting results reported previously, this study does show that, for hydralazine at least, the physical form of nutrients may not play an important role in determining the extent and nature of drug-food interactions.

An interesting result was obtained when plasma levels of levodopa, 3-*O*-methyldopa, and carbidopa were compared in patients with idiopathic Parkinson's disease after taking Sinemet CR, a controlled release formulation containing 200/50 levodopa and carbidopa (ROOS et al. 1993). As might have been expected from previous studies, coadministration with a high-protein meal reduced plasma levels of both levodopa and carbidopa due, amongst other things, to competition for active absorption mechanisms. However, concentrations of the levodopa metabolite 3-*O*-methyldopa increased in the presence of the high-protein meal. Increased metabolite levels may be related to increased gastric residence time in the fed state, giving rise to greater presystemic conversion of levodopa to the metabolite. While levodopa and carbidopa levels were generally reduced by the high-protein meals, the plasma levels yielded a flattened concentration-effect profile. If a "kick" from levodopa is required, then administration on an empty stomach may be appropriate.

In a study in 14 juvenile rheumatoid arthritis patients the mean systemic availability of methotrexate was reduced by 15% and C_{max} by 38% when medication was given immediately after breakfast compared to fasting (DUPUIS et al. 1992). Compared to i.v. administration, the mean absolute bioavailability of methotrexate in these patients was 90% in the fasting state compared to 77% after breakfast. Food has also been shown to reduce systemic availability of the investigational LTD_4 receptor antagonist MK-679 (YOGENDRAN et al. 1992). Administering the drug after a standard meal reduced both C_{max} and AUC(0–12) by approximately 40%.

Reduced nitrendipine absorption due to food is reported in an efficacy study in juvenile patients with severe hypertension (WELLS and SINAIKO 1991).

Nitrendipine failed to lower blood pressure in only 1 of 25 patients. Plasma levels of nitrendipine in 13 patients were variable but opportunity was taken to compare levels in fed and fasted patients. C_{max} values were reduced from 71 to 13 ng/ml and AUC_{ss} values from 158 to 46 ng·h/ml while t_{max} was increased from 1 to 2.4 h by food. Although the food effect probably contributed to variability in nitrendipine levels, this was not further discussed, nor was the basis used to differentiate fed and fasted subjects.

As noted previously with ciprofloxacin (NEUVONEN et al. 1991), co-administration with milk or yoghurt had a marked effect on systemic availability of norfloxacin (KIVISTÖ et al. 1992). In what appears to be emerging as a possible class effect, both the plasma C_{max} and the AUC(0–24) of norfloxacin were reduced by approximately 50% when given with 300 ml milk or yoghurt compared to 300 ml water. Plasma profiles obtained in this study are shown in Fig. 5. The results of this study provide further evidence that, while absorption of oral fluoroquinolones is not affected to a clinically significant extent by solid foods (HOLMES et al. 1985; MONK and CAMPOLI-RICHARDS 1981; VANCE-BRYAN et al. 1990), dairy products containing solubilized calcium may reduce circulating drug levels to an extent that may well be clinically significant depending on the organism(s) being treated.

A further example of the different effects by different diets on absorption is provided by paracetamol, whose absorption was incomplete and abnormally slow in Thai vegetarian compared to nonvegetarian subjects (PRESCOTT et al. 1993). In another study, however, the absorption of tetracycline was similarly depressed by a Mexican meal and also by a conventional Western meal, although the Mexican meal contained double the amount of calcium (COOK et

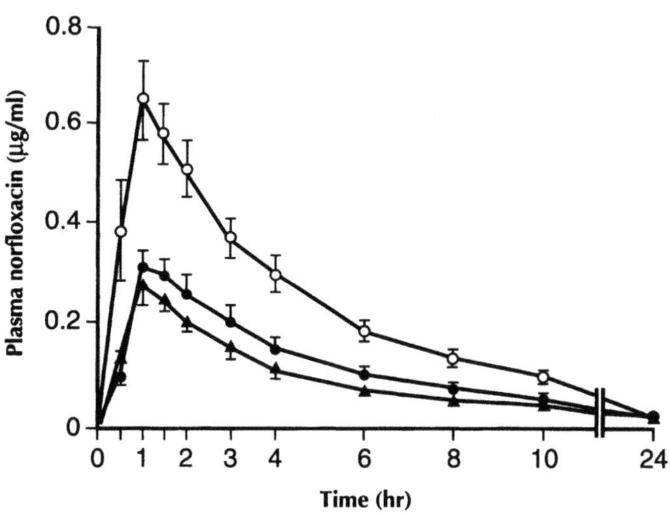

Fig. 5. Mean plasma versus time concentrations (±SE) of norfloxacin in seven subjects following a single 200-mg oral dose of norfloxacin with 300 ml milk (●), yoghurt (▲), or water (○). (From KIVISTÖ et al. 1992)

al. 1993). Clearly both diets in this study contained sufficient free calcium to markedly reduce systemic availability of tetracycline.

Contradictory effects were observed on pravastatin pharmacokinetics and pharmacodynamics when the drug was taken with meals (PAN et al. 1993). In a study conducted in 24 hypercholesterolemic men, mean C_{max} dropped by 49% and AUC by 31% in the presence of food compared to the fasting state. However, reduction in mean total cholesterol and low-density lipoproteins was identical whether pravastatin was taken with or before food. The reduction in pravastatin bioavailability in these hypercholesterolemic patients was similar to that observed in a preliminary study in healthy volunteers. The lack of effect on blood lipids may have been due to food increasing extraction of pravastatin by the liver, which is also the primary site of cholesterol synthesis, LDL/cholesterol clearance, and also pravastatin activity. Other studies within this category have reported 60% reduction in the already poorly absorbed renin inhibitor remikiren (WEBER et al. 1993), a modest reduction in the bioavailability of the leukotriene receptor antagonist SK&F 106203 (NICHOLS et al. 1992), and moderately reduced plasma levels of rufloxacin (SEGRE et al. 1992) in the presence of food. In the last case, however, although mean C_{max} and AUC values were decreased by 16% and 33%, respectively, urinary excretion of unchanged drug increased from 16% to 26% of the administered dose. The increase of urinary excretion suggests that lower plasma levels of rufloxacin may not have been due to reduced drug absorption.

The beta-blocking agent sotalol appears to behave more like atenolol (BARNWELL et al. 1993) than metoprolol or propranolol in that its systemic bioavailability is reduced by approximately 20% by administered food. Like atenolol, sotalol undergoes little or no first-pass metabolism so that any change in hepatic blood flow due to food is not likely to have a positive effect on systemic availability.

Formulation was shown to make a negligible contribution to the nature of drug-food interaction for the neoleptic agent sulpiride (SHINKUMA et al. 1990). Ingestion of a film-coated tablet or a solution with a standard meal resulted in 30% reduction in systemic availability. Further studies with the solution dosage form showed that the food effect increased with increasing meal size, inhibition increasing from 21% with a light meal, to 30% with a medium meal, to 40% with a heavy meal. The systemic availability of the cognition agent tacrine was reduced by approximately 20%, and peak plasma levels by 40% when administered up to 2h following the start of a standard breakfast (WELTY et al. 1994). However, administration with meals reduced gastrointestinal side effects, thus improving patient tolerance. In the case of verapamil, food has been shown not only to reduce systemic availability from a sustained release formulation, but also to alter electrographic effects following a single drug dose. There was no evidence of exaggerated release or "dose dumping" in this study.

A number of studies have examined the effect of food on the absorption and pharmacokinetics of the HIV agent zidovudine (LOTTERER et al. 1991;

UNADKAT et al. 1990; PUHNKE et al. 1993; UEDA et al. 1993). All of these studies, conducted in laboratories in Japan, Germany, and the United States, have reported significantly delayed and reduced absorption with remarkable consistency across studies. Typically, T_{max} is increased from 0.5 to 1.5–2.0h, while C_{max} and AUC are decreased by approximately 50% when zidovudine is administered with a meal compared to fasting. Variability in plasma levels is also increased by food. While no direct relationship has been established between circulating levels and efficacy of zidovudine, doses found to be effective in AIDS patients generally achieve peak serum levels of ≥0.27 μg/ml. However, lower and more prolonged levels, as observed when zidovudine is taken with meals, may maintain efficacy while reducing toxic side effects (UNADKAT et al. 1990).

For the last compound in this category, 882C (1-(β-D-arabinofuranosyl)-5-(1-propynyl)uracil), which is under investigation for treatment of varicella zoster virus infections, food caused a significant reduction in C_{max} and a modest reduction in AUC in healthy volunteers (PECK et al. 1993). The effects observed in this single-dose study led to the recommendation that effective concentrations of this agent can be maintained without food-related restrictions.

E. Interactions Causing Delayed Drug Absorption

Compounds whose absorption is delayed by food are listed in Table 4. The large number of compounds in this category reflects the frequency of this phenomenon, due for the most part to delayed gastric emptying and dissolution.

Delayed absorption may be considered to be unimportant for the majority of drugs, particularly when absolute systemic availability is not altered. This may be particularly true during repeated doses, during which time circulating drug profiles in the nonfasted state would be simply offset in time relative to those in the fasted state, and overall profiles would tend to be somewhat flatter as a result of reduction in peak drug levels. This may be the case with many drugs and formulations included in this category in this chapter. However, as in all the other categories described in this chapter, the importance of an interaction depends on the extent of change compared to the required circulating profile and the therapeutic index of the drug. Peak circulating levels of some of the drugs included in Table 4 are reduced by as much as 60% with concomitant food ingestion. Changes of this magnitude in circulating drug levels are likely to be therapeutically important. Other cases of more moderate delays in absorption are likely to be less important therapeutically. Typically, observed delays in effects or circulating profiles of acetorphan (BERGMANN et al. 1992) and albuterol (HUSSEY et al. 1991) due to food are probably clinically unimportant whereas the 50% reduction in C_{max} reported for aniracetam by food is likely to have clinical consequences (RONCARI 1993;

Table 4. Drug-food interactions resulting in delayed drug absorption

Drug	Dosage form	Dose	Dosage regimen	Subjects	Food details	Fluid volume	Time interval	Sampling duration	Result of food	Reference administration
Acetorphan	–[a]	100 mg	SD[b]	9 healthy subjects	Standard meal, 800 kcal	–	20 min after meal	–	T_{min} for enkephalinase activity 180 min after food compared to 80 min for acetorphan alone	Bergmann et al. (1992)
Albuterol	Volmax and Proventil Repetabs Albuterol sulfate	4 mg	SD	23 healthy male subjects	Standard breakfast[c]	250 ml water	Immediately after meal	30 h plasma	C_{max} reduced 19% for Volmax and 21% for Proventil. T_{max} and t_{lag} delayed	Hussey et al. (1991)
5-Aminosalicylic acid	Tablet	500 mg	MD[d]	12 healthy male subjects	Breakfast, afternoon snack, evening snack,[c] standardized	100 ml water	Immediately after meals	48 h plasma and urine	Systemic absorption delayed but not reduced by food	De Mey and Meineke (1992)
Aniracetam	Tablet	1200 mg	SD	–	–	–	–	–	C_{max} reduced 50%, T_{max} extended from 0.6 to 1.1 h. AUC of metabolites unaffected by food	Roncari (1993) Honma et al. (1986)
Beta-methyldigoxin	Lanirapid	0.2 mg	SD	9 healthy male subjects	Standard breakfast[c] 560 kcal. Other meals also controlled	150 ml water	30 min before or after meal	120 h serum, 240 h urine	Absorption rate significantly reduced after meal. Mean C_{max} reduced 46%, AUC unaffected. Urinary excretion slightly decreased	Tsutsumi et al. (1992)
Cefaclor	Capsules	500 mg	SD	8 healthy male subjects	Two standard breakfasts.[c] One based on rice, the other based on eggs and bread. Six months later repeated, rice meal reduced, bread meal increased	–	30 min after meal	6 h plasma and urine	C_{max} reduced, T_{max} prolonged. Effect increased with larger meal size. Rice meal affected absorption more than bread meal	Oguma et al. (1991)
Cefdinir	10% fine granules	3 mg/kg	SD	8 children	–	–	30 min before or after meal	–	C_{max} reduced 43%. T_{max} increased from 2 h to 4–5 h. AUC reduced by 15%. Urinary recovery marginally reduced	Motohiro et al. (1992)
Cefdinir	5% fine granules	3 mg/kg	SD	20 schoolchildren	–	–	30 min before (14 children) or 30 min after (6 children) meal	–	C_{max} reduced 30%, T_{max} increased from 2 to 3.67 h. Urinary recovery unchanged	Nakamura and Iwai (1992)

Cefdinir	5% fine granules	3 mg/kg	SD	25 younger children	–	–	30 min before (19 children) or 30 min after (6 children) meal	–	C_{max} reduced 52%, T_{max} increased from 2.1 to 3.3 h. Urinary recovery reduced by 48%	NAKAMURA and IWAI (1992)[b]
Cefdinir	5% fine granules	3 mg/kg	SD	15 infants	–	–	30 min before (7 infants) or 30 min after (8 infants) meal	–	No significant differences between groups	NAKAMURA and IWAI (1992)
Cefprozil	Capsule	100 mg	SD	15 healthy male subjects	Standard breakfast[c]	200 ml water	30 min after meal	12 h plasma, 24 h urine	T_{max} increased from 1.5 to 2 h. C_{max}, AUC, urinary excretion not affected	SHUKLA et al. (1992)
Cefprozil	Capsule	250 mg	SD	12 healthy male subjects	Standard breakfast[c]	150 ml water	15 min after meal	8 h plasma	Slight reduction in C_{max}. Mean T_{max} increased from 1.2 to 2 h	BARBHAIYA et al. (1990)
Cefaclor	Capsule	250 mg	SD	12 healthy male subjects	Standard breakfast[c]	150 ml water	15 min after meal	5 h plasma	C_{max} reduced 51%, T_{max} increased from 0.6 to 1.3 h	BARBHAIYA et al. (1990)
CL 275838	Tablet	50 mg	SD	10 healthy male subjects	Standard breakfast[c]	150 ml water	30 min after meal	48 h plasma	T_{max} delayed 2 h. C_{max}, AUC unaffected.	CONFALONIERI et al. (1992)
Diclofenac	Capsule	50 mg	SD	8 healthy male subjects	Standard breakfast	150 ml water	Together with meal	10 h serum	C_{max} reduced 61%, T_{max} increased from 0.8 to 2.4 h. AUC unaffected	TERHAAG et al. (1991)
Diclofenac	Hydrogel bead capsule	150 mg	SD	12 healthy male subjects	Standard breakfast[c]	200 ml water	30 min after meal	24 h plasma	C_{max} reduced 38%, T_{max} increased from 1.85 to 6.25 h. AUC unaffected	THAKKER et al. (1992)
Diltiazem	Pellets in capsule	240 mg	SD	8 healthy subjects 4M, 4F	Standard breakfast, lunch, dinner[c]	100 ml water	After meal	24 h plasma	T_{max} delayed from 7.4 to 9.7 h. C_{max} and AUC unaffected	WILDING et al. (1991a)
Doxycycline	Pellets in capsule	100 mg	SD	20 healthy male subjects	Standard breakfast	180 ml water	With breakfast	48 h plasma and urine	Lag time increased from 0.55 to 1.1 h, T_{max} increased from 1.8 to 3.1 h. C_{max}, AUC, and urinary excretion not significantly affected	WILLIAMS et al. (1990)
Erythromycin acistrate	Enteric-coated tablet	400 mg b.i.d. for 4 days	SD and MD	14 healthy subjects 4M, 10F	Standard light and heavy breakfast[c]	–	Immediately after meal	12 h plasma	After SD, both meals delayed absorption. In seven subjects no absorption occurred in 12 h. After MD food effect diminished	JÄRVINEN et al. (1992)
Fadrozole	Tablet	12 mg	SD	9 healthy Caucasian subjects, 7M, 2F	Standard breakfast[c]	240 ml water	Within 20 min of completing meal	36 h plasma	C_{max} reduced by 15%, T_{max} increased from 1.8 to 2.5 h. AUC not affected	CHOI et al. (1993)

Table 4. *Continued*

Drug	Dosage form	Dose	Dosage regimen	Subjects	Food details	Fluid volume	Time interval	Sampling duration	Result of food	Reference administration
Famotidine	Tablet	40 mg	SD 6 p.m. or 9 p.m.	10 healthy subjects, 4M, 6F	Supper given at 6 p.m. or 9 p.m.	–	With meal	Variable plasma	Absorption delayed somewhat when meal and drug coadministered at 6 p.m. No delay when meal and drug coadministered at 9 p.m.	MARGALITH et al. (1991)
Flurbiprofen	Tablet	100 mg	SD	12 healthy male subjects	30 ml apple juice[c]	180 ml water	After the apple juice	48 h serum	Gastric emptying time almost doubled by meal. Slower initial absorption followed by more rapid second absorption phase	DRESSMAN et al. (1992)
Fluvastatin	Solution or capsule	10 mg	SD	16 healthy male subjects	Standard high-fat breakfast	–	Immediately after meal	12 h blood	C_{max} reduced 69% to 74% and T_{max} increased more than fourfold from solution. C_{max} reduced 50%–60% and T_{max} increased three- to fourfold from capsule. Bioavailability decreased by 15%–25%	SMITH et al. (1993)
Fluvastatin	Capsule	20 mg	SD	22 healthy male subjects	Carbonated beverage	–	With carbonated beverage	–	C_{max} reduced 57%, T_{max} increased from 0.71 to 1.11 h. Bioavailability reduced 28%	SMITH et al. (1993)
Fluvastatin	Capsule	20 mg	SD	Healthy male subjects	Low-fat evening meal (6 p.m.)	–	4 h after meal	12 h plasma	C_{max} reduced 44%, T_{max} increased 57%. Bioavailability reduced 14%	SMITH et al. (1993)
Fusidate sodium	Film-coated tablet	500 mg	SD	–	Standardized breakfast	–	After meal	26 h serum	C_{max} reduced 26%, T_{max} increased from 2.2 to 3.2 h. AUC marginally reduced (16%)	MACGOWAN et al. (1989)
Hydroxychloro-quine	Tablet	155 mg	SD	9 rheumatoid arthritis patients	Light breakfast	–	After meal	–	Absorption delayed due to food to similar extent to healthy subjects	MCLACHLAN and CUTLER (1991)
Isosorbide-5-mononitrate	Controlled release tablet	60 mg	SD	18 subjects	High-fat breakfast	–	–	–	C_{max} reduced 5%, T_{max} increased from 3.1 to 6.5 h. AUC increased 9%	KOSOGLOU et al. (1993)
Lomefloxacin	Capsule	400 mg	SD	12 healthy subjects, 8M, 4F	High-fat or high[c]-carbohydrate breakfast	200 ml water	5 min after meal	48 h plasma, 72 h urine	T_{max} delayed approximately 1 h, C_{max} and AUC unchanged	HOOPER et al. (1990)

Loracarbef	Capsule	400 mg	SD	12 healthy male subjects	Standard breakfast[c]	200 ml water fasting, 100 ml water fed	5 min after meal	24 h serum and urine	C_{max} reduced 29%, T_{max} increased from 68 to 141 min, AUC unchanged	ROLLER et al. (1992)
Loracarbef	Capsule	200 mg	SD	12 subjects	–	–	–	12 h plasma	C_{max} reduced and T_{max} delayed. AUC unchanged	DESANTE and ZECKEL (1992)
Methotrexate	Tablet	7.5 mg	SD	12 healthy male subjects	Standard high-fat breakfast[c]	–	10 min after meal	24 h plasma	C_{max} reduced by 16%, T_{max} delayed from 1.0 to 1.6 h. AUC increased 4.6%	KOZLOSKI et al. (1992b)
Methotrexate	Tablet	15 mg	SD	10 patients, 5M, 5F	Standard French breakfast[c]	–	After meal	24 h plasma	C_{max} reduced 31%, T_{max} increased from 1.3 to 2.0 h. Bioavailability unaffected	OGUEY et al. (1992)
Monofluoro-phosphate	Chewable tablet	10 mg fluoride with 300 mg calcium	SD	8 healthy male subjects	Standard meal[c]	200 ml fluoride-free water	Immediately after meal	48 h plasma and urine	Lag time delayed from 4 to 11 min, C_{max} reduced by 67%. T_{max} increased from 24 min to 2.4 h. AUC and urinary excretion not affected	WARNEKE and SETNIKAR (1993)
Moricizine	Tablet	250 mg	SD	24 healthy male subjects	Standard breakfast[c]	180 ml orange juice	30 min after meal	24 h plasma	C_{max} reduced 24%, T_{max} delayed from 0.9 to 1.2 h. AUC not affected	PIENIASZEK et al. (1991)
Nicorandil	Tablet	10 mg	SD	10 healthy male subject	Standard breakfast[c]	–	10 min after meal	–	C_{max} reduced 13%, T_{max} increased from 0.38 to 0.73 h, AUC increased 24%. Absorption rate consistently decreased by 90%	FRYDMAN (1992)
Nifedipine	–	10 mg	SD	10 healthy subjects	High-fat meal	100 ml water	After meal	24 h plasma	C_{max} reduced by 47%, T_{max} increased from 0.9 to 1.44 h, AUC unaffected. Under same study conditions of nicardipine and nitrendipine unaffected	RAPIN (1992)
Ofloxacin	Tablet	200 mg	SD	7 healthy male subjects	300 ml milk 300 ml yoghurt[c]	300 ml water	Immediately before milk or yoghurt	24 h plasma and urine	C_{max} reduced 18%, T_{max} delayed 1 h by yoghurt. AUC unaffected. Milk had no effect on any parameters	NEUVONEN and KIVISTÖ (1992)
Ofloxacin	Capsule	300 mg	SD	21 healthy male subjects	Standard breakfast[c] or milk	–	30 min after food or milk	24 h plasma	C_{max} reduced 17%, T_{max} delayed 1 h by food, AUC unaffected. Milk had no effect on any parameters	DUDLEY et al. (1991)

Table 4. *Continued*

Drug	Dosage form	Dose	Dosage regimen	Subjects	Food details	Fluid volume	Time interval	Sampling duration	Result of food	Reference administration
Paracetamol	Tablet and solution	500 mg	SD	12 healthy male subjects	200 ml liquid formula enriched with or depleted of dietary fiber[c]	200 ml	With liquid diet	12 h plasma	The two liquid diets did not differentially affect absorption from solution. Fiber-depleted diet delayed absorption from tablets relative to fiber-enriched diet	Walter-Sack et al. (1989)
Paracetamol	Tablet	1000 mg	SD	12 healthy Tswana female volunteers	Cereal, maize, or bacon and egg breakfast[c]	200 ml water	30 min after meal	12 h serum	Absorption delayed to increasing extent by maize, bacon and egg, and cereal breakfasts	Wessels et al. (1992)
Penciclovir (famciclovir prodrug)	–	250 or 500 mg	SD	12 healthy male subjects	–	–	–	24 h plasma	C_{max} reduced 47% and 45%, T_{max} delayed by 1.7 h and 1.8 h from 250- and 500-mg doses, respectively, AUC unaffected	Fowles et al. (1990, 1991)
Penciclovir (famciclovir prodrug)	–	500 mg	SD	18 healthy male subjects	–	–	2 h before or after meal	24 h plasma and urine	C_{max} reduced 18%, T_{max} delayed by 1.1 h	Fowles et al. (1990, 1991)
Rifabutin	Capsule	150 mg	SD	12 healthy male subjects	Standard high-fat breakfast[c]	–	15 min after meal	168 h plasma, 48 h urine	C_{max} reduced 17%, T_{max} delayed 2.4 h, AUC unaffected. Urinary excretion reduced 20%	Narang et al. (1992)
Salsalate	Tablet	1500 mg	SD	17 healthy male subjects	Standard high-fat breakfast[c]	6 ounces water	Immediately after meal	48 h plasma	Salsalate mean C_{max} reduced 17%, T_{max} delayed 1 h, AUC unaffected. Salicylate parameters unaffected	Harrison et al. (1992)
Terazocin	–	2 mg	SD	12 subjects	Standard breakfast[c]	100 ml water	–	30 h plasma	Mean C_{max} reduced 23%, T_{max} delayed 0.9 h, AUC unaffected	McNeal et al. (1991)
Terfenadine	Tablet	120 mg	SD	24 healthy male subjects	Standard breakfast[c]	200 ml water	Immediately after meal	48 h plasma	Mean metabolite 1 C_{max} increased 13%, T_{max} delayed by 0.9 h. AUC unaffected	Eller et al. (1992)
Theophylline	Theo-Dur tablets	400 mg	SD	10 healthy male subjects	High-fat dinner[c]	6 ounces water	Immediately after meal, 8 p.m.	36 h serum	Mean C_{max} increased 33%, T_{max} unchanged, AUC increased 6%. Absorption rate delayed during initial 8 h postdosing	Kann et al. (1989)

Theophylline	Multiparticulate CR pellet	300 mg	SD	12 healthy male subjects	Standard breakfast[c]	150 ml water	Immediately after meal	36 h serum	Delayed drug absorption associated with delayed stomach emptying	Yuen et al. (1993)
Tiagabine	Tablet	8 mg	SD	18 healthy male subjects	–	–	–	–	C_{max} reduced 44%, T_{max} increased from 0.9 to 2.6 h. AUC unaffected	Mengel et al. (1991)
Topiramate	Tablet	100 mg, 400 mg	SD	18 healthy male subjects	Fatty breakfast	–	–	168 h plasma	C_{max} reduced 11% (100 mg) and 13% (400 mg), T_{max} increased by 1.8 h (100 mg) and 2.1 h (400 mg), AUC (0–∞) unaffected	Doose et al. (1992a)
Trazodone	Capsule	100 mg	SD	8 healthy subjects, 4M, 4F	Standard breakfast	–	After meal	26 h serum and urine	C_{max} reduced 22%, T_{max} increased from 1.3 to 2.0 h. Amount of drug absorbed unchanged	Nilsen and Dale (1992)
Valproic acid	Tablet	800 mg	SD	16 healthy male subjects	Light meal, heavy[c] meal	180 ml water	After meal	45 or 48 h plasma	C_{max} reduced 14% and AUC 9% by heavy evening meal, but not affected by light evening meal relative to light breakfast	Ohdo et al. (1992)
Vigabatrin	Tablet	1000 mg	SD	24 healthy male subjects	–	–	–	36 h plasma, 48 h urine	C_{max} reduced 33%, T_{max} increased from 1.0 to 2.14 h. Extent of absorption unaffected	Hoke et al. (1991)
Zalospirone	–	20 mg	SD	24 healthy male subjects	High-fat, low-fat breakfast	–	After meal	24 h plasma	C_{max} reduced 13% by low-fat and 31% by high-fat meals. T_{max} increased 0.5 h and 1.2 h. AUC unaffected	Korth-Bradley et al. (1992)
Zidovudine	–	100 mg	SD	18 asymptomatic HIV-infected subjects	High-fat breakfast	–	30 min or 3 h after meal	10 h plasma	C_{max} reduced 57% (30 min) and 47% (3 h). T_{max} increased by 1 h (30 min) and 0.6 h (3 h). AUC reduced 3% (30 min) and 15% (3 h)	Shelton et al. (1993)
Zidovudine	Capsule	200 mg	SD	11 symptomatic HIV-infected men	Liquid protein meal in 220 ml orange juice[c]	150 ml water	Immediately after meal	8 h serum and urine	C_{max} reduced 32%, T_{max} increased from 49 to 74 min. AUC and renal clearance unaffected	Sahai et al. (1992)

[a] Data not available.
[b] Single dose.
[c] Meal described.
[d] Multiple doses.

HONMA et al. 1986). Delayed absorption of 5-aminosalicylic acid (5-ASA) by food was confirmed in a study in which repeated 500-mg oral doses of 5-ASA were administered to healthy male subjects either 1h before or immediately after meals (DE MEY and MEINEKE 1992). A surprising observation in this study was a marked peak in plasma drug concentrations in the early morning following repeated doses the day before, regardless of dosing time relative to food. Mean values are shown in Fig. 6. This dramatic food-independent effect is postulated to be the result of tablets, or tablet fragments, remaining in the stomach until early morning, at which time they are released quite suddenly as a result of interdigestive phase 3 contractions occurring at that time of day. Lack of adequate interdigestive housekeeping waves during the previous day may have been due to snacks and meals shortening the interdigestive phase. Possible clinical significance is claimed for the delay in beta-methyldigoxin absorption in the presence of food (TSUTSUMI et al. 1992). Mean peak serum beta-methyldigoxin levels were delayed and reduced from 2.10ng/ml and 1.79ng/ml when drug was taken 30min before a standard breakfast and fasting, respectively, to 0.96ng/ml when drug was taken 30min after food. The results obtained with beta-methyldigoxin, which are similar to those reported in some studies for digoxin (SANCHEZ et al. 1973; JOHNSON et al. 1978), are summarized in Fig. 7.

A number of studies have examined and compared the effects of ingested food on circulating levels of oral cephalosporins. Absorption of cefaclor was

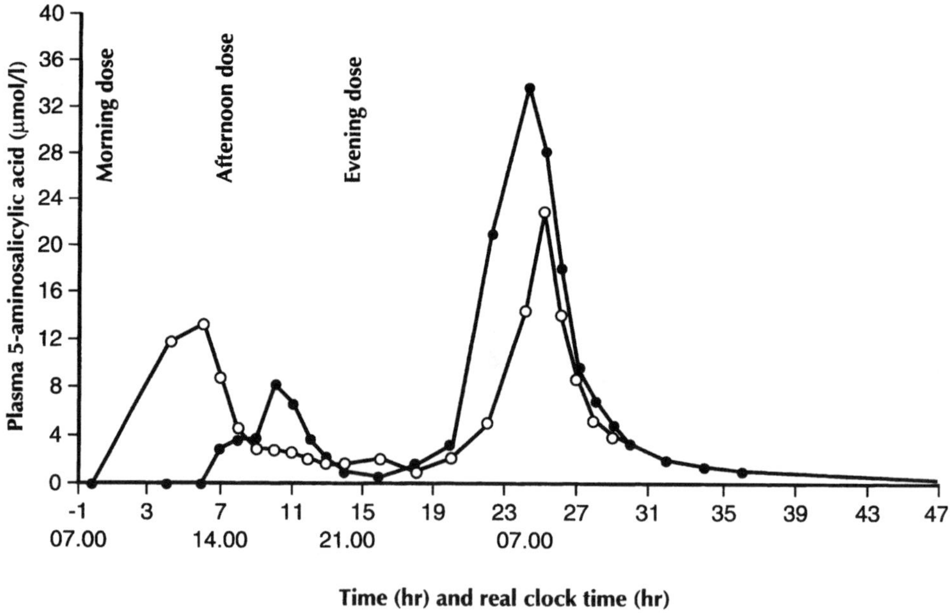

Fig. 6. Median plasma versus time concentrations of 5-aminosalicylic acid in 12 subjects following three oral administrations of 3 × 500-mg 5-aminosalicylic acid tablets either 1h before (○) or at the end of (●) meals. (From DE MEY and MEINEKE 1992)

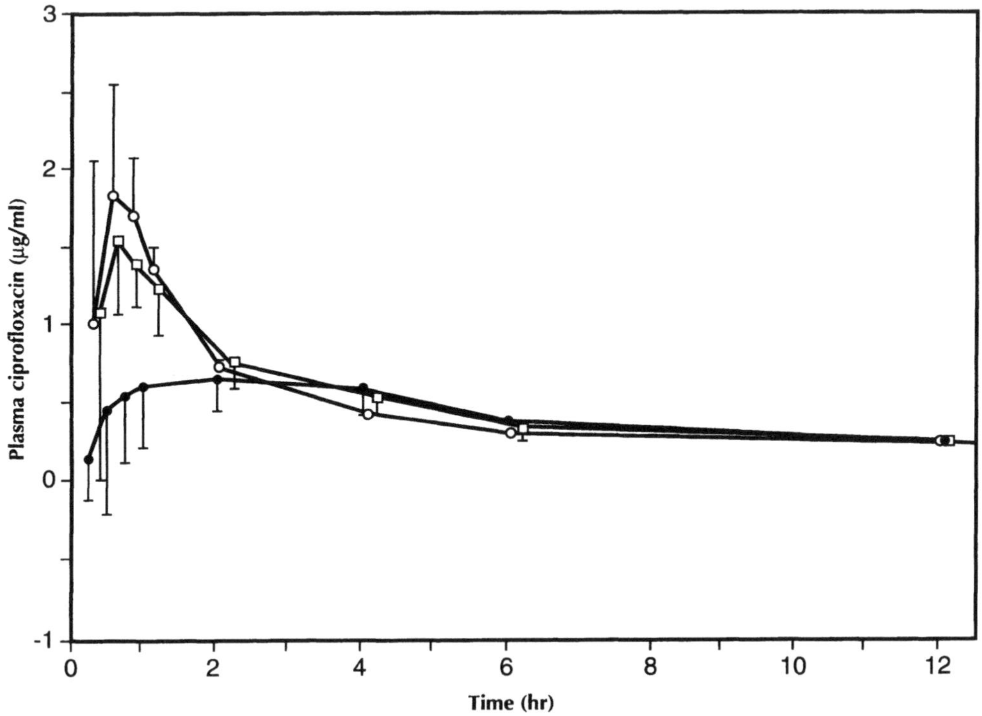

Fig. 7. Mean serum versus time concentrations (±SD) of beta-methyldigoxin in nine subjects following a single 0.2-mg oral dose of beta-methyldigoxin fasting (○), 30 min before (□) and 30 min after (●) a standard breakfast. (From TSUTSUMI et al. 1992)

delayed but not reduced following rice and bread meals, with rice having a greater effect than bread (OGUMA et al. 1991). For both meal types, the effect was greater for a large meal than for a small meal. Similar delays in cefaclor absorption were observed when drug was taken immediately after a standard breakfast, resulting in a 51% reduction in mean peak plasma levels (BARBHAIYA et al. 1990). Under the same study conditions peak levels of cefprozil were reduced by only 14% and plasma profiles were similar after fasting and nonfasting treatments. The different effects on plasma profiles of cefaclor and cefprozil are shown in Fig. 8. It is claimed that because cefaclor is absorbed more rapidly than cefprozil under fasting conditions, slight perturbations in stomach emptying and GI motility are more likely to affect the absorption rate of cefaclor than of cefprozil. Other studies have shown cefadroxyl to be unaffected by food, similarly to cefprozil, while absorption of cephalexin is delayed similarly to cefaclor (LODE et al. 1979). In further studies designed to examine the influence of gastric emptying on cefprozil absorption, food delayed the cefprozil t_{max} in plasma from 1.5 to 2 h but did not affect peak cefprozil plasma levels or absorption efficiency. Pretreatment with metoclopramide reduced the t_{max} and mean residence time while propantheline increased the t_{max} and significantly increased mean residence

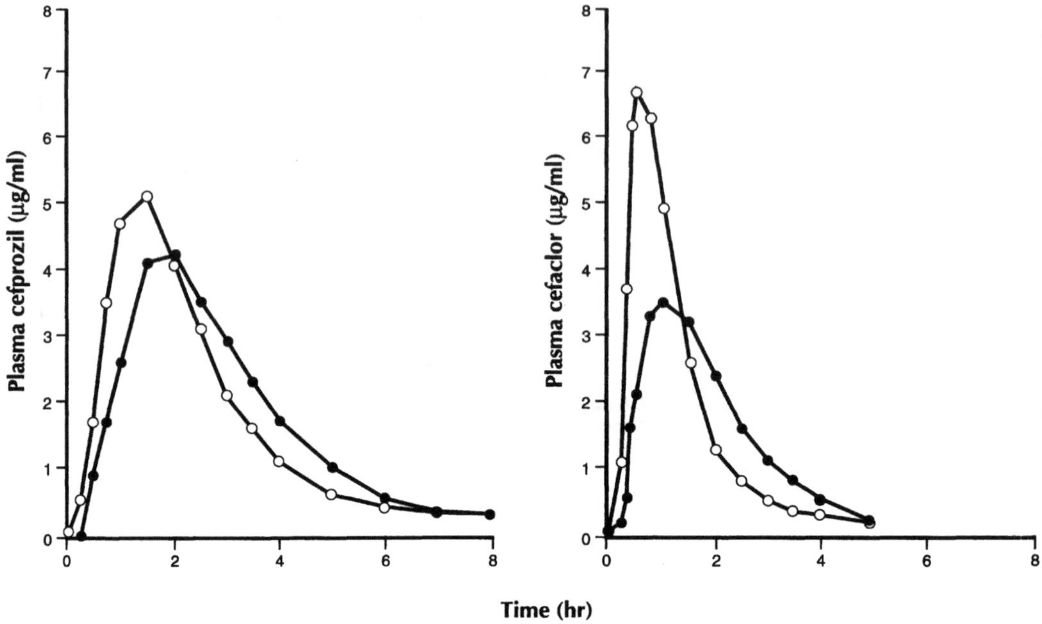

Fig. 8. Mean plasma versus time concentrations of cefprozil and cefaclor in 12 subjects following single 250-mg oral doses of cefprozil or cefaclor under fed (●) and fasting (○) conditions. (From BARBHAIYA et al. 1990)

time. No other pharmacokinetic parameters were affected. Thus, the rate of cefprozil absorption is modestly influenced by gastric emptying rate but the extent of absorption is not affected. Peak plasma levels of cefdinir were reduced 30%–52% by food in children of various ages who received drug as a 5% or 10% fine-granule suspension (MOTOHIRO et al. 1992; NAKAMURA and IWAI 1992b). Interestingly no reduction in C_{max} occurred; in fact the mean C_{max} value increased from 0.61 µg/ml in the fasting state to 0.79 µg/ml in the fed state in infants who received a 5% fine-granule 3-mg/kg pediatric dose.

Coadministered food has been shown to have a marked effect on circulating levels of the nonsteroidal anti-inflammatory agent diclofenac. Absorption from a 150-mg hydrogel bead capsule in healthy male subjects was delayed, yielding a mean C_{max} value of 312 ng/ml at 6.25 h compared to 502 ng/ml at 2 h in the fasted state (THAKKER et al. 1992). Systemic availability was not significantly affected by food but was 50% lower from the capsule dosage than from a solution. In an attempt to elucidate the mechanism of delayed diclofenac absorption, TERHAAG et al. (1991) compared diclofenac absorption in the presence and absence of food using paracetamol as a marker of gastric emptying rate. Under the conditions of this study, food had no effect on the rate or extent of paracetamol absorption, leading to the conclusion that delayed absorption of diclofenac due to food may not be due to delayed stomach emptying alone, but rather some interaction between the drug and food components, possibly adsorption, resulting in delayed stomach emptying of drug together with food. The results observed in these two studies using diclofenac capsules are in contrast to those using diclofenac enteric coated tablets in which peak circulating levels were delayed but not reduced by food (TERHAAG et al. 1990; WILLIS et al. 1981). The capricious relationship between stomach emptying and drug absorption rate was demonstrated with the calcium channel blocking agent diltiazem (WILDING et al. 1991a). Using gamma scintigraphy in healthy subjects, gastric emptying $T_{50\%}$ values of sustained release diltiazem pellets were significantly increased from 91 min fasting to 232 min with food, but the time of peak plasma levels was increased only moderately from a mean value of 7.4 h to 9.7 h. Other pharmacokinetic parameters were unaffected.

Results obtained in a food-effect study with erythromycin acistrate (2′-acetyl erythromycin stearate) appear to be consistent with earlier reports from other studies on erythromycin and its derivatives (JÄRVINEN et al. 1992). Food has been shown to generally reduce the systemic availability of erythromycin and erythromycin stearate (WELLING and TSE 1992) while absorption of erythromycin esters is either unchanged or increased (WELLING et al. 1979). The rate of erythromycin acistrate absorption from single doses was delayed to a small extent following a light meal and to a greater extent following a heavy meal. While the effect was attenuated to some extent after repeated doses, food significantly delayed absorption in some individuals after single and repeated doses, resulting in undetectable levels up to 12 or 24 h postdosing. Data supporting the influence of stomach emptying time on drug absorption rate are provided by a study in which flurbiprofen tablets were administered together with equal volumes of water or calorie-dense apple juice (DRESSMAN et al. 1992). By use of a Heidelberg capsule, apple juice significantly increased mean gastric emptying time from 57 min in the fasted state to 102 min in the fed state. Loo-Reigelman analysis of the resulting serum flurbiprofen data showed a tendency toward lower initial absorption rate followed by a more rapid second phase when subjects received apple juice. Double peaks were observed in plasma profiles from both treatments, but these were less than those observed following a solution flurbiprofen dose. Double peaks are speculated to be related to the existence of two distinct absorption sites separated by a region of relatively lower absorption (PLUSQUELLEC et al. 1987) or the result of biphasic stomach emptying (CLEMENTS et al. 1978). An extensive range of studies conducted in healthy male subjects showed that food can markedly affect circulating levels of the HMG-CoA reductase inhibitor fluvastatin (SMITH et al. 1993). Peak plasma levels were reduced by 69%–74% and T_{max} was increased fourfold by a standard low-fat breakfast following a solution dose of fluvastatin, while peak levels were reduced 50%–60% and t_{max} values increased up to fourfold from a capsule dose. A similar, but somewhat attenuated effect was observed following a low-fat evening meal, with peak levels being reduced by 44% and the time of peak increasing by about 50%. In a third study, administering fluvastatin together with a carbonated beverage reduced peak plasma levels by 57% and t_{max} was increased from 0.8 to 1.1 h. In these studies systemic availability was reduced by 15%–20%. Comparison of efficacy data following the evening dose study showed no significant difference in the reduction in total cholesterol and LDL-cholesterol from baseline between fed and fasted subjects. Moderate food-effects resulting in small delays and/or reductions in peak circulating drug levels have been reported for fusidate sodium (MACGOWAN et al. 1989), hydroxychloroquine (MCLACHLAN and CUTLER 1991), isosorbide-5-mononitrate (KOSOGLOU et al. 1993), and lomefloxacin (HOOPER et al. 1990). A more impressive food effect is reported for the oral beta-lactam antibiotic loracarbef (DESANTE and ZECKEL 1992). While systemic availability was similar in subjects who received 200-mg doses in fasting and nonfasting states, peak plasma levels were reduced over 50% and the time of peak was doubled by food. In the same study, the area under the loracarbef serum curve was approximately doubled by coadministered probenecid due in part to an increase in the elimination $t_{1/2}$ from 1 to 1.5h. A less pronounced effect was obtained in a study in healthy male volunteers in which dosing after a standard meal resulted in a 29% reduction in C_{max} and a doubling of t_{max} compared to the fasting state. It is claimed that, despite the lowering of serum levels of loracarbef by food, levels were nonetheless well above MIC values for susceptible bacteria. Interestingly, administration of food significantly increased urinary excretion of loracarbef so that overall absorption efficiency may have actually been increased.

Two studies examined the effect of food on absorption rate and bioavailability of methotrexate. The first of these examined the effect of a standard high-fat breakfast on absorption of methotrexate from 7.5-mg tablet doses of the sodium salt in healthy male volunteers. Postprandial administration under these conditions resulted in a modest 16% reduction in peak plasma levels and a 30-min delay in t_{max}. The conclusion was drawn that food has no significant effect on the rate and extent of methotrexate absorption. A somewhat greater effect was observed when methotrexate was given fasting or after a standard French breakfast in patients with rheumatoid arthritis (OGUEY et al. 1992). In this case, C_{max} was reduced by 31% and t_{max} increased by 40min. Absolute systemic availability varied from 28% to 94% relative to intravenous methotrexate and was not influenced by food. Despite the somewhat greater reduction in peak levels in this study, it was again concluded that oral methotrexate can be given without regard to food intake. A far more dramatic effect was obtained when fluoride in the form of monofluorophosphate was administered after a standard meal compared to fasting (WARNEKE and SETNIKAR 1993). Food caused a 67% reduction in peak plasma fluoride levels, 369 to 122ng/ml, and increased t_{max} from 1h to over 2–5h. AUC and urinary excretion values were essentially unchanged. Monofluorophosphate is known to pass essentially unchanged through the stomach and to be rapidly absorbed in the upper small intestine. It is hydrolyzed to fluoride in the intestinal mucosa, portal blood, and liver (SETNIKAR and ARIGONI 1988). Thus, slow absorption in this case is attributed to delayed stomach emptying. Plasma fluoride profiles under fed and fasting conditions are shown in Fig. 9.

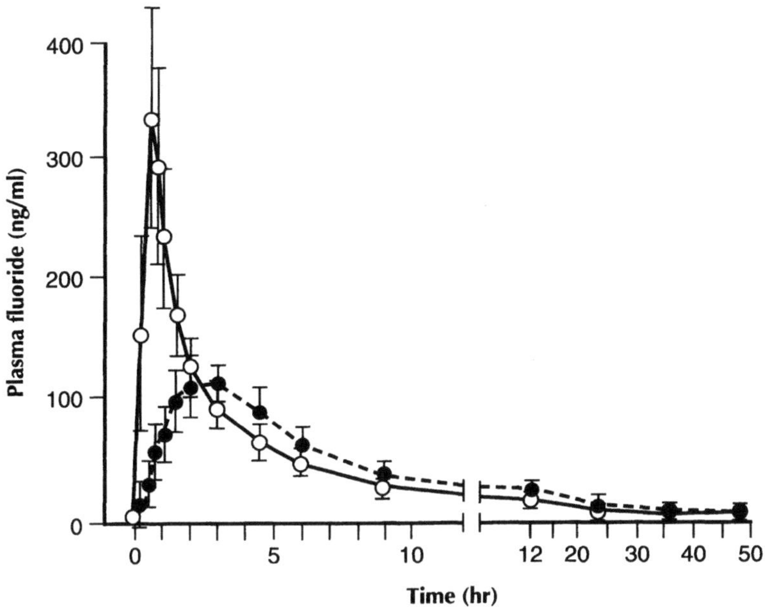

Fig. 9. Mean plasma versus time concentrations (±SD) of fluoride in eight subjects following a single 10-mg oral dose of fluoride as sodium monofluorophosphate fasting (○) or after a standard meal (●). (From WARNEKE and SETNIKAR 1993)

A high-fat meal markedly reduced the absorption rate of the calcium channel blocking agent nifedipine (RAPIN 1992). Peak plasma levels were reduced by 47% and T_{max} increased from 0.9 to 1.4h after a high-fat meal compared to fasting. The related compounds nicardipine and nitrendipine, on the other hand, were unaffected by food. A much smaller food effect was obtained in studies examining the effect of food, milk, and yoghurt on ofloxacin absorption (DUDLEY et al. 1991; NEUVONEN and KIVISTÖ 1992). Milk had no effect, yoghurt caused an 18% reduction and a moderate delay in peak plasma concentrations, while a standard meal gave rise to 20% reduction in peak plasma levels, again with a moderate, approximately 1h, delay. Ofloxacin is thus consistent with other fluoroquinolone antibiotics in that it is only modestly affected by food, at least under the conditions studied, and can generally be administered without regard to meal intake. Meal types have been shown to have different and probably not clinically significant effects on the rate of paracetamol absorption among South African Tswana ethnic groups (WESSELS et al. 1992). Absorption was delayed to an increasing extent by maize, bacon and egg, and cereal breakfasts. The bacon and egg meal reduced the paracetamol C_{max} from 16.9 to 12.6 μg/ml and increased T_{max} from 2.9 to 9.5h following a single 1000-mg paracetamol dose. The greatest delay by the high-fat meal is consistent with the expected impact of this meal on stomach emptying rate.

Food has been shown to decrease the absorption rate of the salicylic acid derivative salsalate (salicylsalicylic acid) (HARRISON et al. 1992). However, the effect is modest, and there was essentially no effect on circulating levels of the major metabolite salicylic acid. A similar modest effect has been reported for terazosin, whose absorption was moderately delayed by food, giving rise to a 23% reduction in C_{max} and an 0.9-h increase in T_{max} (MCNEIL et al. 1991). Consistent with this modest effect, the maximal fall in standing blood pressure after food was similar to that in the fasting state. Similar minor effects were observed with terfenadine, with peak levels of the active metabolite decreasing only 13%, with an 0.9-h delay following a standard high-fat breakfast compared to the fasted state in healthy volunteers (ELLER et al. 1992).

Two studies reporting delayed absorption of theophylline due to food have contributed further to the wide spectrum of food effects that have been reported for this compound and its many formulations. Absorption of theophylline from Theo-Dur tablets was delayed to a small extent, but mean peak levels increased from 4.7 to 6.3 mg/ml from a 400-mg postprandial evening dose compared to the fasting state. While these results are consistent with those reported previously regarding delayed absorption, the marked increase in C_{max}, together with observed intersubject variability in serum profiles obtained following evening doses, may have implications for asthmatic patients with nocturnal asthma who require consistent medication at night. In a second study, the onset and rate of gastric emptying were both delayed by a standard meal and the rate of theophylline absorption from a multiparticular controlled release formulation was also delayed, but to a lesser extent (YUEN et al. 1993). Calculations of the relative amounts of drug absorbed from regions of the GI tract showed that, in fasted individuals, 9% was absorbed from the stomach and 54% from the small intestine. In fed individuals 14% was absorbed from the stomach and 47% from the intestine. Following both treatments 37%–39% of the absorbed dose was absorbed from the colon, showing this to be an important absorption site, at least for sustained-release products of theophylline.

Delayed absorption due to food has been reported for the gamma-aminobutyric acid uptake inhibitor tiagabine (MENGEL et al. 1991). In a study in healthy male volunteers, peak plasma levels were reduced by 44% and T_{max} was increased almost threefold under nonfasting conditions compared to fasting conditions. AUC(0–48) values were similar for the two treatments. Similar observations were made for the structurally novel anticonvulsant topiramate, although peak plasma levels were reduced only 11% and 13% from single 100-mg or 400-mg doses and T_{max} was increased by only 1 h (DOOSE et al. 1992a). Absorption efficacy was unaffected by food. Similar modest delays in absorption have been reported for the serotonin uptake inhibitor trazodone (NILSEN and DALE 1992). A more complex study on valproate pharmacokinetics, on the other hand, shed some light on the influence of food on apparent "circadian rhythm" in absorption of this and possibly other compounds (OHDO et al. 1992). In a study conducted in healthy male subjects, valproate was administered following conventional light breakfast and heavy evening meals and also following identical morning and evening meals. There was no fasting treatment in this study. Following the light breakfast-heavy evening meal doses, absorption of valproate was delayed following the evening dose relative to the morning dose, C_{max} being reduced by 14% and AUC by 9%. When valproate was administered following identical morning and evening meals, there were no differences in the resulting plasma valproate profiles. The different results in the two treatments were attributed to differences in meal sizes and the results of the study led to speculation as to what extent different food effects may have contributed to many "circadian rhythm" effects previously reported for other drugs. Plasma levels obtained in this study are shown in Fig. 10.

Different meals also influenced the extent of drug-food interaction with the serotonin agonist/antagonist zalospirone. In a study in 24 healthy men, administration of a single 20-mg zalospirone dose after a meal containing 43% fat resulted in a 31% reduction in peak circulating drug levels compared to the fasting state. Administration after a meal containing 19% fat, on the other hand, resulted in only a 13% reduction. Peak drug levels were delayed twice as long after the high-fat meal as after the low-fat meal. AUC values and incidences of adverse events were similar for all dosage treatments.

Whereas previously cited reports have claimed delayed and reduced absorption of the AIDS drug zidovudine (see Table 3), other studies have reported only delayed absorption, with no influence by food on overall absorption efficacy (SAHAI et al. 1992; SHELTON et al. 1993). In a study in 18 asymptomatic HIV-infected subjects, C_{max} was reduced 57% and 47% when zidovudine was administered 30 min and 3 h after a high-fat breakfast, with moderate increases in T_{max}, while overall absorption efficacy was unaffected (SHELTON et al. 1993). In a second study (SAHAI et al. 1992), serum zidovudine C_{max} was reduced 32% and T_{max} increased from 49 to 74 min when drug was administered after a liquid protein meal relative to the fasted state in 11 symptomatic HIV-infected men. Again, absolute absorption values were unaffected. The different results obtained in these studies compared to those described in Table 3 present another example of the variable nature of drug-food interactions and the hazards of basing any conclusions on only one study from one laboratory, using a single set of study conditions.

F. Interactions Causing Increased Drug Absorption

Unlike the drugs and dosage forms described so far in this chapter, the rate and extent of systemic availability of the drugs and dosage forms described in this section are increased in the presence of food relative to the fasting state. Publications cited in Table 5 represent a substantial portion of the total number of reports on drug-food interactions during the review period, and reflect

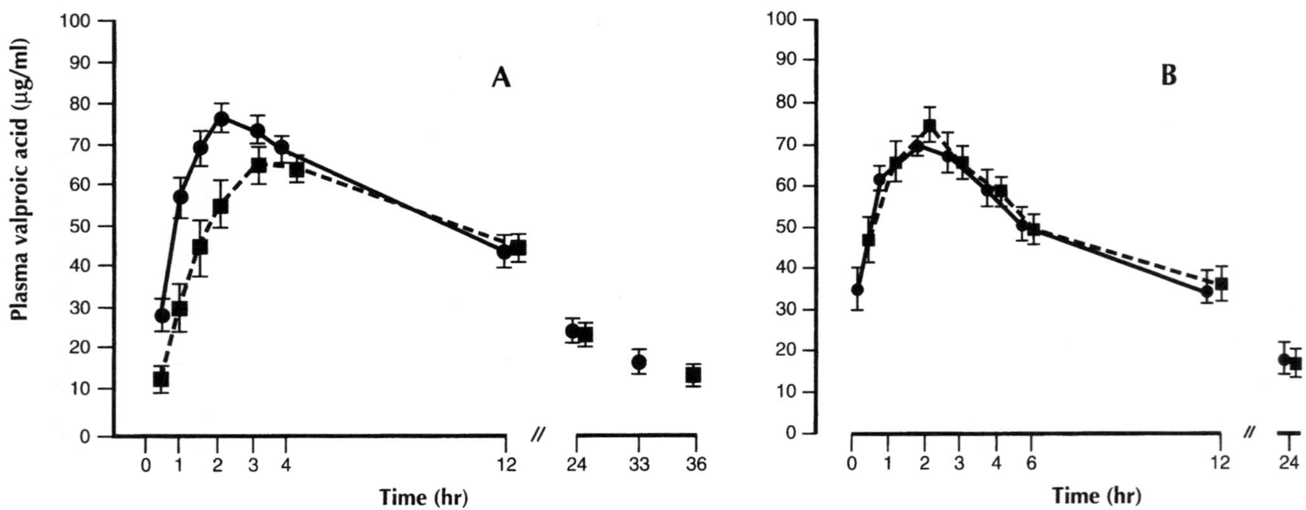

Fig. 10A,B. Mean plasma versus time concentrations (±SD) of valproic acid in 16 subjects following single 800-mg oral doses of valproic acid following a light breakfast (●) or heavy dinner (■) (**A**) or following a light breakfast (●) or light dinner (■) (**B**). (From Ohdo et al. 1992)

Table 5. Drug-food interactions resulting in increased drug absorption

Drug	Dosage form	Dose	Dosage regimen	Subjects	Food details	Fluid volume	Time interval	Sampling duration	Result of food administration	Reference
Alprazolam	Sustained release tablet	3 mg	SD[a]	29 healthy subjects	Standard high-fat meal	–[b]	1 h before, immediately following, 1 h after, and 2 h after meal	48 h blood	C_{max} increased 10%–17% after food. AUC marginally increased (≤6%). Psychomotor performance decrement consistent with changes in blood levles	Wright (1992)
Amiodarone	Solution	300 mg	SD	10 healthy male subjects	Nutrient Realmentyl[c] solution at two dilutions infused 120 min at ligament of Treitz	600 ml water	Drug administered in nutrient solution	24 h plasma, 2 h jejunal aspirations	Amiodarone absorption correlated with lipid absorption. Plasma levels variable	Pfeiffer et al. (1990)
Amocarzine	Film-coated breakable tablet	3 mg/kg	MD[d]	20 male Guatemalan patients infected with *Onchocerca volvulus*	Copious breakfast	100 ml water	1 h after meal	10 h plasma and urine	AUC increased 20%, T_{max} delayed 2 h, C_{max} unaffected. T_{max} of metabolite CGP 13231 delayed 2 h	Lecaillon et al. (1991)
Amocarzine	Film-coated tablet	1200 mg	SD	11 male patients infected with *Onchocerca volvulus*	Standard breakfast[c]	100 ml water	After meal	48 h plasma, 48 or 144 h urine	C_{max} and AUC of parent drug increased threefold. Urinary excretion increased from 0.11% to 0.13% of dose. C_{max}, AUC and urinary excretion of metabolite CGP 13231 increased threefold	Lecaillon et al. (1990)
Astemizole and Pseudoephedrine	Capsule	10 mg astemizole, 140 mg pseudoephedrine	SD	28 healthy subjects	Standard meal	–	With meal	96 h plasma	C_{max} for astemizole and metabolite desmethylastemizole delayed while AUC increased. Pseudoephedrine unaffected	Jallad et al. (1991)
Atovaquone	Tablet	500 mg	SD	18 healthy male subjects	High-fat breakfast[c]	–	45 min after meal	528 h plasma	C_{max} increased fivefold, AUC increased threefold	Rolan et al. (1994)
Atovaquone	Tablet	500 mg	SD	12 healthy male subjects	Toast, toast with low-fat or high-fat butter	–	45 min after meal	336 h plasma	C_{max} and AUC both increased with increasing fat meal. Following fat toast with 56 g butter. C_{max} and AUC increased 5.6- and 4-fold, respectively	Rolan et al. (1994)

Table 5. *Continued*

Drug	Dosage form	Dose	Dosage regimen	Subjects	Food details	Fluid volume	Time interval	Sampling duration	Result of food administration	Reference
Bay-X-1005	Tablet	500 mg	SD	4 healthy subjects	American breakfast	–	–	–	C_{max} and AUC increased 2-fold and 1.4-fold, respectively. T_{max} reduced from 4.4 to 2.2 h	Beckermann et al. (1993)
Brofaromine	Tablet	75 mg	SD	8 healthy male subjects	Standard breakfast	100 ml mineral water	Immediately after food	48 h plasma	C_{max} and AUC increased 1.2-fold. T_{max} reduced from 3 to 2 h	Degen et al. (1993)
Buflomedil	Tablet	600 mg	SD ^{111}In-labelled	7 healthy male subjects	Standard light and heavy breakfasts[c]	100 ml water	After meal	25 h serum, scintigraphy	C_{max} and AUC increased approximately 1.2-fold after heavy meal relative to light meal	Wilson et al. (1991)
Cefetamet pivoxil	Tablet	1000 mg	SD	6 healthy male subjects	–	–	–	–	C_{max} and AUC increased 1.3-fold. T_{max} delayed from 3.0 to 4.8 h	Blouin and Stoeckel (1993)
Cefetamet privoxi	Tablet	1000 mg	SD	16 healthy male subjects	Standard breakfast[c]	Various	With meal	24 h plasma and urine	C_{max} increased 1.2-fold, AUC increased 1.1-fold when taken 1 h after food compared to 1 h before	Tam et al. (1990)
Cefuroxime	Tablet	250 mg	SD	8 or 12 healthy fasting male subjects	Cholecystokinin or hyoscine butylbromide, no food	–	–	–	Hyoscine butylbromide mediated delayed gastric emptying had no effect. Cholecystokinin-mediated increased bile release caused 1.2-fold increases in C_{max} and AUC	MacKay et al. (1991)
CGP 43371	Capsule	800 mg	SD	12 healthy male subjects	Standard breakfast[c]	–	After meal	96 h plasma	C_{max} and AUC increased 11- and 14-fold, respectively. T_{max} unaffected	Sun et al. (1994)
Clarithromycin	Suspension	7.5 mg/kg	SD	24 infants and children with pharyngitis, otitis media, or skin infections	Milk for patients <2 years old, milk and hash brown potatoes for patients ≥2 years old	125 mg/5 ml	–	6 h plasma	C_{max} for parent drug and metabolite increased 1.3- and 1.1-fold. AUC increased 1.4- and 1.1-fold	Gan et al. (1992)

Clarithromycin	Tablet	500 mg	SD	27 healthy male subjects	Standard breakfast	240 ml water	30 min after start of meal	24 h plasma	C_{max} for parent and metabolite increased 1.5- and 1.1-fold, AUC increased 1.25- and 1.1-fold	Chu et al. (1992)
Cyclosporine	Chocolate emulsion	10 mg/kg	SD	8 healthy subjects, 4M, 4F	Low-fat and high-fat breakfast[c]	–	Midway through high-fat breakfast	24 h blood and plasma	Based on blood data, bioavailability was 23% and 42% after low-fat and high-fat meals. Based on plasma data bioavailability was 21% and 79% after low-fat and high-fat meals. Distribution volume and clearance increased with high-fat meals	Gupta et al. (1990)
Danazol	Capsule and lipid emulsion	100 mg	SD	11 healthy female subjects	Standard breakfast	–	15 min after meal	36 h plasma	Emulsion increased bioavailability fourfold relative to capsule, but was unaffected by food. Bioavailability of capsule was increased threefold by food	Charman et al. (1993)
Diltiazem	Extended release capsule	45 mg	SD	16 healthy male subjects	–	–	–	–	C_{max} and AUC increased 37% and 13%, respectively	Frishman (1993)
Encainide	–	35 mg	MD q8h	15 healthy male subjects	Standard breakfast[c]	120 ml water	Mid-meal	24 h serum	Bioavailability of encainide increased 1.8-fold and that of 0-desmethyl encainide 1.3-fold. 3-Methoxy-desmethyl encainide unchanged, as were electrocardiograms	Hilleman et al. (1992)
Felodipine and nifedipine	Tablet	5 mg 10 mg	SD SD	6 healthy male subjects	250 ml grapefruit juice or orange juice	250 ml water	With water or juice	8 h plasma	Felodipine and dehydrofelodipine AUC increased 2.5- and 1.7- fold by grapefruit juice. Nifedipine and dehydronifedipine AUC increased 1.4- and 1.2-fold. Orange juice had negligible effect. Felodipine had greater effects on diastolic blood pressure and heart rate when taken with grapefruit juice	Bailey et al. (1991)

Table 5. *Continued*

Drug	Dosage form	Dose	Dosage regimen	Subjects	Food details	Fluid volume	Time interval	Sampling duration	Result of food administration	Reference
Felodipine	Tablet	5 mg	SD	9 healthy male subjects	200 ml grapefruit juice of different strengths	200 ml water	With water or juice	8 h plasma	C_{max} increased 2.5- and 2.9-fold and AUC increased 2.9- and 3.3-fold by increasing strengths of grapefruit juice. Levels of dehydro-felodipine similarly increased	Edgar et al. (1992)
Fenretidine	Capsule	300 mg	SD	13 healthy male subjects	High-fat breakfast[c]	200 ml water	Immediately after meal	72 h plasma	Fenretidine C_{max} and AUC increased threefold, metabolite C_{max} and AUC increased 2.3- and 2.5-fold	Doose et al. (1992b)
Fenretidine	Capsule	300 mg	SD	15 healthy male subjects	High-fat, -carbohydrate or -protein meals[c]	200 ml water	10 min after meal	72 h plasma	High-fat meal increased bioavailability threefold relative to carbohydrate meal. High-protein meal gave intermediate values	Doose et al. (1992b)
Gepirone	Capsule	20 mg	SD	20 healthy male subjects	Standard breakfast	200 ml water	15 min after meal	48 h plasma	AUC increased 1.4-fold. C_{max} slightly reduced and delayed	Tay et al. (1993)
Itraconazole	Capsule	100 mg	SD	6 healthy male subjects	Standard breakfast[c]	200 ml water	Immediately after meal	96 h plasma	C_{max} increased 3.4-fold, AUC 2.6-fold	Van Peer et al. (1989)
Itraconazole	Capsule	200 mg	SD	28 healthy male subjects	Standard breakfast[c]	200 ml water	Immediately after meal	72 h plasma	C_{max} increased 2- fold, AUC 1.6-fold. Metabolite C_{max} increased 1.4-fold, AUC 1.5-fold	Barone et al. (1993)
Itraconazole and fluconazole	Capsules	100 mg	–	12 healthy subjects, 7M, 5F for each drug	Standard light or full breakfast[c]	100 ml water	5 min after meal	96 h plasma	Itraconazole bioavailability increased 1.4-fold by light meal, and 1.7-fold by heavy meal. Fluconazole unaffected	Zimmermann et al. (1994)
Levodopa	Immediate or controlled release tablet	100 mg IR 200 mg CR	SD	5 healthy male subjects	Light breakfast	–	After meal	8 h plasma	Absolute bioavailability relative to intravenous dose: IR fed > IR fasted > CR fed > CR fasted. Food increased absorption from both formulations	Wilding et al. (1991b)
Levodopa	–	Variable	MD	16 patients with idiopathic Parkinson's disease and severe constipation, 6M, 13F	Diet rich in soluble fiber (DRIF)[c]	–	Few min after fiber supplement	6 h plasma	Total plasma levodopa levels increased 1.3-fold, levels during first h postdose increased 1.5-fold, after 2 weeks on DRIF	Astarloa et al. (1992)

5-Methoxy-psoralen	Tablet	20 mg	SD	9 healthy volunteers, 1M, 8F	Standard breakfast[c]	100 ml water	20–70 min after meal	10 h plasma	C_{max} ranged from 0 to 88 ng/ml fasting and 37–141 ng/ml nonfasting. AUC ranged from 0 to 422 ng·h/ml nonfasting	Ehrsson et al. (1994)
Moclobemide	–	100 mg	SD	8 healthy male subjects	Standard high-protein breakfast[c]	–	–	9 h plasma	Trend toward higher absolute bioavailability after protein meal (67%) compared to fasting (58%)	A.F.D. Cole et al. (1992)
Nifedipine	Extended release tablet	40 mg	–	20 healthy male subjects	Low-fat and high-fat meal	–	45 min after meal	48 h blood	C_{max} increased 1.8-fold and 2.4-fold, AUC increased 1.2-fold and 1.2-fold, by low-fat and high-fat meals, respectively	Kleinbloesem et al. (1993)
Oxcarbazepine	Tablet	600 mg	SD	6 healthy male subjects	Standardized high-fat, high-protein breakfast	100 ml water	Immediately after meal	72 h plasma	C_{max} and AUC of parent drug increased 1.7- and 1.2-fold, respectively. Values for monohydroxy and dihydroxy metabolites also increased	Degen et al. (1994)
Oxybutinin	Solution	15 mg/kg	SD	18 healthy male subjects	High-fat breakfast	–	With meal	–	Bioavailability increased 1.3-fold	Yong et al. (1991)
Phenytoin	Powder	5 mg/kg	–	4 healthy male subjects	Low-fat and high-fat breakfast[c]	200 ml water	Immediately after meal	34 h plasma	C_{max} increased 1.5- and 2.2-fold by low-fat and high-fat meals, respectively. AUC increased 1.5- and 1.9-fold	Hamaguchi et al. (1993)
Progesterone	Capsule	200 mg	MD 5 days	15 healthy postmenopausal women	Standard breakfast[c]	–	Immediately after or 2 h before	24 h plasma on day 1, 10 h plasma on day 5	On day 1, C_{max} increased 5-fold and AUC 2-fold. Similar effects on day 5	Simon et al. (1993)
Repirinast	Tablet	300 mg	SD	12 healthy male subjects	High-carbohydrate or high-fat breakfast[c]	100 ml water	Immediately after meal	72 h plasma	C_{max} and AUC increased 1.6- and 1.9-fold by high-carbohydrate breakfast and 3.2- and 2.4-fold by high-fat breakfast	Schaefer et al. (1993)
Sparfloxacin	–	300 mg	SD	3 healthy subjects	–	–	–	–	C_{max} increased 1.5-fold, AUC 1.2-fold, and urinary excretion 1.4-fold	Tanimura et al. (1991)
S-1108	–	200 mg	SD	6 healthy male subjcets	Japanese breakfast ~500 Kcal	100 ml water	Immediately after meal	10 h blood, 24 h urine	C_{max} increased 1.1-fold, AUC 1.4-fold	Saito (1993)
S-1108	–	100 mg	SD	Healthy male subjects	–	–	–	–	C_{max} increased 1.4-fold, AUC 1.5-fold. T_{max} increased from 1.4 to 2.5 h	Nakashima et al. (1993)

Table 5. *Continued*

Drug	Dosage form	Dose	Dosage regimen	Subjects	Food details	Fluid volume	Time interval	Sampling duration	Result of food administration	Reference
Theophylline	Capsule	900 mg	SD	20 healthy subjects	High-fat breakfast[c]	–	15 min after meal	72 h serum	C_{max} increased 1.6-fold, AUC 1.2-fold	Cook et al. (1990)
Ticlopidine	Tablet	250 mg	SD	12 healthy male subjects	High-fat breakfast[c]	200 ml water	30 min after meal	24 h plasma	C_{max} and AUC increased 1.2-fold	Shah et al. (1990)
Tramadol	Tablet	100 mg	SD	18 healthy male subjects	High-fat breakfast	–	Immediately after meal	36 h plasma	C_{max} and AUC of tramadol increased 1.2- and 1.1-fold, respectively. Metabolite levels only moderately affected	Liao et al. (1992)
Vanoxerine	Tablet	100 mg	SD	12 healthy male subjects	High-fat or low-fat breakfast[c]	100 ml water	After meal	–	C_{max} increased 1.5-fold by high-fat meal, unaffected by low-fat meal. AUC increased 1.8-fold and 3.6-fold by low-fat and high-fat meals, respectively	Ingwersen et al. (1993)
Vinpocetine	Tablet	40 mg	SD	8 healthy subjects, 4M, 4F	Standard breakfast, 1350 Kcal	150 ml mineral water	10 min before start of meal, 10 min after start of meal, 20 min after termination of meal	12 h serum	C_{max} increased 1.6-, 2.3-, and 1.9-fold while AUC increased 1.6-, 1.7-, and 2.0-fold when drug was taken 10 min before, 10 min after starting, and 30 min after finishing meal	Lohmann et al. (1992)
Zalospirone	–	10 mg	MD q 8 h	24 18- to 45-year-old and 24 ≥ 65-year-old male and female subjects	Medium-fat meal	–	30 min after meal	–	C_{max} and AUC increased twofold. Elderly subjects gave higher levels than younger subjects	Klamerus et al. (1993)
566C80	Tablet	500 mg	SD	18 healthy male subjects	–	–	45 min after meal	–	C_{max} increased 5.4-fold, AUC 3.3-fold	Rolan et al. (1992)
566C80	Tablet	500 mg	SD	12 healthy male subjects	Toast, toast plus 28 g butter (LOFAT), toast plus 56 g butter (HIFAT)	–	–	–	C_{max} increased 2.5- and 3.2-fold after LOFAT and HIFAT, respectively, compared to plain toast.	Rolan et al. (1992)

[a] Data not available.
[b] Single dose.
[c] Not relevant.
[d] Meal described.
[e] Multiple doses.

again, not only the broad spectrum of effects that may result from drug-food interactions, but also their frequent unpredictability. The compounds in this section tend to be lipophilic and poorly water soluble, but this is not always the case. Similarly some interactions are relatively trivial and do not warrant further discussion, while others are substantial and likely to be clinically significant.

Amiodarone is an amphiphilic substance and its bioavailability is extremely variable (LATINI et al. 1984; RIVA et al. 1982). In order to examine its absorption in the presence of food, amiodarone was administered into the jejunum together with two nutrient solutions, one 3.3 Kcal/min and the other 1.3 Kcal/min, with the former containing total lipid and caloric load 2.5 times greater than the latter (PFEIFFER et al. 1990). Based on intestinal testing, absorption of amiodarone correlated significantly with lipid absorption rate. However, plasma C_{max} and AUC values were extremely variable and tended to be higher from the 1.3-Kcal/min infusion. These seemingly paradoxical results were attributed to wide fluctuations in amiodarone pharmacokinetics, distribution, and metabolism.

Conflicting results from drug-food interactions have been reported with the onchocerciasis agent amocarzine (CGP 6140) (LECAILLON et al. 1991). In 20 male Guatemalan patients who received a 3-mg/kg oral dose, systemic availability was increased 20% when drug was taken with a "copious" breakfast compared to fasting. In a similar study, when the dose was increased to 1200 mg, both the peak plasma levels and systemic availability of amocarzine were increased approximately threefold when drug was given after a meat and noodle standard breakfast, relative to fasting (LECAILLON et al. 1990). The AUC of the metabolite CGP 13231 also increased threefold in this study, although it was unaffected at the lower dose. Mean plasma levels of amocarzine and its metabolite CGP 13231 are shown in Fig. 11.

The considerable increase in absorption due to food after the high dose of amocarzine may be related to a greater degree of solubilization after the meal or to decreased presystemic metabolism. The nature of the food in the lower-dose study was not described, but is likely to have been similar to that used in the high-dose study. A controlled high-dose, low-dose study would provide useful mechanistic information on this interaction. Substantially increased absorption due to food has been reported for the lipophilic antiprotozoal agent atovaquone (ROLAN et al. 1994). In healthy male volunteers, peak atovaquone plasma levels increased over fivefold while systemic bioavailability increased over threefold when drug was given 45 min after a Western-style high-fat breakfast compared to fasting. Mean plasma profiles obtained over 48 and 528 h are shown in Fig. 12.

Complementary studies using meals with different fat content, aqueous suspensions, and oily emulsion vehicles, and also predosing with a cholecystokinin octapeptide, led to the conclusion that the observed food effect with atovaquone was probably due to the combined effects of bile release and also increased solubility due to a direct effect of the fatty meal.

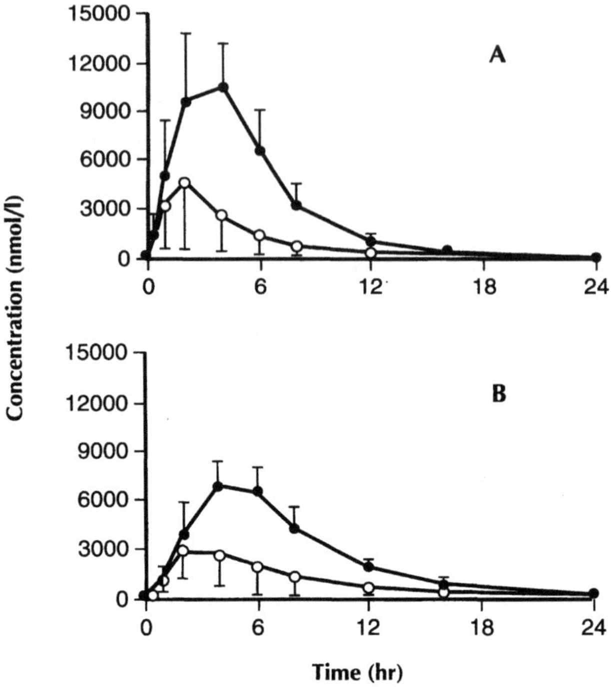

Fig. 11A,B. Mean plasma versus time concentrations (±SD) of CGP 6140 (**A**) and its *N*-oxide metabolite CGP 13231 (**B**) in 11 patients following a single 1200-mg oral dose of CGP 6140 fasting (○) or following a large breakfast (●). (From LECAILLON et al. 1990)

More modest drug-food interactions are reported for the 5-lipoxygenase inhibitor BAY-X-1005 (BECKERMANN et al. 1993), the selected monoamine oxidase-A inhibitor brofaromine (DEGEN et al. 1993), and the vasoactive agent buflomedil (WILSON et al. 1991). With all three compounds, administration after a standard breakfast resulted in 1.2-fold to 2-fold increases in absorption. In the case of buflomedil, the extent of absorption was increased 1.2-fold after a heavy meal compared to a light meal, while increases for the other agents were observed relative to the fasting state.

Consistent with some previous observations on other cephalosporins (WELLING 1977), food has been shown to delay but also to moderately increase absorption of cefetamet pivoxil. Absorption of cefetamet pivoxil, the pivaloyloxymethyl ester prodrug of cefetamet, was delayed by food, and mean peak plasma levels occurred at 4.8h compared to 3h fasting. Overall bioavailability and peak plasma levels increased approximately 25%–30% (BLOUIN and STOECKEL 1993). A similar effect to this was observed when cefetamet pivoxil was administered 1h after a standard breakfast, although plasma profiles were similar when drug was administered either with or 1h before the standard breakfast (TAM et al. 1990).

Increasing fluid volume had no effect on absorption. Plasma profiles ob-

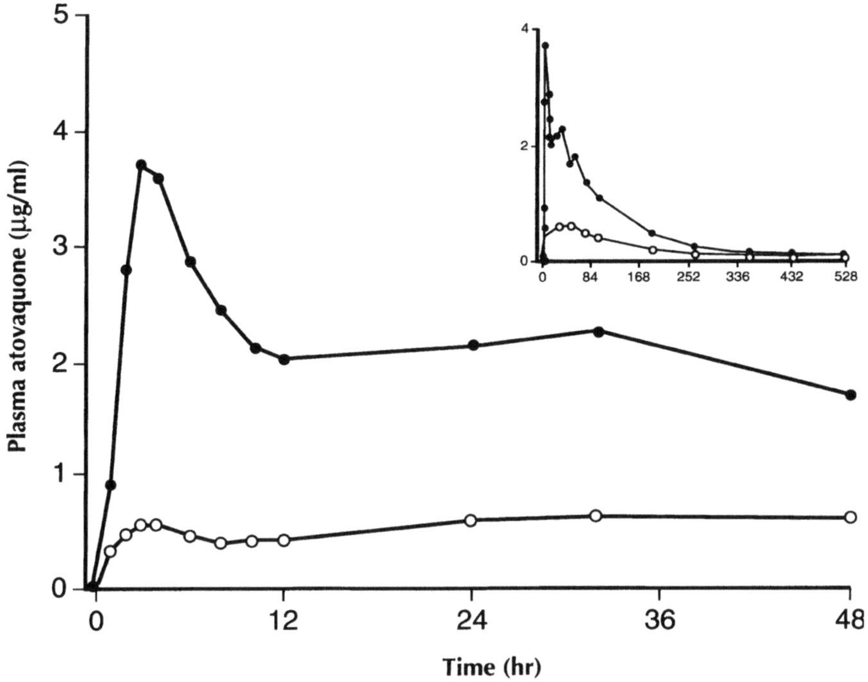

Fig. 12. Mean plasma versus time concentrations of atovaquone in 18 subjects following a single 500-mg oral dose of atovaquone fasted (○) or 45 min after a high-fat meal (●). (From Rolan et al. 1994)

tained with cefetamet pivoxil administered under fasting and nonfasting conditions, and with different fluid volumes, are shown in Fig. 13. The lack of effect of fluid volume in this study may have been due to the relatively large fluid volumes used, 250 ml and 450 ml. However, the lack of a food effect when drug was administered directly with the meal was unexpected. A proposed explanation is that while drug was taken with water 1 h before and 1 h after breakfast, it was taken with the tea or coffee provided for the group that received the drug with food. Previous studies have shown that bioavailability of cefuroxime is increased when taken with food (Williams and Harding 1984). In an attempt to identify the mechanism of this interaction, cefuroxime was administered to healthy male subjects 30 min before intravenous hyoscine butylbromide or immediately before intravenous cholecystokinin 8 (MacKay et al. 1991). Hyoscine butylbromide had no effect on cefuroxime absorption while cholecystokinin resulted in a 20% increase in cefuroxime C_{max} and AUC values. These results led to the conclusion that bile release, but not gastric emptying, may be at least partially responsible for increased cefuroxime absorption in the presence of food.

Possibly the most remarkable food effect reported during the review period involved the lipophilic hypolipidemic compound CGP 43371 (Sun et al. 1994). Administration of single 800-mg capsule doses of CGP 43371 after a

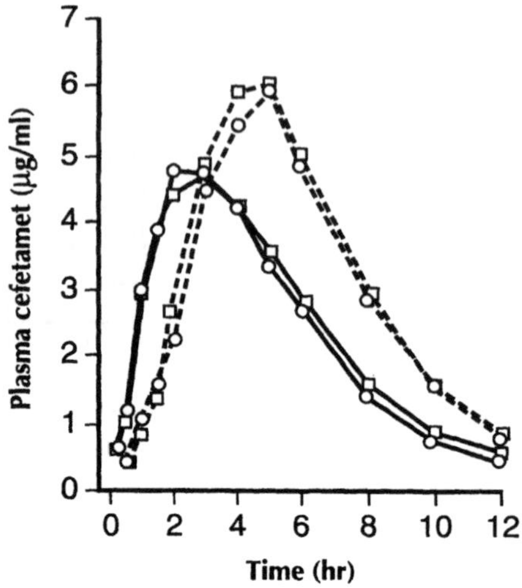

Fig. 13. Mean plasma versus time concentrations of cefetamet in 12 subjects following a single 1000-mg oral dose of cefetamet pivoxil under fasted (—) or fed (---) conditions with 250 ml (○) or 450 ml (□) water. (From TAM et al. 1990)

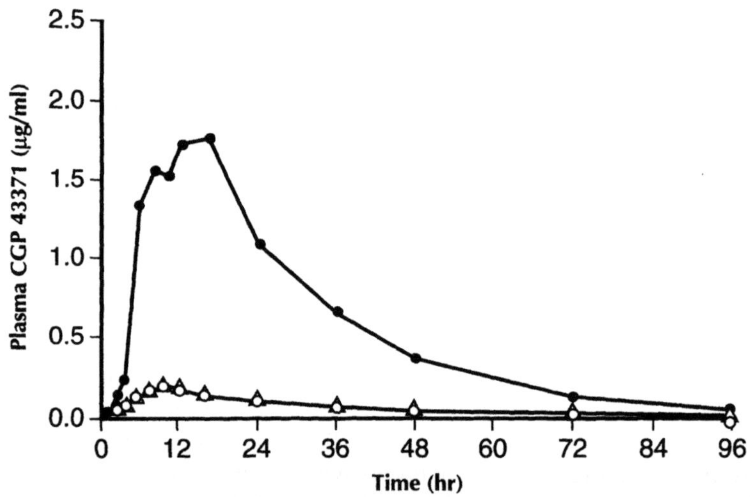

Fig. 14. Mean plasma versus time concentrations of CGP 43371 in 12 subjects following a single 800-mg oral dose of CGP 43371 as a dispersion (△) or capsule (○) under fasting conditions or as a capsule after a standard meal (●). (From SUN et al. 1994)

standard breakfast caused an 11-fold increase in peak plasma drug levels and a 13-fold increase in overall bioavailability relative to the fasting state. Plasma levels from this study are shown in Fig. 14.

It is proposed that, as CGP 43371 is absorbed mainly from the ileum, delayed gastric emptying would enable more compound to disintegrate and dissolve before reaching this absorption site. It is further proposed that CGP 43371 dosage should be modified depending on dosing relative to food intake. This study provides a truly dramatic example of the extent to which drug-food interactions can influence circulating drug profiles, with obvious clinical implications.

Similar food effects were observed in infants and adults for the new macrolide antibiotic clarithromycin (CHU et al. 1992; GAN et al. 1992). Following a 7.5-mg/kg dose to infants and children aged 6 months to 10 years with various infections, either fasting or after milk and/or hash brown potatoes, peak plasma levels were 4.6 μg/ml and 3.6 μg/ml after nonfasting and fasting doses, respectively (GAN et al. 1992). Although these increases were modest, systemic availability increased by 40%, indicating better overall absorption with food. In the study in adults, food taken immediately before a 500-mg clarithromycin dose increased the extent of absorption by approximately 25% (CHU et al. 1992). In both of these studies plasma levels of the major active metabolite 14-hydroxyclarithromycin were moderately increased by approximately 10%. Different results were obtained from blood and plasma analysis regarding the effect of low-fat and high-fat meals on cyclosporine absorption (GUPTA et al. 1990). Based on plasma data, cyclosporine systemic availability was 23% and 42% after low-fat and high-fat meals, respectively. Based on blood data, relative values were 21% and 79%. Thus, the apparent increase in cyclosporine bioavailability appears to depend on the sample matrix. This in turn is probably related to differential penetration of cyclosporine into red cells depending on the amount of absorbed fat. Plasma to blood cyclosporine clearance ratios were 1.8 and 1.3 after high-fat and low-fat meals, respectively. Other studies have shown that high-fat meals increased cyclosporine clearance, but not mean residence time, based on either blood or plasma measurements (GUPTA and BENET 1990).

Absorption of the heterocyclic steroid derivative danazol (CHARMAN et al. 1993) and also the retinoid fenretinide (DOOSE et al. 1992b) is substantially increased by food. In the case of danazol, systemic availability from a capsule dose was increased over threefold by food in healthy female subjects, resulting in mean peak plasma levels of 37 and 101 ng/ml, after fasting and fed doses, respectively. In the case of fenretidine, bioavailability and peak plasma levels both increased threefold following a high-fat meal compared to fasting (DOOSE et al. 1992). Administration of fenretidine in an oil suspension to fasting subjects yielded intermediate values. Further examination of the effect of meal composition showed that a high-fat meal resulted in plasma fenretinide bioavailability three times greater than a carbohydrate meal, with a high-protein meal yielding intermediate results. Mean plasma profiles obtained in these studies are shown in Figs. 15 and 16. These results give rise to the recommendation that fenretinide be given with food in order to optimize potential therapeutic benefit.

A specific drug-food interaction has recently been identified between a component in grapefruit juice and the calcium antagonist felodipine (BAILEY

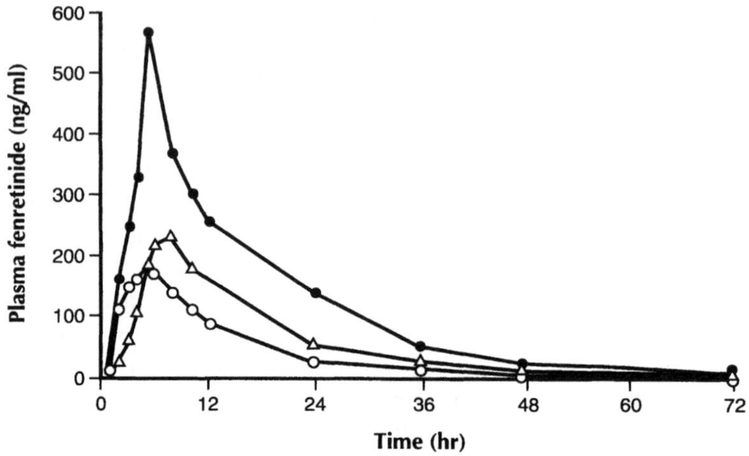

Fig. 15. Mean plasma versus time concentrations of fenretinide in 13 subjects following a single 300-mg oral dose of fenretidine fasting (○), after a meal (●), and as a 20-ml neutral oil suspension administered fasting (△). (From DOOSE et al. 1992b)

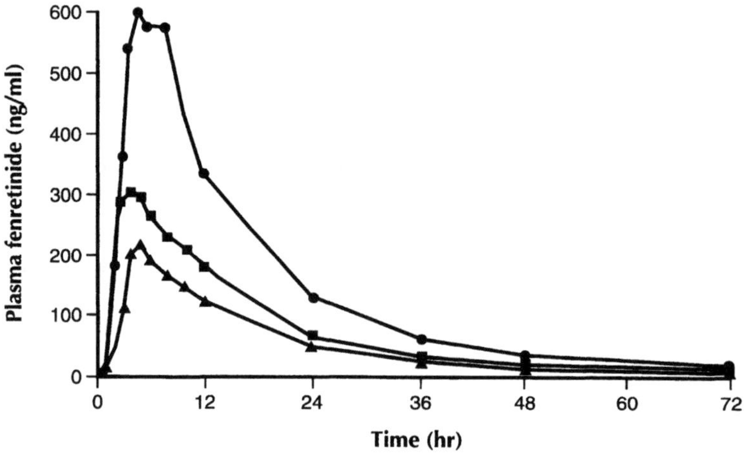

Fig. 16. Mean plasma versus time concentrations of fenretinide in 15 subjects following a single 300-mg oral dose of fenretidine after a high-fat (●), high-protein (□), or high-carbohydrate (▲) meal. (From DOOSE et al. 1992b)

et al. 1991; EDGAR et al. 1992). In a study in six men with borderline hypertension, felodipine and dehydrofelodipine systemic availability increased 2.5- and 1.7-fold, respectively, when felodipine was taken with two 250-ml double-strength grapefruit juices, relative to water (BAILEY et al. 1991). Under the same conditions, plasma levels of nifedipine and dehydronorfedipine increased 1.4- and 1.2-fold. The results with felodipine were reproduced in another study in nine healthy middle-aged men (EDGAR et al. 1992). The interaction with grapefruit juice, which is believed to be a class effect for the dihydropyridines, is thought to be due to inhibition of first-pass oxidative metabolism by flavonoids in the grapefruit juice. However, the precise mechanism of interaction has yet to be identified.

Food has been shown to increase the systemic availability of the antifungal agent itraconazole under a variety of conditions. Peak plasma itraconazole levels increased 3.4-fold, and systemic availability 2.6-fold, in six healthy male subjects following a standard breakfast compared to fasting (VAN PEER et al. 1989). Values for these parameters were increased 2- and 1.6-fold following a standard breakfast in an expanded study in 28 male volunteers (BARONE et al. 1993). In contrast to itraconazole, absorption of the closely related compound fluconazole was relatively insensitive to food, both C_{max} and AUC being slightly reduced by a full meal and essentially unchanged by a light meal, relative to fasting. While these divergent results are consistent with previous data on these agents, there is no mechanistic explanation for their different behavior. Food effects on absorption of levodopa were influenced by the type of meal but to a lesser extent by the dosage form. In six healthy volunteers levodopa absolute bioavailability from an immediate release dosage form was 86.4% and 80.4% from fed and fasting treatments, respectively. Levodopa availability from a controlled release dosage form was 71% and 63.6% from fed and fasted treatments (WILDING et al. 1991b). Thus, although the controlled release dosage form yielded lower absolute bioavailability, the food effect was similar for both formulations, increasing bioavailability by approximately 10%. On the other hand, a diet rich in insoluble fiber (DRIF) was shown to increase levodopa plasma levels by 30%, and levels during the 1st h postdose by 50%, in patients after 2 weeks on a DRIF diet compared to baseline (ASTARLOA et al. 1992). Thus, the DRIF may serve a useful dual purpose of relieving constipation and also increasing plasma levels and presumably effectiveness of levodopa. As the effect on levodopa levels occurred mainly during early periods after drug administration, it appears that levodopa absorption was accelerated due to increased gastrointestinal motility and shorter gastric emptying time.

While the effect of food on nifedipine absorption from conventional tablets is somewhat variable, food appears to give rise to a moderate but significant increase in nifedipine absorption from an extended release tablet (KLEINBLOESEM et al. 1993). Peak plasma nifedipine concentrations were increased 1.8- and 2.4-fold by low-fat and high-fat meals, respectively. Overall systemic availability was increased 1.2-fold by both treatments. Increased nifedipine plasma levels appeared to be without effect on blood pressure relative to the fasting state, but mean heart rate increased by 10 bpm after both postprandial doses compared to 5 bpm fasting. Peak plasma levels and systemic availability of the major monohydroxy metabolite of the antiepileptic compound oxcarbazepine were increased 1.2- and 1.5-fold when a 600-mg tablet dose was taken after a high-fat, high-protein breakfast (DEGEN et al. 1994). Plasma concentrations of the dihydroxy metabolite were similarly affected by food intake. Powders of different particle size have been shown to give rise to different food-drug interactions for the anticonvulsant agent phenytoin. Using a commercial Japanese powder of mean particle size 190 μM, systemic availability was increased 1.5- and 2.2-fold by low-fat and high-fat meals (Hamaguchi et al. 1993). Plasma profiles in four subjects are shown in Fig. 17. The marked increase in plasma phenytoin levels due to food in this study differs from results obtained earlier by Sekikawa et al. (1980), who used a Japanese commercial powder of mean particle size 47.1 μM. The different results obtained in the two studies are probably related to slower dissolution from the larger particle size formulation in the fasted state and a greater, positive effect on the systemic availability from this formulation due to the direct influence of food and also delayed gastric emptying. A far more dramatic food effect occurred with oral micronized progesterone (Simon et al. 1993). Repeated doses of micronized progesterone were administered in capsules to 15 healthy postmenopausal women, either 2h before or immediately after a standard breakfast for 5 days. Plasma profiles were obtained on days 1 and 5. On day 1, peak plasma levels of progesterone, obtained by radioimmunoassay, were increased fivefold and systemic availability twofold by food. Similar results were obtained on day 5. Increased progesterone absorption due to food was attributed either to a direct drug-food interaction in the gastrointestinal tract or to increased blood flow to the liver causing decreased presystemic clearance. No attempt was made to further examine or differentiate these possible mechanisms. A further example of differential food effects

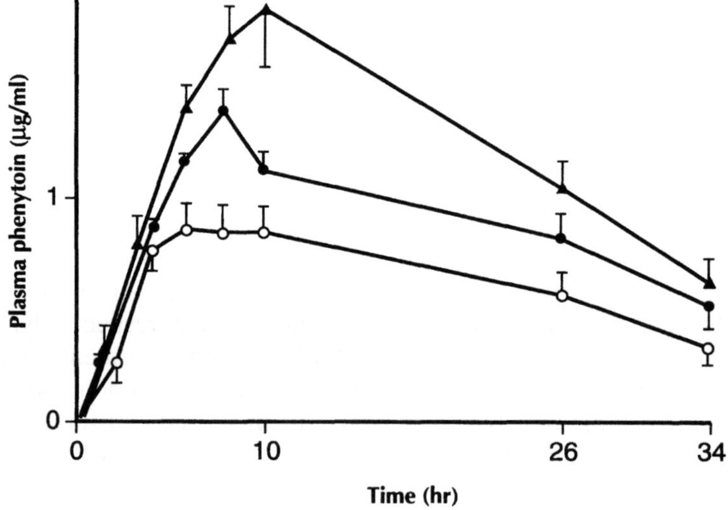

Fig. 17. Mean plasma versus time concentrations (±SE) of phenytoin in four subjects following a single 5-mg/kg oral dose of free acid phenytoin powder with large particle size fasting (○), and after low-fat (●) and high-fat (▲) meals. (From Hamaguchi et al. 1993)

on the extent of drug-food interaction is provided by the antiallergy agent repirinast (SCHAEFER et al. 1993). In a study conducted in 12 healthy men, peak plasma levels and systemic availability of the active de-esterified metabolite were increased 1.6- and 1.9-fold by a high-carbohydrate continental breakfast, and 3.2- and 2.4-fold by a high-fat American breakfast. The terminal elimination half-life of the active metabolite from plasma was considerably prolonged by both meals (12.7 h vs 2.5 h for the high-carbohydrate meal and 15.4 h vs 7.1 h for the high-fat meal), suggesting prolonged absorption in the nonfasting state.

Similar to observations with cefetamet pivoxil (BLOUIN and STOECKEL 1993; TAM et al. 1990), food had a positive effect on the new ester-type oral cephalosporin S-1108. In two separate studies conducted in Japan the systemic availability of S-1108 was increased approximately 1.5-fold, and peak plasma levels 1.1- to 1.4-fold, following a Japanese-style breakfast (NAKASHIMA et al. 1993; SAITO 1993). Urinary recovery was increased by both treatments in both studies. S-1108 was also administered to patients fasting 10 min before intramuscular ceruletide diethylamine, or 90 min after oral ranitidine hydrochloride. Ceruletide diethylamine had no effect on S-1108 absorption, but T_{max} was delayed. Ranitidine, on the other hand, had a negative effect on S-1108 absorption. Thus, neither increased bile flow nor increased gastric pH seemed to contribute to any food-related increase in S-1108 absorption. Delayed gastric emptying appears to be a reasonable alternative.

Further studies with theophylline demonstrated increased absorption due to food from single oral doses of Theo-24 capsules to healthy subjects (COOK et al. 1990). Mean peak plasma theophylline levels increased 1.6-fold and systemic availability 1.2-fold when this theophylline formulation was taken immediately after a high-fat breakfast compared to the fasting state. Additional experiments comparing dog and man showed that absorption of theophylline in the two species followed the same pattern regarding food effects and also for different theophylline formulations, thus providing support, for these theophylline products and study conditions at least, for predictability of dog bioavailability data in comparison to man for commercial oral formulations. Modest food-related increases have been reported for the platelet inhibitor ticlopidine (SHAH et al. 1990) and the analgesic agent tramadol (LIAO et al. 1992). A more dramatic interaction was observed with the piperazine derivative dopamine reuptake inhibitor vanoxerine (INGWERSEN et al. 1993). Administration of vanoxerine as a 100-mg dose in 4 × 25-mg tablets to 12 healthy men after low-fat and high-fat breakfasts increased systemic availability 1.8-fold and 3.6-fold, respectively. Despite the considerable increase in systemic availability after the high-fat meal, C_{max} was increased less than twofold due to delayed absorption. Peak drug plasma levels occurred at 2.5 h after the high-fat meal compared to 45 min for fasting. The mechanism of increased absorption in this study was not addressed. However, one subject who was virtually unaffected by food intake was a "poor metabolizer" of debrisoquine, which suggested that decreased first-pass metabolism, possibly related to increased splanchnic blood flow, may have contributed to the food effect in the other subjects.

Absorption of the nootropic agent vinpocetine and also the 5-HT_{1a} partial antagonist zalospirone is modestly increased by food (LOHMANN et al. 1992; KLAMERUS et al. 1993). Administration of vinpocetine tablets 10min before and 10min after starting and 30min after a standard 1350-Kcal breakfast increased systemic availability 1.6-, 1.7-, and 2.0-fold relative to fasting. Peak plasma levels were affected similarly. Mean plasma profiles in eight healthy male and female subjects in this study are shown in Fig. 18. Peak plasma levels and areas under plasma curves of zalospirone were increased approximately 1.4-fold by food in both young (18–45 years old) and elderly (≥65 years old) subjects (KLAMERUS et al. 1993). In addition, plasma levels were almost doubled in elderly subjects relative to young subjects. The results of these studies led to the recommendation that both vinpocetine and zalospirine be taken with or after meals.

The last drug considered in this category illustrates again the dramatic positive effect that food can have on circulating drug profiles. In a study in 18 healthy men, the systemic availability of a novel antiprotozoal agent 566C80 was increased 3.3-fold, while C_{max} was increased 5.4-fold when administered

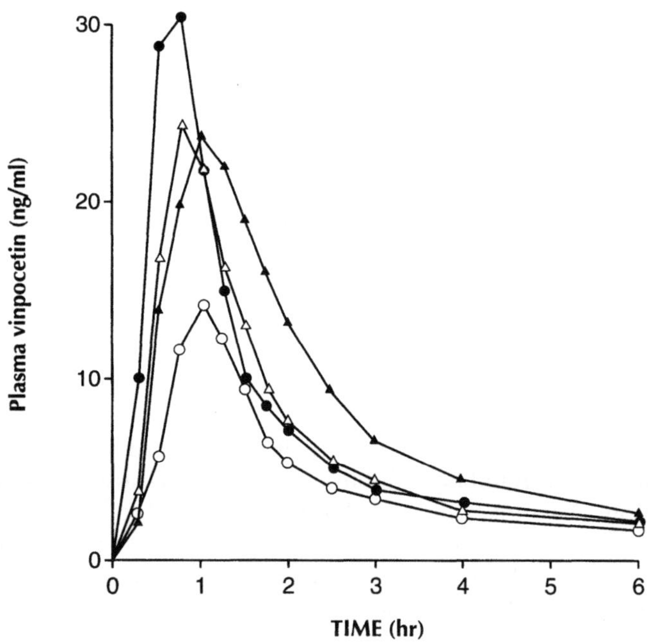

Fig. 18. Mean plasma versus time concentrations of vinpocetine in eight subjects following a single 10-mg oral dose of vinpocetine fasting (○) and 10min before (△), 10min after (●), and 30min after (▲) starting a standard breakfast. (From LOHMANN et al. 1992)

after food (HUGHES et al. 1991). In attempts to elucidate the mechanism of this interaction, 566C80 was given fasted and with meals of varying fat content, as an aqueous suspension, and as a oily emulsion. In a related study 566C80 was given after an infusion of cholecystokinin octapeptide (CCK-OP) (ROLAN et al. 1992). Pharmacokinetic results from these studies led to the conclusion that increased absorption of 566C80 after food could be quantitatively accounted for by dietary fat and that stimulation of bile secretion may be a small component of the food effect.

G. Interactions Causing Accelerated Drug Absorption

Although, as described in the previous section, food may often increase drug absorption and this may be accompanied by faster absorption rates, there have been few reports of food increasing the rate of drug absorption but having no effect on overall systemic bioavailability. Three examples of this phenomenon are shown in Table 6 and are described here.

The nonacidic, nonsteroidal anti-inflammatory agent nabumetone is converted in the liver to the active metabolite 6-methoxy-2-naphthylacetic acid (6MNA). This is the predominant form circulating in plasma. Administration of nabumetone to healthy individuals with food resulted in a 1.5-fold increase in peak 6MNA plasma levels, but overall systemic availability was unaffected, relative to fasting (HYNECK 1992). Administration of nabumetone with milk or antacid had a similar effect on 6MNA plasma profiles. The lipophilic character of nabumetone may account for increased absorption rates in the presence of food and milk, but an alternative explanation is necessary to explain the antacid effect. Increased dissolution is unlikely for this nonacidic molecule. 6MNA profiles in fasted individuals and the effects of food, milk, and antacid are shown in Fig. 19.

Increased absorption rate of naproxen with food, resulting in a modest reduction in T_{max} and a greater increase in the 0- to 2-h area under the naproxen plasma curve, may have been due, in part at least, to a fluid volume effect (LAFONTAINE et al. 1990). The treatment scheme for naproxen with food in healthy subjects included 675 ml drinking water, while the fasted treatments provided only 300 ml water. Ingested food may nonetheless have contributed to faster naproxen absorption causing a transient increase in gastric pH, thereby increasing the dissolution rate of naproxen, which is poorly soluble in acidic media. Increased biliary excretion may have also contributed to accelerated naproxen dissolution and absorption. Although absorption of valproate from conventional tablets is delayed by food (OHDO et al. 1992), absorption from a sustained release formulation containing a mixture of sodium valproate and valproic acid appears to be accelerated (ROYER-MORROT et al. 1993). In a study using a Depakine Chrono 500-mg sustained release valproate/valproic

Table 6. Drug food interactions resulting in accelerated drug absorption

Drug	Dosage form	Dose	Dosage regimen	Subjects	Food details	Fluid volume	Time interval	Sampling duration	Result of food administration	Reference
Nabumetone	–[b]	1000 mg	SD[a]	12 healthy subjects	High-fat meal	–	–	96 h plasma	C_{max} increased 1.5-fold, AUC unchanged	HYNECK (1992)
Naproxen	Tablet	500 mg	SD	6 healthy subjects	Low-fat meal[c]	–	At beginning of meal	48 h plasma	AUC(0–2 h) increased. Fluid volume may have contributed	LAFONTAINE et al. (1990)
Valproate sodium and valproic acid	Sustained release formulation	333 mg valproate sodium, 145 mg valproic acid	SD	12 healthy female subjects	Standard breakfast[c]	30 ml water	Midpoint of meal	72 h plasma	C_{max} and 6 h absorption increased 1.2-fold. Systemic availability unchanged	ROYER-MORROTT et al. (1993)

[a] Single dose.
[b] Data not available.
[c] Meal described.

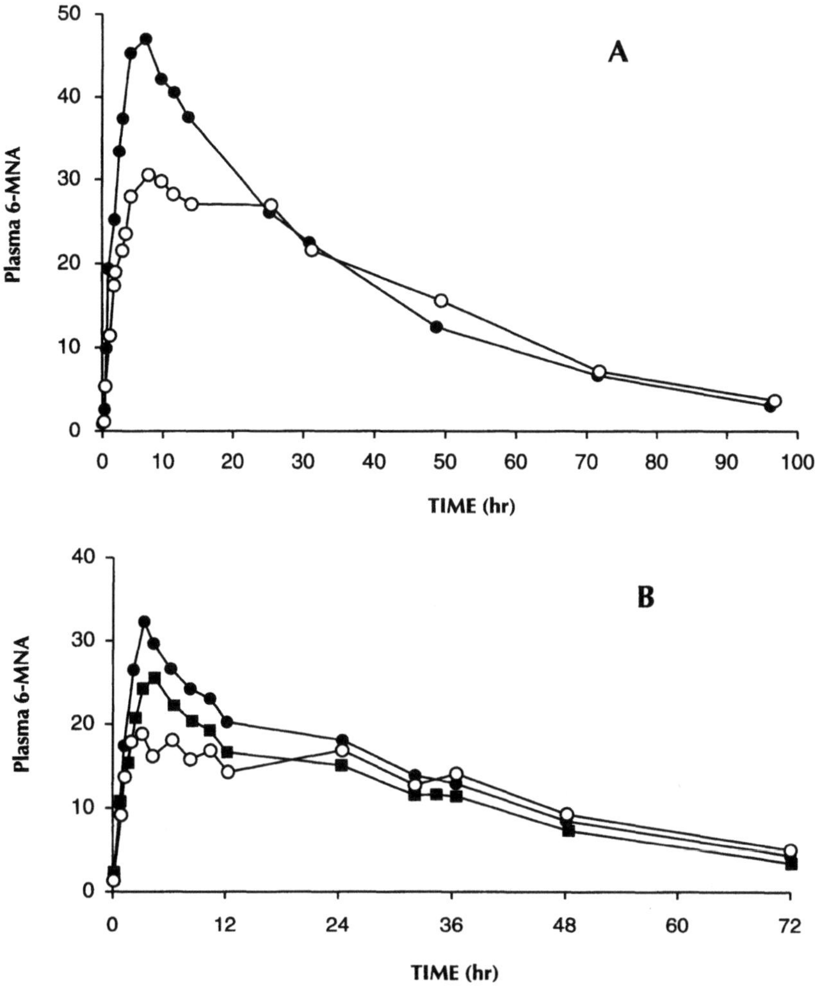

Fig. 19A,B. Mean plasma versus time concentrations of the active metabolite of nabumetone in **A** 12 subjects following a single 1000-mg oral dose of nabumetone fasted (○) or after a meal (●), and in **B** 15 patients following a single 1000-mg oral dose of nabumetone administered with water (○), milk (●), and aluminum hydroxide (■). (From HYNECK 1992).

acid formulation in healthy young women, ingestion at the mid-point of a standard breakfast caused a 1.2-fold increase in both the peak plasma valproate levels and 6h absorption, while overall systemic availability was unchanged, relative to fasting. While increased bile flow and faster dissolution, may have contributed to the faster absorption rate, the divergent results obtained in the two studies suggest that the latter result may be a formulation effect rather than a direct drug effect.

H. Cases in Which Food Has No Effect on Drug Absorption

The reports summarized in Table 7 describe studies in which food had no effect, or an insignificant effect, on the rate and extent of drug absorption. In some cases compounds appearing in this table have already been cited under previous categories in Tables 3–6 and this again reflects the varied results that may be obtained under different study conditions, from different laboratories, or from different formulations of the same drug. Many of the results described in Table 7 are results obtained from routine screening for drug-food interactions, while others arise from deliberate strategies to minimize the influence of factors such as coadministered food on the rate and extent of drug absorption.

This may have been the case with the triazolobenzodiazepine alprazolam, whose absorption was essentially unchanged by food when it was administered in a prototype mixed polymeric controlled release tablet formulation. Mean peak circulating drug levels increased by 12% with food, but other pharmacokinetic parameters were unchanged (ELLER and DELLA-COLETTA 1990). Lack of food effect in this case is not surprising as in vitro dissolution rate for this formulation was shown to be pH-independent and in vivo plasma clearance of alprazolam is low so that metabolism is primarily determined by hepatic metabolic capacity rather than by blood flow. In the case of the new dihydropyridine calcium channel antagonist amlodipine, plasma profiles of unchanged drug in fasting and fed subjects were identical, as were profiles from capsule and solution doses. Plasma profiles obtained in this study are shown in Fig. 20. In the case of bambuterol, food had no effect on circulating levels of either bambuterol or its active metabolite terbutaline (ROSENBORG et al. 1991).

While the absorption of hydrochlorothiazide has previously been reported to be both increased (BEERMAN and GROSCHINSKY-GRIND 1978) and decreased (BARBHAIYA et al. 1982) by food from conventional single-drug formulations, absorption of both hydrochlorthiazide and bisoprolol was unaffected by food when these drugs were administered in a combination tablet to healthy volunteers (MURALIDHARAN et al. 1992). No significant differences were observed in plasma pharmacokinetic parameters, except that the hydrochlorothiazide T_{max} was increased 26%, or in the percentage of hydrochlorothiazide excreted in urine.

Improved safety of the selective monoamine oxidase A inhibitors with respect to the classical "cheese" effect is demonstrated by the compound brofaramine. Ingestion of brofaramine together with cheese containing the equivalent of the PD_{30} dose of tyramine resulted in no change in blood pressure in one subject, and a maximum change of only 20 mmHg in three subjects. The mean increase in blood pressure was only 11 mmHg compared to 40 mmHg from an equivalent dose of tyramine. It is claimed that the lack of interaction with tyramine-rich foods greatly increases the benefit-risk ratio of these MAO-A inhibitors (BIECK et al. 1993). Neither food nor

Table 7. Cases in which food had no effect on drug absorption

Drug	Dosage from	Dose	Dosage regimen	Subjects	Food details	Fluid volume	Time interval	Sampling duration	Result of food administration	Reference
Alprazolam	Matrix SR tablet	1 mg	SD[a]	21 healthy subjects, 12M, 9F	Standard high-fat breakfast[b]	170 ml water	30 min after starting meal	36 h plasma	C_{max} increased 12%, AUC and t_{max} unchanged	ELLER and DELLA-COLETTA (1990)
Amlodipine	Capsule	10 mg	SD	12 healthy male subjects	Standard breakfast[b]	150 ml water	Within 30 min after completing meal	192 h plasma	No effect on plasma levels	FAULKNER et al. (1989)
Bambuterol	–	20 mg	MD[c] 1 tablet each evening, 14 days	22 healthy subjects	Dinner	–	Immediately after meal	–	No effect on plasma levels or urinary excretion of bambuterol or active metabolite terbutaline	ROSENBORG et al. (1991)
Bisoprolol and hydrochlorothiazide	Combination tablet	10 mg bisoprolol, 6.25 mg hydrochlorothiazide	SD	16 healthy subjects	–	–	With meal	48 h plasma and urine	T_{max} for hydrochlorothiazide increased 20% otherwise no effect on plasma or urine parameters	MURALIDHARAN et al. (1992)
Brofaramine	–	150 mg/day, 14 days	MD	10 subjects	Cheese or 25–75 mg tyramine	–	–	Blood pressure determination	Tyramine caused 40 mm increase in blood pressure. Mean increase only 11 mm with cheese	BIECK et al. (1993)
Bromocriptine	Tablet	7.5 mg	SD	7 healthy male subjects	Standard breakfast[b]	100 ml water	After meal	14 h plasma	C_{max} decreased 10%, AUC unchanged	KOPITAR et al. (1991)
Carbamazepine	Tablet	200 mg	SD	12 patients with mild to advanced chronic renal failure	Very low protein diet	–	–	–	AUC decreased 10%, T_{max} reduced from 8.5 to 6 h, C_{max} unaffected	BANNWARTH et al. (1992)
Cardizem	Capsule	360 mg	SD	18 healthy subjects	–	–	–	36 h plasma	Diltiazem C_{max} reduced 12%, AUC unchanged. Deacetyldiltiazem C_{max} increased 11%, AUC unchanged	YU et al. (1992)
Cefetamet pivoxil	Syrup	385 mg	SD	18 healthy male subjects	Standard breakfast[b]	10 ml syrup	With meal	15 h plasma, 24 h urine	Plasma and urine parameters for syrup unaffected. Bioavailability from syrup lower than from tablet	DUCHARME et al. (1993)

Table 7. *Continued*

Drug	Dosage from	Dose	Dosage regimen	Subjects	Food details	Fluid volume	Time interval	Sampling duration	Result of food administration	Reference
Cimetitine and ranitidine	Tablet	400 mg cimetidine 150 mg ranitidine	SD	18 healthy male subjects	Standard breakfast[b]	100 ml water	15 min after starting meal	8 h plasma	No significant effect on plasma levels of either compound. In fasted subjects antacid treatment reduced plasma levels of cimetidine and ranitidine. In fed subjects antacid had no effect on either drug	DESMOND et al. (1990)
Cyclosporine	Solution	3.51 ± 1.35 mg/kg/day	MD (b.i.d.)	14 renal transplant recipients, 9M, 5F	Moderate and trace-fat breakfast[b]	125 ml orange juice	Immediately after meal	12 h plasma	C_{max} and AUC unaffected by trace-fat and moderate-fat meals. T_{max} prolonged from 3.3 to 5.1 h by moderate-fat meal	HONCHARIK et al. (1991)
Cyclosporine	Capsules	~6 mg/kg	SD	11 healthy male subjects	Standard light breakfast[b], also bile acid tablet	150 ml water	With meal	32 h plasma	Systemic availability unaffected by food but C_{max} reduced 17%. Addition of bile acid tablets increased availability 25%	LINDHOLM et al. (1990)
Diazepam, ethinyl estradiol, norethindrone, propranolol	–	5 mg 0.07 mg 1.0 mg 20 mg	SD	8–10 healthy subjects	Olestra or triglyceride on[b]	6 ounces water	With diet	Plasma up to 48 h	No significant differences in absorption when administered with olestra, triglyceride oil, or water, except T_{max} for diazepam prolonged with triglyceride oil	ROBERTS and LETT (1989)
Diazepam	Conventional tablet and slow-release capsule	60 mg CR 120 mg SR	SD	24 healthy male subjects	Standard breakfast	200 ml water	30 min after starting meal	24 h plasma	Food had no effect on plasma levels of diazepam and its metabolites from either formulation, but SR formulation, increased systemic availabilty 69% relative to conventional tablet	DU SOUICH et al. (1990)

E2020	–	2 mg	SD	12 healthy male subjects	Standard breakfast	–	Within 30 min after meal	168 h plasma	Rate and extent of absorption unaffected	Mihara et al. (1993)
Fluvoxamine	Tablet	50 mg	SD	12 healthy subjects, 8M, 4F	Standard breakfast[b]	240 ml water or orange juice	15 min after starting meal	72 h plasma	Rate and extent of absorption unaffected	Van Harten et al. (1991)
Ibuprofen	–	600 mg	SD	11 healthy male subjects	Standard breakfast[b]	200 ml water	30 min after starting meal	10 h plasma	Plasma levels of S(+) ibufrofen higher than those of R(–) ibuprofen under fasting and nonfasting conditions. Food modestly reduced C_{max} of S(+) and R(–), but systemic availability, and isomer ratios unaffected	Levine et al. (1992)
Levodopa	Solution	125 mg	SD	8 healthy male subjects	10.5 g or 30.5 g protein meals[b]	100 ml water	15 min after meal	8 h plasma	Meal with 30.5 g protein had no effect on levodopa absorption. Absorption was reduced 10%, and C_{max} by 26% by low-protein meal	Robertson et al. (1991)
Methotrexate	–	9.5 ± 5.8 mg/week	SD	11 patients with rheumatoid arthritis	Standard meal	–	–	24 h plasma	Rate and extent of absorption unaffected	Phelan et al. (1991)
Metoprolol succinate	Controlled release tablet	400 mg	SD	18 healthy male subjects	Standard high-fat breakfast	–	–	–	Tendency to increase C_{max} and AUC, but not significant. Similar C_{max} values	Sandberg et al. (1990)
Metoprolol succinate	Controlled release tablet	50 mg	SD	12 healthy subjects	Standard carbohydrate breakfast[b]	–	–	–	Plasma levels unaffected by food	Sandberg et al. (1988)
Morphine sulfate	Sustained release tablet	30 mg 9.12 h	MD	24 healthy male subjects	Standard caffeine-free meals	180 ml water	Immediately after meals	84 h plasma	Rate and extent of absorption unaffected	Bass et al. (1992)
Mosapride citrate	Tablet	10 mg	SD	10 healthy male subjects	Sandwiches and 200 ml orange juice	200 ml water	30 min after meal	24 h plasma	C_{max} and AUC not significantly affected. T_{max} increased from 0.6 to 0.9 h	Sakashita et al. (1993)
Moxonidine	Tablet	0.2 mg	SD	18 healthy male subjects	Standard breakfast	200 ml water	30 min after meal	24 h plasma and urine	C_{max} decreased 14%, T_{max} marginally increased. Systemic availability and urinary excretion unchanged	Theodor et al. (1992)

Table 7. *Continued*

Drug	Dosage form	Dose	Dosage regimen	Subjects	Food details	Fluid volume	Time interval	Sampling duration	Result of food administration	Reference
Moxonidine	Tablet	0.2 mg	SD	15 healthy male subjects	–	–	–	–	C_{max} decreased 14%, T_{max} marginally increased. Systemic availability unchanged	WEIMANN and RUDOLPH (1992)
Nefiracetam	Tablet	100 mg	SD	39 healthy Japanese male subjects	Standard light breakfast	100 ml water	30 min after meal	24 h plasma and urine	Rate and extent of absorption not affected	FUJIMAKI et al. (1992)
Paroxetine	–	30 mg	SD	41 healthy male subjects	Standard continental breakfast, low-fat or high-fat diet, or milk	–	Immediately after meal	–	Rate and extent of absorption not affected by standard breakfast, nor between low-fat meals. AUC reduced 42% by 1000 ml milk	GREB et al. (1989)
Piroximone	–	25 mg or 50 mg	SD	12 healthy male subjects	Standard breakfast[b]	–	Immediately before meal	24 h plasma and urine	Following the 25-mg dose, rate and extent of absorption, unaffected. Following the 50-mg dose, extent of absorption marginally increased	HAEGELE et al. (1991)
Procainamide	Slow release tablet	–	SD	Healthy subjects	High-fat meal	–	During meal	–	Negligible effect on rate and extent of absorption	DEVRIES et al. (1991)
Pseudoephedrine and brompheniramine	GITS combination tablet	180 mg pseudoephedrine 10 mg brompheniramine	SD	12 healthy male subjects	High-fat breakfast	–	After meal	96 h plasma	Rate and extent of absorption of both drugs unaffected	CHAO et al. (1991)
Rifabutin	Capsule	150 mg	SD	12 healthy male subjects	Standard breakfast[b]	–	15 min after starting meal	168 h plasma, 48 h urine	C_{max} reduced 16% but other parameters unchanged	NARANG et al. (1992)
Sparfloxacin	–	400 mg	SD	10 healthy subjects	Standard breakfast	–	Concomitant intake	96 h plasma	Rate and extent of absorption unaffected	THEBAULT et al. (1990)
Sparfloxacin	–	200 mg	SD	6 healthy subjects	–	–	–	48 h plasma	Rate and extent of absorption unaffected	SAKASHITA et al. (1991)

Temafloxacin	Crushed tablet via nasogastric tube	600 mg	SD	18 healthy male subjects	Osmolite enteral feeding solution	200 ml water	Concomitant with enteral feeding solution	48 h plasma	Rate and extent of absorption unaffected	Lubowski et al. (1992)
Temafloxacin	Tablet	600 mg	SD	Healthy subjects	Standard breakfast[b]	240 ml water	20 min after meal	48 h plasma	T_{max} delayed from 1.8 to 3.0 h. Other parameters unaffected	Granneman and Mukherjee (1992)
Theophylline	Monospan capsule	450 mg	SD	22 healthy male subjects	High-fat breakfast[b]	6 ounces water	Immediately after meal	72 h plasma	Rate and extent of absorption unaffected	Harrison et al. (1993)
Theophylline	Monospan capsule	900 mg		29 healthy male subjects	High-fat breakfast[b]	6 ounces water	Immediately after meal	72 h plasma	Rate and extent of absorption unaffected	Harrison et al. (1993)
Theophylline	Theo-24 Capsules	6 mg/kg	SD	6 healthy male subjects	Ensure enteral feeding, as ten boluses	100 ml water	One hour after third enteral feeding bolus	72 h plasma	Rate and extent of absorption unaffected by enteral feeding	Plezia et al. (1990)
Theophylline	Uniphyl tablets	Variable, in 200-mg increments	MD 1900 h each day	22 patients (14M, 8F) with chronic airways obstruction	Evening meal	–	45 min after meal	Peak and trough plasma levels, together with pulmonary function tests	Marginal increases in C_{max} and C_{min} values. No differences in pulmonary function test results	Arkinstall et al. (1991)
Tiaprofenic acid	Sustained release tablet with ^{152}Sm	600 mg	SD	7 healthy subjects, 5M, 2F	Light breakfast or heavy breakfast[b]	100 ml water	After meal	24 h plasma	Rate and extent of absorption unaffected by meals despite slower gastric emptying	Wilding et al. (1992)
Trimetazidine	Sustained release formulation	80 mg	SD	12 healthy male subjects	Standard breakfast	–	With meal	48 h plasma	Rate and extent of absorption unaffected	Chaufour et al. (1992)
Verapamil	Sustained release formulation	240 mg	SD	12 healthy male subjects	Standard breakfast[b]	–	10 min after meal	36 h plasma	No significant differences in plasma profiles of verapamil or nor-verapamil	Devane et al. (1990) Devane and Kelly (1991)
Verapamil	Pellet filled capsule, or pellets sprinkled on food	240 mg	SD	32 healthy male volunteers	Apple sauce	150 ml water	With meal	48 h plasma	No significant differences in plasma profiles of verapamil or nor-verapail	Kozloski et al. (1992a)

[a] Single dose.
[b] Meal described.
[c] Multiple dose.

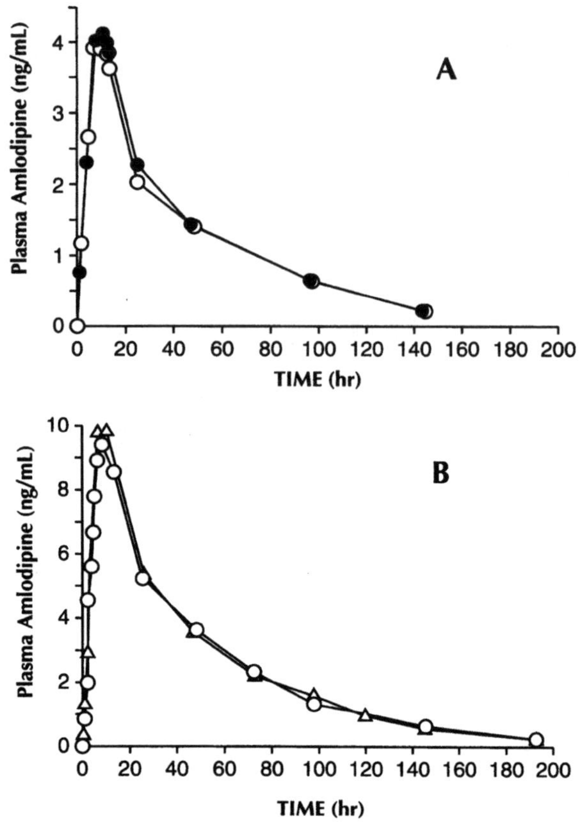

Fig. 20A,B. Mean plasma versus time concentrations of amlodipine in 12 subjects following **A** a single 10-mg oral dose of amlodipine fasting (○) and following a standard breakfast (●) and **B** a single 20-mg oral dose of amlodipine as a solution (○) and capsule (△) dose fasting. (From FAULKNER et al. 1989)

metoclopramide had a significant effect on plasma profiles of the ergot derivative bromocriptine (KOPITAR et al. 1991). Consistent with its effects on gastric emptying, food caused a slight delay in bromocriptine absorption, while metoclopramide had the opposite effect. However, both changes were trivial and overall bioavailability was unaffected by either treatment.

While previously cited studies showed increases in absorption of the cephalosporin prodrug cefetamet pivoxil from tablets (BLOUIN and STOECKEL 1993; TAM et al. 1990), a study using an oral syrup formulation of cefetamet pivoxil showed no food effect when a 385-mg (cefetamet free acid equivalents) dose of cefatemet pivoxil was administered together with a standard breakfast (DUCHARME et al. 1993). Mean C_{max} values in plasma were 2.7 μg/ml with food compared to 2.9 μg/ml fasting and absolute bioavailability was 38% and 34% after fed and fasting doses, respectively. Interestingly, the syrup had significantly lower absolute systemic bioavailability than that of a tablet administered under fed conditions. This is attributed to the lack of effect by food on the bioavailability of cefetamet pivoxil from the syrup formulation. While the mechanism for the different responses of cefetamet pivoxil tablets and syrup to food is unknown, it is claimed that plasma profiles of unbound cefetamet are adequate to treat susceptible organisms (DUCHARME et al. 1993). Absorption of both cimetidine and ranitidine was unaffected when they were administered to healthy male subjects after a standard breakfast compared to fasting (DESMOND et al. 1990). In the fasted state, absorption of cimetidine was decreased 24%, and ranitidine 59%, when these compounds were taken together with an antacid. In the fed state, however, coadministered antacid did not have the same effect on cimetidine or ranitidine absorption. It is proposed that the antacid effect seen in the fasting state is related to impaired tablet dissolution and to drug binding to unabsorbed antacid. Abolition of this effect by food may be due to competition for drug-binding sites on the antacid. However, this binding must be weak because of the lack of effect by food on drug absorption in the absence of antacid.

Data on the effect of food on cyclosporine absorption are conflicting. In a study cited in Table 5, bioavailability from a chocolate emulsion was 23% and 42% after low-fat and high-fat meals, respectively (GUPTA et al. 1990). Other studies have reported no significant change (LINDHOLM et al. 1990) or decreased cyclosporine absorption with food (KEOWN et al. 1982). Significant increases in cyclosporine absorption with food have been reported in renal transplant patients (PTACHCINSKI et al. 1985). Two further studies have reported little effect by food on cyclosporine absorption (HONCHARIK et al. 1991). In the first study, cyclosporine was administered to 14 renal transplant patients immediately following a moderate- or trace-fat breakfast. Neither meal had any significant effect on cyclosporine pharmacokinetic parameters. Mean C_{max} values were 410, 346, and 365 ng/ml, and mean AUC values were 2115, 2085, and 2145 ng·h/ml following fasting, moderate-fat, and high-fat treatments at an average cyclosporine dose of 3.51 mg/kg per day. In the second study, conducted in healthy male subjects, a standard light breakfast caused a 17% reduction in mean peak plasma cyclosporine levels, but had no effect on area under plasma profiles, with mean values of 7283 and 7453 ng·h/ml from a 6-mg/kg dose to fasted and fed subjects, respectively (LINDHOLM et al. 1990). Addition of bile salts to the nonfasted treatment, on the other hand, caused mean C_{max} values to increase 1.1-fold, and overall bioavailability 1.2-fold, relative to fasting. These results suggest that, in this study at least, bile acid formation is an important determinant of cyclosporine absorption. This observation does not explain the various outcomes of cyclosporine-food interactions obtained in other studies.

In an attempt to further elucidate the mechanism of drug-food interactions, and to identify an appropriate surrogate test meal, the absorption of some widely divergent compounds, diazepam, propranolol, norethindrone, and ethinyl estradiol, was examined in the presence of water, triglyceride oil, and a sucrose polyester, Olestra (ROBERTS and LEFF 1989). The study was conducted in healthy subjects. The absorption of both diazepam (GREENBLATT et al. 1978) and propranolol (MELANDER et al. 1977) has been previously shown to be increased by solid meals, although a more recent study with diazepam showed that neither the rate nor extent of absorption was altered by food, either from a conventional formulation or a slow-release formulation (DU SOUICH et al. 1990). The liquid preparations had no significant effect on the absorption of any of the four compound studied, except that the time of peak plasma diazepam levels was increased by triglyceride oil, relative to the Olestra and water treatments. The inconsistency between results obtained from the two earlier diazepam studies makes it difficult to assess the overall predictability of the liquid formulation for actual diazepam-food interactions in a clinical environment.

Food was shown to have no effect on the rate and extent of absorption of the new cholinesterase inhibitor E2020 (MIHARA et al. 1993). Following single 2-mg p.o. doses of E2020 to healthy male volunteers, mean peak plasma E2020 levels of 3.3 and 3.2 ng/ml and AUC (0–168 h) values of 166.5 and 172.8 ng·h/ml were obtained under fasting and fed conditions. The nature of the meal in this case was not described but was presumably a standard Japanese-style breakfast. Similar lack of effects are reported for the selective serotonin reuptake inhibitor fluvoxamine (VAN HARTEN et al. 1991) and also the systemic availability of the S(+) and R(−) enantiomers of ibuprofen (LEVINE et al. 1992). On the other hand, sucralfate, often administered to reduce gastrointestinal side effects of nonsteroidal anti-inflammatory agents, reduced peak concentrations of both enantiomers of ibuprofen. Neither food nor sucralfate had any influence on R(−) to S(+) enantiomer inversion. A report on the lack of food effect on levodopa absorption illustrates, again, the complexity and unpredictability of drug-food interactions. An earlier report described inhibition of levodopa absorption, based on clinical results, by high-protein diets (GILLESPIE et al. 1973). Later studies, included in this review, described reduced levodopa absorption after a standard luncheon (ROOS et al. 1993) and increased absorption from both immediate and controlled release formulations following a light breakfast (WILDING et al. 1991b). In a further study conducted in healthy volunteers, a meal containing 30.5 g protein had no effect on levodopa absorption from a 150-mg solution dose, whereas absorption was reduced by 10%, and peak plasma levodopa levels by 26%, when drug was ingested following a meal containing only 10.5 g protein (ROBERTSON et al. 1991). The poor bioavailability of levodopa following the low-protein meal, relative to the fasting state, suggests that low-protein diets do not increase levodopa absorption. It is speculated that any beneficial effects of a low-protein diet on levodopa efficacy for treatment of Parkinson's disease may be related to reduced competition for transport across the blood-brain barrier rather than increased systemic availability.

Absorption of the beta-adrenoreceptor antagonist metoprolol has previously been shown to be significantly increased by food (PHELAN et al. 1991). The effect of food was further examined on a new controlled release formulation of metoprolol succinate. Following a standard high-fat breakfast, peak plasma levels and overall bioavailability of metoprolol from a single 400-mg dose increased by only a small extent (SANDBERG et al. 1988). Following a standard high-carbohydrate breakfast, plasma metoprolol levels from a 50-mg dose were unaffected. Thus, it appears that absorption of metoprolol from this controlled release formulation is affected far less by food than the conventional release formulation studied previously.

Previous studies using single doses of controlled release formulations of morphine have reported increased and also unchanged absorption in the presence of food (KAIKO et al. 1990; HUNT and KAIKO 1991). In order to determine the effect of food on controlled release morphine under clinical dosing conditions, morphine sulfate was administered in sustained release tablets at a dose level of 30 mg every 12 h for seven doses either immediately after or 2 h before standard caffeine-free meals (BASS et al. 1992). Plasma morphine levels monitored during the repeated dosing and profiled during the 12 h following the last dose were similar from fasted and fed treatments. Profiles obtained during the period following the last morphine sulfate dose are shown in Fig. 21. The AUC for fed subjects during the 12-h period was 9.8% less than that for fasted subjects. The lack of food effect observed in this repeated dose study was observed using the same sustained release formulation to that associated with increased absorption in the presence of food following single doses (HUNT and KAIKO 1991).

Two separate studies showed that food does not affect the rate or extent of absorption of the new imidazoline antihypertensive agent moxonidine.

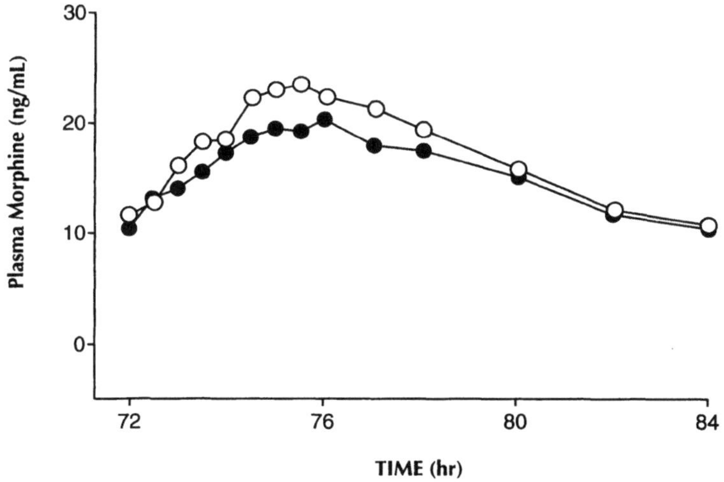

Fig. 21. Mean plasma versus time concentrations of morphine during the 72- to 84-h dosing interval in 24 subjects following repeated 30-mg q 12 h oral doses of morphine sulfate as sustained release tablets administered under fasting (○) and fed (●) conditions. (From BASS et al. 1992)

Following single 0.2-mg doses of moxonidine tablets to healthy male subjects, mean peak plasma levels of moxonidine decrease 14% while systemic availability and urine excretion were essentially unchanged by food (THEODOR et al. 1992). In a second study in healthy male subjects, almost identical results were obtained, with modest, nonsignificant, increases in peak moxonidine plasma levels and no effect on absorption in the presence of food (WEIMANN and RUDOLPH 1992). Similar to other quinolone antibiotics, absorption of sparfloxacin appears to be essentially unaffected by food. Following single 400-mg doses to healthy subjects, the time of peak sparfloxacin plasma levels was increased from 3.1 to 4.7 h by food, but peak levels and systemic availability were unaffected (THEBAULT et al. 1990). In a similar study conducted in healthy individuals, plasma sparfloxcin levels following a single 200-mg dose under fed and fasted conditions were almost superimposable (SAKASHITA et al. 1991). Similarly, the rate and extent of temafloxacin absorption were essentially unchanged when administered by nasogastric tube with or without enteral feeding (LUBOWSKI et al. 1992), and from conventional tablets administered as a single 200-mg dose under fasting and fed conditions (GRANNEMAN and MUKHERJEE 1992). Plasma profiles of temafloxacin from 2 × 200-mg tablets and 1 × 400-mg tablet fasting, and 1 × 400-mg tablet nonfasting, are shown in Fig. 22.

Incorporating the nonsteroidal-anti-inflammatory agent tiaprofenic acid into a novel sustained release system comprising discrete sustained release

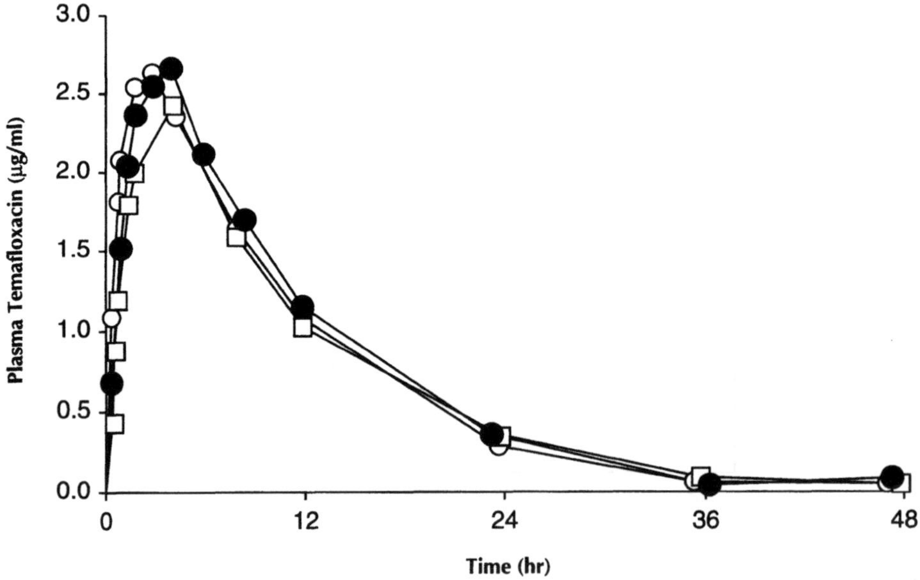

Fig. 22. Mean plasma versus time concentrations of temafloxacin in 18 subjects following a single oral dose of 1 × 400-mg tablet fasting (○), 2 × 200-mg tablets fasting (□), and 1 × 400-mg tablet after food (●). (From GRANNEMAN and MUKHERJEE 1992)

pellets in a single tablet gave rise to absorption profiles that were unaffected by food (WILDING et al. 1992). Healthy volunteers received 600 mg drug in the fasting state and also after standard light and heavy breakfasts. Gamma scintigraphy measurements showed that stomach emptying of ^{152}Sm-labeled pellets was delayed by both light and heavy meals. Despite the delayed stomach emptying, plasma profiles of tiaprofenic acid were almost superimposable after fasting and fed treatments.

Although food has recently been shown to delay (KANN et al. 1989; YUEN et al. 1993) and also to increase (COOK et al. 1990) absorption of theophylline from controlled release formulations, other studies have shown little effect. The rate and extend of theophylline absorption from Monospan capsules, in which drug is contained in 1-mm-diameter beadlets, were almost identical when administered to healthy volunteers fasting or immediately after a high-fat breakfast (HARRISON et al. 1993). Similarly, the systemic availability of theophylline from Theo-24 capsules was unaffected by enteral liquid feeding in healthy male subjects (PLEZIA et al. 1990). Earlier reports on the lack of effect by food on theophylline absorption from Uniphyl tablets during chronic evening dosing were confirmed in a subsequent study which monitored both the serum drug levels and also clinical efficacy in patients with chronic airways obstruction (ARKINSTALL et al. 1991). Following mean daily (evening) theophylline doses of 818 mg, peak plasma theophylline concentrations were 14.4 μg/ml with food and 13.1 μg/ml fasting. Mean trough levels were 7.4 and 6.9 μg/ml. There were no treatment-related differences in pulmonary function tests including spirometry, asthma symptom scores, side effects, or the need for beta-agonist inhalers.

The last compound to be discussed in this section is of interest largely because of the divergent results obtained regarding food interactions with controlled release preparations. Verapamil absorption was previously shown to be reduced by 30% and peak serum levels by 48% when a sustained release tablet was taken with food (HOON et al. 1992). However, different results were obtained when verapamil was administered in a different sustained release formulation. Twelve healthy male volunteers received single 200-mg verapamil doses in a newly marketed sustained release formulation employing rate-controlling technology in the form of a capsule, either fasting or 10 min after a standard breakfast (DEVANE et al. 1990; DEVANE and KELLY 1991). Plasma profiles from the fed and fasting treatments, together with verapamil profiles from a conventional 80-mg dose administered every 8 h are shown in Fig. 23. Plasma profiles of verapamil following the sustained release capsule were superimposable in the fed and fasting states. Plasma profiles of the metabolite norverapamil were similarly unaffected by food. Mean peak plasma verapamil plasma levels were 77 ng/ml following both sustained release treatments, compared to 95 ng/ml following the conventional tablet. Areas under plasma verapamil concentration versus time curves from zero to infinite time were 1541 and 1387 ng·h/ml from fasted and fed sustained release doses, respectively, compared to 1370 ng·h/ml from the conventional tablet. The

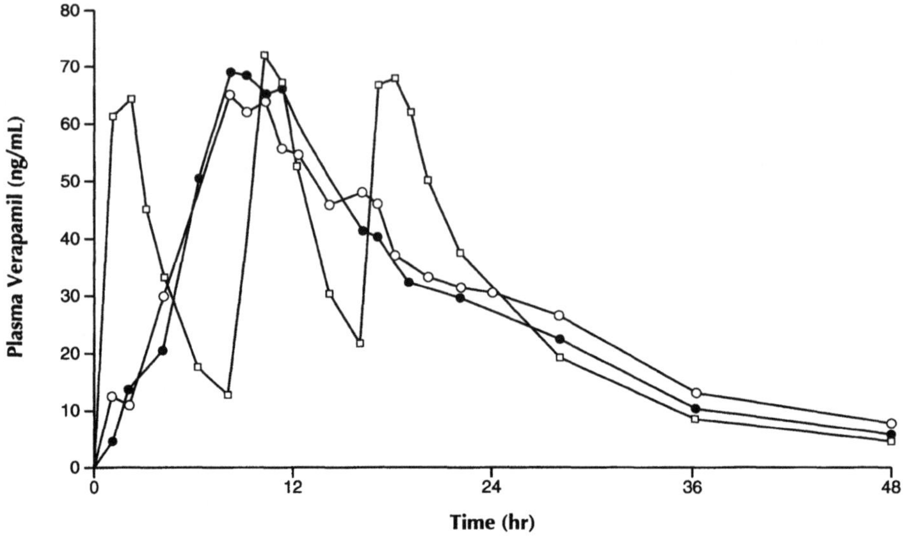

Fig. 23. Mean plasma versus time concentrations of verapamil in 12 subjects following once daily oral doses of a 240-mg sustained release verapamil capsule under fasting (○) and fed (●) conditions and after three times daily oral doses of an 80-mg immediate release tablet fasting (□). (From DEVANE and KELLY 1991)

divergent results obtained from these studies are most likely related to the formulations used. Release of drug from membrane controlled release formulations, including those using osmotic pump technology, are generally less sensitive to the gastrointestinal environment, including the presence of food, than other types of controlled release formulations. The advantage of a rate-controlling membrane dosage form for verapamil was further demonstrated in a study in which drug was administered as extended release pellets in the fasting state and also sprinkled on apple sauce. There were no significant differences in plasma verapamil or norverapamil profiles between the fasting and apple sauce treatments (KOZLOSKI et al. 1992a).

I. Conclusions

The number of articles published during the 5 years covered by this review illustrates the high level of interest, from scientific, regulatory, and clinical perspectives, on interactions between drugs and ingested food influencing drug absorption. As in the initial review on this topic (WELLING 1977), drugs and drug formulations continue to fall naturally into four major categories of those whose absorption is reduced, delayed, increased, or not affected by food. In this review, a fifth category of drugs whose absorption is accelerated but not increased by food was identified.

Grouping of drugs and formulations in this way is entirely empirical. The interactions could have been grouped differently, for example, among therapeutic classes or according to physicochemical characteristics. Whichever way the interactions are sorted, one is nonetheless always confronted with the tremendous divergence of results that occur, largely independent of therapeutic class and of physicochemical properties. This is the conundrum that presents to those who concern themselves with drug absorption, circulating drug profiles, and, inevitably, clinical efficacy consequences. Among the reports summarized in this review, approximately 57 reported delayed drug absorption as a result of food interactions, 40 reported reduced absorption, 49 reported increased absorption, 3 reported accelerated absorption, and 46 reported no interaction. Many of the drugs whose absorptin was reduced or increased also exhibited delayed absorption to varying extents.

The last decade may be considered to have been a data-collecting period. Altogether there have been approximately 400 studies reported in the literature describing drug-food interactions, and the patterns of interaction have not changed significantly with time. No new mechanisms or pivotal relationships have been described, and one can only conclude that, of all the possible interactions that may occur and which are summarized in Tables 1 and 2, any interaction that predominates will determine the final observed effect. Results of various reported studies have indicated that the nature of a drug-food interaction may be at least partially predictable from a physicochemical perspective, but accurate predictability is plagued by many exceptions, which are often spectacular in nature. The lack of predictability of the outcome of drug-food interactions is such that this phenomenon must always be considered, and is in fact becoming mandatory in all drug development programs, and in clinical use.

The clinical impact of a drug-food interaction, as stated earlier, depends on the nature and degree of change in circulating drug levels, the margin of safety, and the slope of the drug concentration-response curve. A small change in circulating drug levels for a drug with a relatively flat dose response curve is likely to be of little clinical consequence. On the other hand, a large change in circulating drug levels for a drug with a relatively steep dose response curve, or a relative narrow safety margin, may be critical.

While it continues to be necessary to explore drug-food interactions for drugs and drug formulations, additional avenues need to be explored to seek mechanistic patterns that may lead to better prediction of the nature and extent of changes in circulating drug levels due to the presence of food, and their possible clinical impact.

References

Albertioni F, Juliusson G, Liliemark J (1993) On the bioavailability of 2-chloro-2′-deoxyadenosine (CdA); the influence of food and omeprazole. Eur J Clin Pharmacol 44:579–582

Arkinstall WW, Brar PS, Stewart JH (1991) Repeated dosing of Uniphyl tablets under fed and fasting conditions: comparison of serum theophylline levels, pulmonary function and asthma symptoms. Ann Allergy 67:583–587

Astarloa R, Mena MA, Sánchez V, De La Vega L, De Yébenes JG (1992) Clinical and pharmacokinetic effects of a diet rich in insoluble fiber on Parkinson's disease. Clin Neuropharmacol 15:375–380
Axelson JE, Chan GLY, Kirsten EB, Mason WD, Lanman EC, Kerr CE (1987) Enhancement of the bioavailability of propafenone by food. Br J Clin Pharmacol 23:735–741
Bailey DG, Spence JD, Munoz C, Arnold JMO (1991) Interaction of citrus juices with felodipine and nifedipine. Lancet 337:268–269
Bannwarth B, Combes C, Vinçon G, Bouchet JL, Aparicio M, Bégaud B (1992) Influence of low protein diet on the pharmacokinetics of carbamazepine in chronic renal failure. Fundam Clin Pharmacol 6:222
Barbhaiya RH, Patel RB, Corrick-West HP, Joslin RS, Welling PG (1982) Comparative bioavailability of hydrochlorothiazide from oral tablet dosage forms, determined by plasma and urinary excretion models. Biopharm Drug Dispos 3:329–336
Barbhaiya RH, Shukla UA, Gleason CR, Shyu WC, Pittman KA (1990) Comparison of the effects of food on the pharmacokinetics of cefprozil and cefaclor. Antimicrob Agents Chemother 34:1210–1213
Barnwell SG, Laudanski T, Dwyer M, Story MJ, Guard P, Cole S, Attwood D (1993) Reduced bioavailability of atenolol in man: the role of bile acids. Int J Pharm 89:245–250
Barone JA, Koh JG, Bierman RH, Colaizzi JL, Swanson KA, Gaffar MC, Moskovitz BL, Mechlinski W, Van de Velde V (1993) Food interactions and steady state pharmacokinetics of itraconazole capsules in healthy male volunteers. Antimicrob Agents Chemother 37:778–784
Bass J, Shepard KV, Lee JW, Hulse J (1992) An evaluation of the effect of food on the oral bioavailability of sustained-release morphine sulfate tablets (Oramorph SR) after multiple doses. J Clin Pharmacol 32:1003–1007
Basu TK (1988) Drug-nutrient interactions. Croom Helm, London
Beckermann B, Beneke M, Böttcher M, Dietrich H, Horstmann R, Seitz I (1993) Influence of formulation, food or antacids on the pharmacokinetics of BAY X 1005 in human volunteers. Arch Pharmacol 347 [Suppl]:R27
Beerman B, Groschinsky-Grind M (1978) Enhancement of the gastrointestinal absorption of hydrochlorothiazide by propantheline. Eur J Clin Pharmacol 13:385–387
Belch JJF, McLaren M, Lau CS, MacKay IR, Bancroft A, McEwen J, Thompson JM (1993) Cicaprost, an orally active prostacyclin analogue: its effects on platelet aggregation and skin blood flow in normal volunteers. Br J Clin Pharmacol 35:643–647
Bergmann JH, Simonneau G, Chassany O, Caulin C (1992) Effect de la prise d'un repas et d'un anti acide sur la cinétique de l'acetorphan. Therapie 47:229
Bieck PR, Antonin K-H, Schmidt E (1993) Clinical pharmacology of reversible monoamine oxidase-A inhibitors. Clin Neuropharmacol 16 [Suppl 2]:S34–S41
Blouin RA, Stoeckel K (1993) Cefetamet pivoxil clinical pharmacokinetics. Drug Dispos 25:172–188
Brookes LG, Sambol NC, Lin ET, Gee W, Benet LZ (1991) Effect of dosage form, dose and food on the pharmacokinetics of metformin. Pharm Res 8 [Suppl]:S320
Chao ST, Prather D, Pinson D, Coen P, Pruitt B, Knowles M, Place V (1991) Effect of food on bioavailability of pseudoephedrine and brompheniramine administered from a gastrointestinal therapeutic system. J Pharm Sci 80:432–435
Charman WN, Rogge MC, Boddy AW, Berger BM (1993) Effect of food and a monoglyceride emulsion formulation on danazol bioavailability. J Clin Pharmacol 33:381–386
Chaufour S, Funck-Brentano C, Jaillon P (1992) Pharmacokinetics of trimetazidine SR in healthy volunteers: influence of food on oral drug disposition. Fundam Clin Pharmacol 6:225
Choi RL, Sun JX, Kochak GM (1993) The effect of food on the relative bioavailability of fadrazole hydrochloride. Biopharm Drug Dispos 14:779–784

Chu S-Y, Park Y, Locke C, Wilson DS, Cavanaugh JC (1992) Drug-food interaction potential of clarithromycin, a new macrolide antimicrobial. J Clin Pharmacol 32:32–36

Clements JA, Heading RC, Nimmo WS, Prescott CF (1978) Kinetics of acetaminophen absorption and gastric emptying in man. Clin Pharmacol Ther 24:420–431

Cole AFD, Baxter JG, Jackson BJ, Hew-Wing P, Güntert TW, Lalka D (1992) Pharmacokinetic and metabolic aspects of the moclobemide-food interaction. Psychopharmacology 106:S37–S39

Cole SK, Story MJ, Laudanski T, Dwyer M, Attwood D, Robertson J, Barnwell SG (1992) Targeting drugs to the enterohepatic circulation: a potential drug delivery system designed to enhance the bioavailability of indomethacin. Int J Pharm 80:63–73

Confalonieri S, Cosmi G, Guiso G, Gherardi S, Guido M, Caccia S (1992) The pharmacokinetics of a potential memory-enhancing compound, CL 275838, in fasted and fed volunteers. Drug Invest 4:322–328

Conway EL, McNeil JJ, Hurley J, Jackman GP, Krum H, Howes LG, Louis WJ (1993) The effects of food on the oral bioavailability of doxazosin in hypertensive subjects. Drug Invest 6:90–95

Cook CS, Hauswald CL, Grahn AY, Kowalski K, Karim A, Koch R, Schoenhard GL, Oppermann JA (1990) Suitability of the dog as an animal model for evaluating theophylline absorption and food effects from different formulations. Int J Pharm 60:125–132

Cook HJ, Mundo CR, Fonseca L, Gasque L, Moreno-Esparza R (1993) Influence of the diet on bioavailability of tetracycline. Biopharm Drug Dispos 14:549–553

De Mey C, Meineke I (1992) Prandial and diurnal effects on the absorption of orally administered enteric coated 5-aminosalicylic acid (5-ASA). Br J Clin Pharmacol 33:179–182

Degen PH, Cardot JM, Czendlik C, Dieterle W (1993) Influence of food on the disposition of the monoamine oxidase-A inhibitor brofaromine in healthy volunteers. Biopharm Drug Dispos 14:209–215

Degen PH, Flesch G, Cardot J-M, Czendlik C, Dieterle W (1994) The influence of food on the disposition of the antiepileptic oxcarbazepine and its major metabolites in healthy volunteers. Biopharm Drug Dispos 15:519–526

Desante KA, Zeckel ML (1992) Pharmacokinetic profile of loracarbef. Am J Med 92:16S–19S

Desmond PV, Harman PJ, Gannoulis N, Kamm M, Mashford ML (1990) The effect of an antacid and food on the absorption of cimetidine and ranitidine. J Pharm Pharmacol 42:352–354

Devane JG, Kelly JG (1991) Effect of food on the bioavailability of a multiparticulate sustained-release verapamil formulation. Adv Ther 8:48–53

Devane JG, Kelly JG, Geoghegan B (1990) Pharmacokinetic and in vitro characteristics of sustained release verapamil products. Drug Dev Ind Pharm 16:1233–1248

DeVries TM, Voigtman RE, Posvar EL, Nesbitt RU, Forgue ST (1991) Pharmacokinetic characterization of a procainamide tablet suitable for twice daily dosing. Pharm Res 8 [Suppl]:S315

Doose DR, Gisclon LG, Stellar SM, Riffits JM, Hills JF (1992a) The effect of food on the bioavailability of topiramate from 100- and 400-mg tablets in healthy male subjects. Epilepsia 33 [Suppl 3]:105

Doose DR, Minn FL, Stellar S, Nayak RK (1992b) Effects of meals and meal composition on the bioavailability of fenretidine. J Clin Pharmacol 32:1089–1095

Dressman JB, Berardi RR, Elta GH, Gray TM, Montgomery PA, Lau HS, Pelekoudas KL, Szpunar GL, Wagner JG (1992) Absorption of flurbiprofen in the fed and fasted states. Pharm Res 9:901–907

Drew RH, Gallis HA (1992) Azithromycin-spectrum of activity, pharmacokinetics, and clinical applications. Pharmacotherapy 12:161–173

Du Souich P, Lery N, Lery L, Varin F, Boucher S, Vezina M, Pilon D, Spenard J, Caillé G (1990) Influence of food on the bioavailability of diltiazem and two of its metabolites following the administration of conventional tablets and slow-release capsules. Biopharm Drug Dispos 11:137–147

Ducharme MP, Edwards DJ, McNamara PJ, Stoeckel K (1993) Bioavailability of syrup and tablet formulations of cefetamet pivoxil. Antimicrob Agents Chemother 37:2706–2709

Dudley MN, Marchbanks CR, Flor SC, Beals B (1991) The effect of food or milk on the absorption kinetics of ofloxacin. Eur J Clin Pharmacol 41:569–571

Dupuis LL, Koren G, Silverman ED, Laxer RM (1992) Influence of food on the bioavailability of oral methotrexate (MTX). Arthritis Rheum 35 [Suppl]:S57

Edgar B, Bailey D, Bergstrand R, Johnsson G, Regårdh CG (1992) Acute effects of drinking grapefruit juice on the pharmacokinetics and dynamics of felodopine and its potential clinical relevance. Eur J Clin Pharmacol 42:313–317

Ehrsson H, Wallin I, Ros AM, Eksborg S, Berg M (1994) Food-induced increase in bioavailability of 5-methoxypsoralen. Eur J Clin Pharmacol 46:375–377

Eller MG, Della-Coletta AA (1990) Absence of effect of food on alprazolam absorption from sustained release tablets. Biopharm Drug Dispos 11:31–37

Eller MG, Walker BJ, Yuh L, Antony KK, McNutt BE, Okerholm RA (1992) Absence of food effects on the pharmacokinetics of terfenadine. Biopharm Drug Dispos 13:171–177

Ewe K, Press AG, Dederer W (1989) Gastrointestinal transit of undigestible solids measured by metal detector EAS II. Eur J Clin Invest 19:291–297

Faulkner JK, Hayden ML, Chasseaud LF, Taylor T (1989) Absorption of amlodipine unaffected by food. Arzneimittelforschung 39:799–801

Fiese EF, Steffen SH (1990) Comparison of the acid stability of azithromycin and erythromycin A. J Antimicrob Chemother 25 [Suppl A]:39–47

Fowles SE, Fairless AJ, Pierce DM, Prince WT (1991) A further study of the effect of food on the bioavailability and pharmacokinetics of penciclovir after oral administration of famciclovir. Br J Clin Pharmacol 30:657P–658P

Fowles SE, Pierce DM, Prince WT, Thow JC (1990b) Effect of food on the bioavailability and pharmacokinetics of penciclovir, a novel antiherpes agent, following oral administration of the prodrug, famciclovir. Br J Clin Pharmacol 29:620P–621P

Frishman WH (1993) A new extended-release formulation of diltiazem HCl for the treatment of mild-to-moderate hypertension. J Clin Pharmacol 33:612–622

Frost RW, Carlson JD, Dietz AJK, Heyd A, Lettieri JT (1989) Ciprofloxacin pharmacokinetics after a standard or high-fat/high-calcium breakfast. J Clin Pharmacol 29:953–955

Frydman A (1992) Pharmacokinetic profile of nicorandil in humans: an overview. J Cardiovasc Pharmacol 20 [Suppl 3]:S34–S44

Fujimaki Y, Sudo K, Hakusui H, Tachizawa H, Murasaki M (1992) Single- and multiple-dose pharmacokinetics of nefiracetam, a new nootropic agent, in healthy volunteers. J Pharm Pharmacol 44:750–754

Gan V, Chu S-Y, Kusmiesz HT, Craft JC (1992) Pharmacokinetics of a clarithromycin suspension in infants and children. Antimicrob Agents Chemother 36:2478–2480

Gertz BJ, Holland SD, Kline WF, Matuszewski BK, Porras AG (1993) Clinical pharmacology of alendronate sodium. Osteoporosis Int [Suppl 3]:S13–S16

Gillespie NG, Mena I, Cotzias GC, Bell MA (1973) Diets affecting treatment of parkinsonism with levodopa. J Am Diet Assoc 62:525–528

Granneman GR, Mukherjee D (1992) The effect of food on the bioavailability of temafloxacin. A review of 3 studies. Clin Pharmacokinet 22 [Suppl 1]:48–56

Greb WH, Brett MA, Buscher G, Dierdorf H-D, von Schrader HW, Wolf D, Mellows G, Zussman BD (1989) Absorption of paroxetine under various dietary conditions and following antacid intake. Acta Psychiatr Scand 80 [Suppl 350]:99–101

Greenblatt DJ, Allen MD, MacLaughlin DS, Harmatz JS, Shader RI (1978) Diazepam absorption: effect of antacids and food. Clin Pharmacol Ther 24:600–609

Gupta SK, Benet LZ (1990) High-fat meals increase the clearance of cyclosporine. Pharm Res 7:46–48
Gupta SK, Manfro RC, Tomlanovich SJ, Gambertoglio JG, Garovoy MR, Benet LZ (1990) Effect of food on the pharmacokinetics of cyclosporine in healthy subjects following oral and intravenous administration. J Clin Pharmacol 30:643–653
Haegele KD, Hinze C, Jode-Ohlenbusch A-M, Creamer G, Borlack J (1991) Effects of a standardized meal on the pharmacokinetics of the new cardiotonic agent piroimone. Arzneimittelforschung 41:1225–1229
Hamaguchi T, Shinkuma D, Irie T, Yamanaka Y, Morita Y, Iwamoto B, Miyoshi K, Mizuno N (1993) Effect of a high-fat meal on the bioavailability of phenytoin in a commercial powder with a large particle size. Int J Clin Pharmacol 31:326–330
Hanyok JJ (1993) Clinical pharmacokinetics of sotalol. Am J Cardiol 12:19A–26A
Harrison LI, Riedel DJ, Armstrong KE, Goldlust MB, Ekholm BP (1992) Effect of food on salsalate absorption. Ther Drug Monit 14:87–91
Harrison LI, Mitra AK, Kehe CR, Klinger NM, Wick KA, McCarville SE, Cooper KM, Chang SF, Roddy PJ, Berge SM, Kisicki JC, Dockhorn R (1993) Kinetics of absorption of a new once-a-day formulation of theophylline in the presence and absence of food. J Pharm Sci 82:644–648
Hilleman DE, Mohiuddin SM, Destache CJ, Stoysich AM, Nipper HC, Maleskar MA (1992) Impact of food on the bioavailability of encainide. J Clin Pharmacol 32:833–837
Hoke JF, Chi EM, Anthony KK, Kulmala HK, Sussman NM, Okerholm RA (1991) Effect of food on the bioavailability of vigabatrin tablets. Epilepsia 32 [Suppl 3]:6–7
Holmes B, Brogden RN, Richards DM (1985) Norfloxacin: a review of its antimicrobial activity, pharmacokinetic properties and therapeutic use. Drugs 30:482–513
Honcharik N, Yatscoff RW, Jeffery JR, Rush DN (1991) The effect of meal composition on cyclosporine absorption. Transplantation 52:1087–1089
Honma A, Ikeda K, Udo N, Sdamori M, Hasegawa K, Kuruma I, Uto M, Nakahara T, Fujita T (1986) Aniracetam: clinical phase I study of aniracetam. J Clin Ther Med 2:929–952
Hoon TJ, McCollam PL, Beckman KJ, Hariman RJ, Bauman JL (1992) Impact of food on the pharmacokinetics and electrocardiographic effects of sustained release verapamil in normal subjects. Am J Cardiol 70:1072–1076
Hooper WD, Dickinson RG, Eadie MJ (1990) Effect of food on absorption of lomefloxacin. Antimicrob Agents Chemother 34:1797–1799
Hopkins A (1966) The pattern of gastric emptying: a new review of old results. J Physiol (Lond) 182:144–149
Hughes WT, Kennedy W, Shenep JL, Flynn PM, Hetherington SV, Fullen G, Lancaster DJ, Stein DS, Palte S, Rosenbaum D, Lian SHT, Blum MR, Rugers M (1991) Safety and pharmacokinetics of 566C80, a hydroxynaphthoquinone with anti-Pneumocystis carinii activity: a phase I study in human immunodeficiency virus (HIV)-infected man. J Infect Dis 163:843–848
Hunt TL, Kaiko RF (1991) Comparison of the pharmacokinetic profiles of two oral controlled-release morphine formulations in healthy young adults. Clin Ther 13:482–488
Hussey EK, Donn KH, Powell JR, Lahey AP, Pakes GE (1991) Albuterol extended-release products: effect of food on the pharmacokinetics of single oral doses of Volmax® and Proventil® repetabs in healthy male volunteers. J Clin Pharmacol 31:561–564
Hyneck MA (1992) An overview of the clinical pharmacokinetics of nabumetone. J Rheumatol 19 [Suppl 36]:20–24
Ingwersen SH, Mant TGK, Larson JJ (1993) Food intake increases the relative oral bioavailability of vanoxerine. Br J Clin Pharmacol 35:308–310
Iseki K, Satoh Y, Sugawara M, Miyazaki K (1994) Effect of Enterued® administration on the intestinal absorption of orally active cefem antibiotics. J Pharm Soc Jpn 114:233–240

Jallad NS, Callejas HJ, Woo-Ming RB, Weidler DJ (1991) The pharmacokinetics of astemizole plus pseudoephedrine when taken with a standardized meal. J Clin Pharmacol 31:863
Järvinen A, Nykänen S, Mattila J, Haataja H (1992) Effect of food on absorption and hydrolysis of erythromycin acistrate. Arzneimittelforschung 42:73–76
Johnson BF, O'Grady J, Sabey GA, Bye C (1978) Effect of a standard breakfast on digoxin absorption in normal subjects. Clin Pharmacol Ther 23:315–319
Kaiko R, Grandy R, Thomas G, Goldenheim P (1990) A single-dose study of the effect of food ingestion and timing of dose administration on the pharmacokinetic profile of 30mg sustained-release morphine sulfate tablets. Curr Ther Res 47:869–878
Kann J, Levitt MJ, Horodniak JW, Pav JW (1989) Food effects on the nighttime pharmacokinetics of Theo-Dur tablets. Ann Allergy 63:282–286
Keown PA, Stiller CR, Laupacis AL, Howson W, Coles R, Stawecki M, Koegler J, Carruthers G, McKenzie N, Sinclair NR (1982) The effects and side effects of cyclosporine: relationships to drug pharmacokinetics. Transplant Proc 14:659–661
Kivistö KT, Ojala-Karlsson P, Neuvonen PJ (1992) Inhibition of norfloxacin absorption by dairy products. Antimicrob Agents Chemother 36:489–491
Klamerus KJ, Troy SM, Ben-Maimon CS, Chiang ST (1993) Effect of age, gender, and food on zalospirone disposition. Clin Pharmacol Ther 53:193
Kleinbloesem CH, Ouwerkerk M, Brödenfeldt R (1993) Food effect with extended release formulations: nifedipine. Clin Pharmacol Ther 53:207
Knupp CA, Milbrath R, Barbhaiya RH (1993) Effect of time of food administration on the bioavailability of didanosine from a chewable tablet formulation. J Clin Pharmacol 33:568–573
Kopitar Z, Vrhovac B, Povšič L, Plavšič F, Francetić I, Urbančič J (1991) The effect of food and metoclopramide on the pharmacokinetics and side effects of bromocriptine. Eur J Drug Metab Pharmacokinet 16:177–181
Korth-Bradey JM, Fruncillo RJ, Klamerus K-J, Chiang ST, Conrad KA (1992) The influence of high and low fat meals on single dose pharmacokinetics of zalospirone. Pharm Res 9 [Suppl]:S326
Kosoglou T, Kazierad D, Schentag JJ, Patrick JE, Heimark L, Flannery BE, Affrime MB (1993) Effect of food and gastric emptying rate (GER) on the bioavailability (BA) of 5-ISMN. Clin Pharm Ther 53:206
Kozloski GD, de Vito JM, Johnson JB, Holmes GB, Adams MA, Hunt TL (1992a) Bioequivalence of verapamil hydrochloride extended-release pellet-filled capsules when opened and sprinkled on food and when swallowed intact. Clin Pharm 11:539–542
Kozloski GD, De Vito JM, Kisicki JC, Johnson JB (1992b) The effect of food on the absorption of methotrexate sodium tablets in healthy volunteers. Arthritis Rheum 35:761–764
Kuhn M (1993) Drug interactions and their nursing implications. J NY State Nurses Assoc 24:10–16
Lafontaine D, Mailhot C, Vermeulen M, Bissonnette B, Lambert C (1990) Influence of chewable sucraflate or a standard meal on the bioavailability of naproxen. Clin Pharm 9:773–777
Lasswell AB, Loreck ES (1992) Development of a program in accord with JACAHO standards for counseling on potential drug-food interactions. J Am Diet Assoc 92:1124–1125
Latini R, Tognoni G, Kates RE (1984) Clinical pharmacokinetics of amiodarone. Clin Pharmacokinet 9:136–156
Lecaillon JB, Dubois JP, Soula G, Pichard E, Poltera AA, Ginger CD (1990) The influence of food on the pharmacokinetics of CGP 6140 (amocarzine) after oral administration of a 1200mg single dose to patients with onchocerciasis. Br J Clin Pharmacol 30:629–633
Lecaillon JB, Poltera AA, Zea-Flores G, De Ramirez I, Nowell De Arevalo A (1991) Influence of food related to dose on the pharmacokinetics of amocarzine and of its

N-oxide metabolite CGP 13231, after oral administration to 20 onchocerciasis male patients from Guatemala. Trop Med Parasitol 42:286–290
Levine MAH, Walker SE, Paton TW (1992) The effect of food or sucralfate on the bioavailability of S(+) and R(–) enantiomers of ibuprofen. J Clin Pharmacol 32:1110–1114
Liao S, Hills J, Stubbs RJ, Nayak RK (1992) The effect of food on the bioavailability of tramadol. Pharm Res 9 [Suppl]:S308
Liedholm H, Wåhlin-Boll E, Melander A (1990) Mechanisms and variations in the food effect on propranolol bioavailability. Eur J Clin Pharmacol 38:460–475
Lindholm A, Henricsson S, Dahlqvist R (1990) The effect of food and bile acid administration on the relative bioavailability of cyclosporin. Br J Clin Pharmacol 29:541–548
Lode H, Stahlmann R, Koeppe P (1979) Comparative pharmacokinetics of cephalexin, cefaclor, cefadroxil, and CGP 9000. Antimicrob Agents Chemother 16:1–6
Lohmann A, Dingler E, Sommer W, Schaffler K, Wober W, Schmidt W (1992) Bioavailability of vinpocetine and interference of the time of application with food intake. Arzneimittelforschung 42:914–917
Lotterer E, Ruhnke M, Trauemann M, Beyer R, Bauer FE (1991) Decreased and variable systemic availability of zidovudine in patients with AIDS if administered with a meal. Eur J Clin Pharmacol 40:305–308
Lubowski TJ, Nightingale CH, Sweeney K, Quintiliani R (1992) The relative bioavailability of temafloxacin administered through a nasogastric tube with and without enteral feeding. Clin Pharmacokinet 22 [Suppl 1]:43–47
Lucas S, Rowinsky E, Wargin W, Hohneker J, Hsieh A, Donebower R (1993) Results of a study of the effect of food on the bioavailability and pharmacokinetics of Navelbine® liquid-filled soft gelatin capsules. Proc Am Soc Clin Pharmacol 12:160
MacGowan AP, Greig MA, Andrews JM, Reeves DS, Wise R (1989) Pharmacokinetics and tolerance of a new film-coated tablet of sodium fusidate administered as a single oral dose to healthy volunteers. J Antimicrob Chemother 23:409–415
MacKay J, Mackie AE, Palmer JL, Moult A, Baber NS (1991) Investigation into the mechanism for the improved oral systemic bioavailability of cefuroxime from cefuroxime axetil when taken after food. Br J Clin Pharmacol 33:226P–227P
Margalith D, Duroux P, Bauerfeind P, Emde C, Koelz H-R, Biollaz J, Klotz U, Armstrong D, Blum AL (1991) Famotidine should be taken with supper: the effect of drug-meal interactions on gastric acidity and plasma famotidine levels. Eur J Gastroenterol Hepatol 3:405–412
Mattila MJ, Neuvonen PJ, Gothoni G, Hackman R (1972) Interference of iron preparations and milk with the absorption of tetracyclines. In: Backer SB, Neuhaus GA (eds) Toxicological problems of drug combinations. Excerpta Medica, Amsterdam, pp 128–133
McLachlan AJ, Cutler DJ (1991) Bioavailability of hydroxychloroquine in rheumatoid arthritis patients using deconvolution techniques. Aust J Hosp Pharm 21:333
McLean AJ, McNamara PJ, DuSouich P, Gibaldi M, Lalka D (1978) Food, splanchnic blood flow and bioavailability of drugs subject to first-pass metabolism. Clin Pharmacol Ther 24:5–10
McNeil JJ, Drummer OH, Raymond K, Conway EL, Louis WJ (1991) The influence of food on the oral bioavailability of terazosin. Br J Clin Pharmacol 32:775–776
Melander A, McLean A (1983) Influence of food intake on presystemic clearance of drugs. Clin Pharmacokinet 8:286–296
Melander A, Danielson K, Schersten B, Wahlin E (1977) Enhancement of the bioavailability of propranolol and metoprolol by food. Clin Pharmacol Ther 22:108–112
Melander A, Stenberg P, Liedholm H, Scherston B, Wahlin-Boll E (1979) Food-induced reduction in bioavailability of atenolol. Eur J Clin Pharmacol 16:327–330
Mengel H, Gustavson LE, Soerensen HJ, McKelvy JF, Pierce MW, Houston A (1991) Effect of food on the bioavailability of tiagabine HCl. Epilepsia 32 [Suppl 3]:6

Mihara M, Ohnishi A, Tomono Y, Hasegawa J, Shimamura Y, Yamazaki K, Morishita N (1993) Pharmacokinetics of E2020 a new compound for Alzheimer's disease in healthy male volunteers. Int J Clin Pharmacol 31:223–229
Monk JP, Campoli-Richards DM (1987) Ofloxacin: a review of its antibacterial activity, pharmacokinetic properties and therapeutic use. Drugs 33:346–391
Motohiro T, Handa S, Yamada S, Oki S, Tsumura N, Yoshinaga Y, Sasaki H, Aramaki M, Oda K, Kawakami A, Koga T, Sakata Y, Nishiyama T, Yamashita F, Hayashi M, Amamoto M, Murakami T, Ono E, Shimizu T, Fukuda T, Watanabe J, Nagai K, Sueyoshi K, Morita J, Hashimoto N, Yuge K, Kuda N, Ando H, Dai S, Tominaga K (1992) Pharmacokinetics and clinical effects of cefdinir 10% fine granules in pediatrics. Jpn J Antibiot 45:74–86
Muralidharan G, Yacobi A, Blotner S, Bryzinski B, Carver A, Simon J, Faulkner R (1992) Pharmacokinetics of bisoprolol (B) and hydrochlorothiazide (HCTZ) after administration of a combination tablet of B/HCTZ (10/6.25 mg) in fasted and nonfasted healthy volunteers. Pharm Res 9 [Suppl]:S321
Nakamura H, Iwai N (1992a) Pharmacokinetic study on oral antibiotics in pediatrics. III. A pharmacokinetic study on cefprozil in pediatrics. Jpn J Antibiot 45:1489–1504
Nakamura H, Iwai N (1992b) Pharmacokinetic studies on oral antibiotics in pediatrics. V. A pharmacokinetic study on cefdinir in pediatrics. Jpn J Antibiot 45:12–27
Nakashima M, Uematsu T, Oguma T, Yoshida T, Kimura Y, Konishi M (1993) Phase I study of S-1108, a new ester-type oral cephem antibiotic. Chemotherapy 41 [Suppl 1]:109–125
Narang PK, Lewis RC, Bianchine JR (1992) Rifabutin absorption in humans: relative bioavailability and food effect. Clin Pharmacol Ther 52:335–341
Nazareno LA, Holazo AA, Limjuco R, Massarella JW, Koss-Twardy S, Min B (1992) The effect of food on absorption of dideoxycytidine (ddC) in HIV-positive patients. Pharm Res 9:S321
Neuvonen PJ, Kivistö KT (1989) The clinical significance of food-drug interactions: a review. Med J Aust 150:36–40
Neuvonen PJ, Kivistö KT (1992) Milk and yoghurt do not impair the absorption of ofloxacin. Br J Clin Pharmacol 33:346–348
Neuvonen PJ, Kivistö KT, Lehto P (1991) Interference of dairy products with the absorption of ciprofloxacin. Clin Pharmacol Ther 50:498–502
Nichols AI, Everitt D, Portelli S, Moore N, Law D, Yeulet S, Howland K, Audet P, Rizzo S, Ilson B, Jorkasky D (1992) The effect of food on the single dose pharmacokinetics of the leukotriene receptor antagonist SK&F 106203. Pharm Res 9 [Suppl]:S309
Nilsen OG, Dale O (1992) Single dose pharmacokinetics of trazodone in healthy subjects. Pharmacol Toxicol 71:150–153
Oguey D, Kölliker F, Gerber NJ, Reichen J (1992) Effect of food on the bioavailability of low-dose methotrexate in patients with rheumatoid arthritis. Arthritis Rheum 35:611–614
Oguma T, Yamada H, Sawaki M, Narita N (1991) Pharmacokinetic analysis of the effects of different foods on absorption of cefaclor. Antimicrob Agents Chemother 35:1729–1735
Ohdo S, Nakano S, Ogawa N (1992) Circadian changes of valproate kinetics depending on meal condition in humans. J Clin Pharmacol 32:822–826
Ohtsubo K, Fujii N, Higuch S, Aoyama T, Goto I, Tatsuhara T (1992) Influence of food on serum ambenonium concentration in patients with myasthenia gravis. Eur J Clin Pharmacol 42:371–374
Olanoff LS, Walle T, Cowart TD, Walle KU, Oexmann MJ, Conradi EC (1986) Food effects on propranolol systemic and oral clearance: support for a blood flow hypothesis. Clin Pharmacol Ther 40:408–414
Oosterhuis B, Jonkman JHG (1993) Pharmacokinetic studies in healthy volunteers in the context of in vitro/in vivo correlations. Eur J Drug Metab Pharmacokinet 18:19–30

Pan HY, Devault AR, Brescia D, Willard DA, McGovern ME, Whigan DB, Ivashkiv E (1993) Effect of food on pravastatin pharmacokinetics and pharmacodynamics. Int J Clin Pharmacol 31:291–294

Palazzini E, Cristoformi M, Babbini M (1992) Bioavailability of a new controlled-release oral naproxen formulation with and without food. Int J Clin Pharmacal Res 12:179–184

Peck RW, Weatherley BC, Posner J (1993) The pharmacokientics of 882C, a thymidine analogue with potent anti-VZV activity in healthy volunteers. Antiviral Res 20 [Suppl 1]:132

Pfeifer S (1993) Einfluß von Nahrung auf die Pharmakokinetik von Arzneistoffen. Pharmazie 48:3–16

Pfeiffer A, Vidon N, Bovet M, Rongier M, Bernier JJ (1990) Intestinal absorption of amiodarone in man. J Clin Pharmacol 30:615–620

Phelan MJI, Orme M, Williams E, Thompson RN (1991) Food has no effect on the pharmacokinetics of methotrexate in patients with rheumatoid arthritis. Arthritis Rheum 34 [Suppl]:S91

Pieniaszek HJ, Rakestraw DC, Schary WL, William RL (1991) Influence of food on the oral absorption and bioavailability of moricizine. J Clin Pharmacol 31:792–795

Plezia PM, Thornley SM, Kramer TH, Armstrong EP (1990) The influence of enteral feedings on sustained-release theophylline absorption. Pharmacotherapy 10:356–361

Plusquellec Y, Compistron G, Stevens S, Barre J, Jung J, Tillement JP, Houin G (1987) A double-peak phenomenon in the pharmacokinetics of veralipride after oral administration: a double site model for drug absorption. J Pharmacokinet Biopharm 15:225–239

Prescott LF, Yoovathaworn K, Makarananda K, Saivises R, Sriwatanakul K (1993) Impaired absorption of paracetamol in vegetarians. Br J Clin Pharmacol 36:237–240

Ptachcinski RJ, Venkataramanan R, Rosenthal JT, Burckart GJ, Taylor RJ, Hakala TR (1985) The effect of food on cyclosporine absorption. Transplantation 40:174–176

Rapin J-R (1992) Effect of food on calcium antagonists pharmacokinetics. Pharm Weekbl 14 [Suppl E]:E5

Riva E, Gerna M, Latini R, Giani P, Volpi A, Moggioni A (1982) Pharmacokinetics of amiodarone in man. J Cardiovasc Pharmacol 4:264–269

Roberts RJ, Leff RD (1989) Influence of absorbable and nonabsorbable lipids and lipidlike substances on drug bioavailability. Clin Pharmacol 45:299–304

Robertson DRC, Higginson I, MacKlin BS, Renwick AG, Waller DG, George CF (1991) The influence of protein containing meals on the pharmacokinetics of levadopa in healthy volunteers. Br J Clin Pharmacol 31:413–417

Roe AR (1989) Diet and drug interactions. Van Nostrand Reinhold, New York

Rolan PE, Mercer AJ, Westherley BC, Holdich T, Ridout G, Meire H, Posner J (1992) Investigation of the factors responsible for a food-induced increase in absorption of a novel antiprotozoal drug 566C80. Br J Clin Pharmacol 33:226P–227P

Rolan PE, Mercer AJ, Weatherley BC, Holdich T, Meire H, Peck RW, Ridout G, Posner J (1994) Examination of some factors responsible for food-induced increase in absorption of atovaquone. Br J Clin Pharmacol 37:13–20

Roller S, Lode H, Stelzer I, Deppermann KM, Boeckh M, Koeppe P (1992) Pharmacokinetics of loracarbef and interaction with acetylcysteine. Eur J Clin Microbiol Infect Dis 11:851–855

Roncari G (1993) Human pharmacokinetics of aniracetam. Drug Invest 5 [Suppl 1]:68–72

Roos RAC, Tijssen MAJ, Van der Velde EA, Breimer DD (1993) The influence of a standard meal on Sinemet CR absorption in patients with Parkinson's disease. Clin Neurol Surg 95:215–219

Rosenborg J, Nyberg L, Delén A-M (1991) Dietary habits do not influence the dosage of Bambec® tablets. Eur Respir J 4 [Suppl 14]:434S

Royer-Morrot M-J, Zhiri A, Jacob F, Necciari J, Lascombes F, Royer RJ (1993) Influence of food intake on the pharmacokinetics of a sustained release formulation of sodium valproate. Biopharm Drug Dispos 14:511–518
Ruhnke M, Bauer FE, Seifert M, Trautmann M, Hille H, Koeppe P (1993) Effects of standard breakfast on pharmacokinetics of oral zidovudine in patients with AIDS. Antimcrob Agents Chemother 34:2153–2158
Russell GAB, Martin RP (1989) Flecainide toxicity. Arch Dis Child 64:860–862
Sahai J, Gallicano K, Garber G, McGilvery I, Hawley-Foss N, Turgeon N, Cameron DW (1992) The effect of a protein meal on zidovudine pharmacokinetics in HIV-infected patients. Br J Clin Pharmacol 33:657–660
Saito A (1993) Effects of food, ceruletide and ranitidine on the pharmacokinetics of the oral cephalosporin S-1108 in humans. Chemotherapy 39:374–385
Sakashita M, Yokogawa M, Yamaguchi T, Sekine Y (1991) Pharmacokinetics of sparfloxacin in man. Yakubutsu Dotai 6:43–51
Sakashita M, Yamaguchi T, Miyazaki H, Sekine Y, Nomiyama T, Tanaka S, Miwa T, Harasawa S (1993) Pharmacokinetics of the gastrokinetic agent mosapride citrate after single and multiple oral administration in healthy subjects. Arzneimelforschung 43:867–872
Sanchez N, Sheiner LB, Halkin H, Melmon KL (1973) Pharmacokinetics of digoxin: interpreting bioavailability. Br Med J 20:132–134
Sandberg A, Ragnarsson G, Johnson UE, Sjögren J (1988) Design of a new multiple-unit controlled-release formulation of metoprolol – Metoprolol CR. Eur J Clin Pharmacol 33 [Suppl]:S3–S7
Sandberg A, Abrahamsson B, Regårdh C-G, Wieselgren I, Bergstrand R (1990) Pharmacokinetic and biopharmaceutic aspects of once daily treatment with metoprolol CR/ZOK: a review article. J Clin Pharmacol 30:S2–S16
Schaefer HG, Beermann D, Horstmann R, Wargenau M, Heibel BA, Kuhlmann J (1993) Effect of food on the pharmacokinetics of the active metabolite of the prodrug repirinast. J Pharm Sci 82:107–109
Segre G, Cerretani D, Moltoni L, Urso R (1992) Pharmacokinetics of rufloxacin in healthy volunteers. Eur J Clin Pharmacol 42:101–105
Sekikawa H, Nakano M, Takada M, Arita T (1980) Influence of dietary components on the bioavailability of phenytoin. Chem Pharm Bull (Tokyo) 28:2443–2449
Semple HA, Koo W, Tam YK, Ngo L-Y, Coutts RT (1991) Interactions between hydralazine and oral nutrients in humans. Ther Drug Monit 13:304–308
Setnikar I, Arigoni R (1988) Chemical stability and mode of gastrointestinal absorption of sodium monofluorophosphate. Arzneimittelforschung 38:45–49
Shah J, Fratis A, Ellis D, Murakami S, Teitelbaum P (1990) Effect of food and antacid on absorption of orally administered ticlopidine hydrochloride. J Clin Pharmacol 30:733–736
Shelton MJ, Morse GD, Portmore A, Blum R, Saddler B, Reichman RC (1993) Prolonged, but not diminished zidovudine (ZDV) absorption by food. Clin Pharmacol Ther 53:207
Shinkuma D, Hamaguchi T, Kobayashi M, Yamanaka Y, Mizuno N (1990) Effects of food intake and meal size on the bioavailability of sulpiride in two dosage forms. Int J Clin Pharmacol 28:440–442
Shukla VA, Pittman KA, Barbhaiya RH (1992) Pharmacokinetics interactions of cefprozil with food, propantheline, metoclopramide, and probenecid in healthy volunteers. J Clin Pharmacol 32:725–731
Shyu WC, Knupp CA, Pittman KA, Dunkle L, Barbhaiya RH (1991) Food-induced reduction in bioavailability of didanosine. Clin Pharmacol Ther 50:503–507
Simon JA, Robinson DE, Andrews MC, Hildebrand JR III, Rocci ML, Blake RE, Hogden GD (1993) The absorption of micronized progesterone: the effect of food, dose proportionality, and comparison with intramuscular progesterone. Fertil Steril 60:26–33
Smith HT, Jokubaitis LA, Troendle AJ, Hwang DS, Robinson WT (1993) Pharmacokinetics of fluvastatin and specific drug interactions. Am J Hypertens 6:375S–382S

Sörgel F, Kinzig M (1993) Pharmacokinetics of gyrase inhibitors. I. Basic chemistry and gastrointestinal disposition. Am J Med 94:44S–55S
Sun JX, Cipriano A, Chan K, Klibaner M, John VA (1994) Effect of food on the relative bioavailability of a hypolipidemic agent (CGP 43371) in healthy subjects. J Pharm Sci 83:264–266
Svensson CK, Edwards DJ, Mauriello PM, Barde SH, Foster AC, Lanc RA, Middleton E Jr, Lalka D (1983) Effect of food on hepatic blood flow: implication in the food effect phenomenon. Clin Pharmacol Ther 34:316–323
Tam YK, Kneer J, Dubach UC, Stoeckel K (1990) Effects of timing of food and fluid volume on cefetamet pivoxil absorption in healthy normal volunteers. Antimicrob Agents Chemother 34:1556–1559
Tanimura M, Kataoka S, Fujita Y (1991) Basic and clinical studies of sparfloxacin in urinary tract infections. Chemotherapy 39 [Suppl 4]:523–530
Tay LK, Sciacca MA, Sostrin MB, Farmen RH, Pittman KA (1993) Effect of food on the bioavailability of gepirone in humans. J Clin Pharmacol 33:631–635
Taylor DM, Massey CA, Willson WG, Dhillon SA (1993) Lowered phenytoin serum concentrations during therapy with liquid food concentrates. Ann Pharmacother 27:369
Terhaag B, Hrdlicka P, Gramatte T, Richter K, Feller K (1990) Zum Einfluß der Nahrung auf die Pharmakokinetik von Diclofenac aus Rewodina-25-dragees. Z Klin Med 45:443–446
Terhaag B, Gramatte T, Hrdlicka P, Richter K, Feller K (1991) The influence of food on the absorption of diclofenac as a pure substance. Int J Clin Pharmacol 29:418–421
Thakker KM, Mangat S, Wagner W, Castellana J, Kochak GM (1992) Effect of food and relative bioavailability following single doses of diclofenac 150 mg hydrogel bead (HGB) capsules in healthy humans. Biopharm Drug Dispos 13:327–335
Thebault JJ, Montay G, Ebmeier M, Douin MJ, Millerioux L, Chassard D, Mignot A (1990) Effect of food on the bioavailability of the new quinolone sparfloxacin. Proc 30th Interscience Conference on Antimicrob Agents Chemotherapy, Atlanta, Oct. 21–24, pp 294
Theodor RA, Weimann H-J, Weber W, Müller M, Michaelis K (1992) Influence of food on the oral bioavailability of moxonidine. Eur J Drug Metab Pharmacokinet 17:61–66
Trovato A, Nuhlicek DN, Midtling JE (1991) Drug-nutrient interactions. Am Fam Physician 44:1651–1658
Tsutsumi K, Nakashima H, Kotegawa T, Nakano S (1992) Influence of food on the absorption of beta-methyldigoxin. J Clin Pharmacol 32:157–162
Ueda Y, Matsumoto F, Imai T, Sakurai I, Takahashi T, Morita M (1993) Effect of probenecid and a meal load on the pharmacokinetics of zidovudine. Chemotherapy (Tokyo) 41:499–503
Unadkat JD, Collier AC, Crosby SS, Cummings D, Opheim KE, Corey L (1990) Pharmacokinetics of oral zidovudine (azidothymidine) in patients with AIDS when administered with and without a high-fat meal. AIDS 4:299–232
Van Harten J, Van Bemmel P, Dobrinska MR, Ferguson RK, Raghoebar M (1991) Bioavailability of fluvoxamine given with and without food. Biopharm Drug Dispos 12:571–576
Van Peer A, Woestenborghs R, Heykants J, Gasparini R, Gauwenbergh G (1989) The effects of food and dose on the oral systemic availability of itraconazole in healthy subjects. Eur J Clin Pharmacol 36:423–426
Vance-Bryan K, Guay DRP, Rotschafer JC (1990) Clinical pharmacokinetics of ciprofloxacin. Clin Pharmacokinet 19:434–461
Walter-Sack IE (1992) What is "fasting" drug administration? Eur J Clin Pharmcol 42:11–13
Walter-Sack IE, De Vries JX, Nickel B, Stenzhorn G, Weber E (1989) The influence of different formula diets and different pharmaceutical formulations on the systemic

availability of paracetamol, gallbladder size, and plasma glucose. Int J Clin Pharmacol 27:544–550
Warneke G, Setnikar I (1993) Effects of a meal on the pharmacokinetics of fluoride from oral monofluorophosphate. Arzneimittelforschung 43:590–595
Weber C, Roos B, Birnboeck H, Van Brummelen P (1993) Effect of food on the oral bioavailability of the renin inhibitor remikirin (RO 42-5892). Br J Clin Pharmacol 36:177P–178P
Weimann H-J, Rudolph M (1992) Clinical pharmacokinetics of moxonidine. J Cardiovasc Pharmac 20 [Suppl] 4:S37–S41
Welling PG (1977) Influence of food and diet on gastrointestinal drug absorption: a review. J Pharmacokinet Biopharm 5:291–334
Welling PG (1989) Effects of food on drug absorption. Pharmacol Ther 43:425–441
Welling PG (1993) Necessity of food studies: implications of food effects. In: Midha KK, Blume HH (eds) Bio-international: bioavailability, bioequivalence and pharmacokinetics. Medpharm, Stuttgart, pp 211–221
Welling PG, Tse FLS (1982) The influence of food on the absorption of antimicrobial agents. J Antimicrob Chemother 9:7–27
Welling PG, Elliott RL, Pitterle ME, Corrick-West HP, Lyons LL (1979) Plasma levels following single and repeated doses of erythromycin estolate and erythromycin stearate. J Pharm Sci 68:150–155
Wells TG, Sinaiko AR (1991) Antihypertensive effect and pharmacokinetics of nitrendipine in children. J Pediatr 118:638–843
Welty DF, Siedlick PH, Posvar EL, Selen A, Sedman AJ (1994) The temporal effect of food on tacrine bioavailability. J Clin Pharmacol 34:985–988
Wessels JC, Koeleman HA, Boneschans B, Steyn HS (1992) The influence of different types of breakfast on the absorption of paracetamol among members of an ethnic group. Int J Clin Pharmacol 30:208–213
Wilding IR, Hardy JC, Maccari M, Ravelli V, Davis SS (1991a) Scintigraphic and pharmacokinetic assessment of a multiparticulate sustained release formulation of diltiazem. Int J Pharm 76:133–143
Wilding IR, Hardy JG, Davis SS, Melia CD, Evans DF, Short AH, Sparrow RA, Yeh KC (1991b) Characterization of the in vivo behavior of a controlled-release formulation of levodopa (Sinemet CR). Clin Neuropharmacol 14:305–321
Wilding IR, Davis SS, Sparrow RA, Bloor JR, Hayes G, Ward GT (1992) The effect of food on the in vivo behavior of a novel sustained release formulation of tiaprofenic acid. Int J Pharm 83:155–161
Williams DB, O'Reilly WJ, Boehm G, Story MJ (1990) Absorption of doxycycline from a controlled release pellet formulation: the influence of food on bioavailability. Biopharm Drug Dispos 11:93–105
Williams L, Davis JA, Lowenthal DT (1993) The influence of food on the absorption and metabolism of drugs. Clin Nutr 77:815–829
Williams PE, Harding SM (1984) The absolute bioavailability of oral cefuroxime axetil in male and female volunteers after fasting and after food. J Antimicrob Chemother 13:191–196
Willis JV, Kendall MJ, Jack DB (1981) The influence of food on the absorption of diclofenac after single and multiple oral doses. Eur J Clin Pharmacol 19:33–37
Wilson CG, Washington N, Greaves JL, Washington C, Wilding IR, Hoadley T, Simms EE (1991) Predictive modelling of the behavior of a controlled release buflomedil HC1 formulation using scintographic and pharmacokinetic data. Int J Pharm 72:79–86
Wix AR, Doering PL, Hatton RC (1992) Drug-food interaction counseling programs at teaching hospitals. Am J Hosp Pharm 49:855–860
Wright CE (1992) The influence of meal timing on the kinetics (PK) and dynamics (PD) of alprazolam sustained release tablets (ASR). Pharm Res 9 [Suppl]:S285
Yamaguchi T, Ikeda C, Sekine Y (1986) Intestinal absorption of a β-adrenergic blocking agent Nadolol. II. Mechanism of the inhibitory effect on the intestinal absorption of Nadolol by sodium cholate in rats. Chem Pharm Bull (Tokyo) 34:3836–3843

Yogendran L, McLaughlin K, Vadas EB, Margolskee DJ, Bechard S, Hseih JY-K, Lin C, Rogers JD, Sciberras DG, James IM (1992) Comparative bioavailability of different formulations of MK-679 an orally active LTD_4 receptor antagonist. Br J Clin Pharmacol 34:169P–170P

Yong C-L, Yu D, Eden L, Eichmeyer L, Giessing D (1991) Effect of food on the pharmacokinetics of oxybutinin in normal subjects. Pharm Res 8 [Suppl]:S320

Yu DK, Morrill B, Bhargava VO, Giesing DH, Weir SJ (1992) Effect of food coadministration on Cardizem CD™ capsule bioavailability. Pharm Res 9 [Suppl]: S323

Yuen KH, Deshmukh AA, Newton JM, Short M, Melchor R (1993) Gastrointestinal transit and absorption of theophylline from a multiparticulate controlled release formulation. Int J Pharm 97:61–77

Zimmermann T, Yeates RA, Laufen H, Pfaff G, Wildfeuer A (1994) Influence of concomitant food intake on the oral absorption of two triazole antifungal agents, itraconazole and fluconazole. Eur J Clin Pharmacol 46:147–150

CHAPTER 4

Drug Interactions at Plasma and Tissue Binding Sites

J.C. McELNAY

A. Introduction

The binding of drugs to plasma and tissue proteins is a major determinant of drug distribution throughout the body. This binding has also a very important effect on drug dynamics since it is only the free (unbound) drug which can diffuse to, and interact with, receptor sites, i.e., bound drug is pharmacologically inactive. The relative ease of study in vitro of the binding of drugs in human plasma and in separate protein fractions (mainly albumin) has led to an extensive literature on the subject, with the influence of other drugs and/or endogenous materials on drug binding being a common topic for investigation. The relative difficulty in obtaining samples of viable human tissues has meant that work on the binding of drugs to human tissues has been very limited by comparison, although this may change with the more widespread use of tissue culture techniques. It is as a result of this large body of largely in vitro data that the topic of protein binding displacement interactions has gained prominence as a possible mechanism of drug-drug interactions; however, as mentioned in Chap. 1 of this volume, the importance of plasma protein binding displacement interactions has been overestimated and overstated and only in very specific cases do they result in adverse outcomes. In this chapter the influence of plasma and tissue binding on drug kinetics is discussed to facilitate consideration of the consequences of drug displacement from binding sites and their possible clinical significance. Those cases in which displacement can lead to clinically significant outcomes, which are much more restrictive than much of the literature and popular belief amongst practitioners would suggest, are considered in some detail. Before moving on the these issues, the proteins involved in drug binding are described.

B. Proteins Involved in Drug Binding

The main binding proteins in plasma are albumin and α_1-acid glycoprotein. Albumin consists of a single polypeptide chain and is present in plasma at a concentration of 35–45 g/l (approximates to 0.6 mM) in normal healthy individuals. Despite its large molecular weight, albumin is not retained exclusively in the plasma compartment but is distributed extravascularly. Most of the albumin that moves out of the vascular system is returned to the bloodstream via the thoracic duct (Rothschild et al. 1973). Approximately 13g albumin is synthesised (liver) and catabolised daily. Lowered serum albumin concentrations can be precipitated by many factors, e.g., neoplastic disease, burns, and lowered albumin concentrations are often seen in the elderly, particularly in those elderly patients who are suffering from chronic disease. The binding of albumin can be compromised by the presence of increased concentrations of endogenous displacing agents, e.g., in kidney disease. Albumin, unlike α_1-acid glycoprotein, which only binds basic drugs, can bind both acidic and basic drugs. The binding of drugs by these plasma proteins is of course dependent on the concentration of the protein in the plasma. Diseases/events that can increase the concentration of α_1-acid glycoprotein will therefore result in increased drug binding, e.g., myocardial infarction and Crohn's disease (Piafsky et al. 1978).

As well as protein concentration, the other main determinants of the amount of protein binding in plasma are the number of binding sites per molecule of protein and the strength of binding between the drug and its binding site. Two major binding areas have been identified on albumin, e.g., piroxicam has been shown to bind to the apazone locus (site I area) and to a lesser extent to the diazepam site (site II) (Bree et al. 1990). The strength of binding is measured in terms of the association constant between drug and protein; the larger the association constant the more avid the binding (D'Arcy and McElnay 1982). A drug may have different classes of binding sites with different affinities, e.g., tolbutamide has been estimated to have 2.27 primary binding sites (with an association constant of $21.86 \times 10^4 M^{-1}$) and 8.21 secondary binding sites (with an association constant of $1.71 \times 10^2 M^{-1}$) per molecule of albumin (Crooks and Brown 1974). If a drug has only one class of binding site, a straight line plot is achieved in the graphical representation devised by Scatchard (Fig. 1), which, having binding data over a range of protein and/or drug concentrations, can be used to determine n (number of binding sites per molecule of protein) and K_a (association constant). The graph becomes curved if more than one class of binding site is involved and interpretation becomes more difficult. The difficulties involved in interpreting binding data have been addressed by Plumbridge et al. (1978).

There is much current interest in chirality of drugs in relation to interaction with receptor sites and the production of different pharmacological effects by different optical isomers of the same compound. Binding of drugs to plasma proteins can also take place in a stereoselective manner. A recent paper, for example, examined the protein binding of propranolol and verapamil enantiomers in maternal and foetal serum. It was found that for propranolol in maternal but not in foetal serum, the difference in binding of the R- and S-isomers was significant, with the R/S ratio being significantly larger in the mother than in the foetus. For verapamil, the difference in binding of the R- and S-enantiomers was significant in both the mother and the infant, but the R/S ratio was similar in mother and foetus (Belpaire et al. 1995).

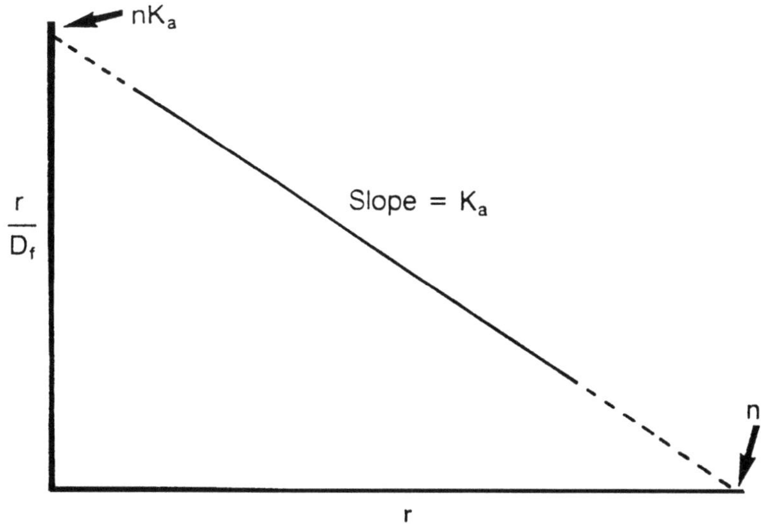

Fig. 1. Plot devised by SCATCHARD (1949) for drug/macromolecule interaction for a single class of binding site (r moles of drug bound per mole protein; D_f molar free drug concentration; n number of primary binding sites per molecule protein; K_a association constant of drug for binding sites). If more than a single class of binding site is involved in the binding interaction, the plot will be curved

The protein binding of drugs in tissues also primarily involves albumin (JUSKO and GRETCH 1976). Other agents are, however, involved, e.g., digoxin binds strongly to cardiac muscle (DOHERTY et al. 1967), bilirubin binds with ligandin (DAVIS and YEARY 1975), certain cytotoxics bind with DNA (ALLEN and CREAVEN 1974) while many drugs are bound by melanin (SALAZAR-BOOKAMAN 1994). In addition, some drugs are bound to red blood cells. Recent in vitro binding studies, for example, have indicated the presence of at least three kinds of binding sites on red blood cells for dorzolamide, a novel topical carbonic anhydrase inhibitor (HASEGAWA et al. 1994).

C. Influence of Plasma and Tissue Binding on Drug Kinetics

Both plasma and tissue binding of drugs, via their influence of maintaining a high concentration gradient of free drug between the lumen of the intestine and plasma post administration of a drug orally, will favour a more rapid absorption of drugs from the intestine (or indeed injection site). These effects are, however, very marginal in comparison to the influence of protein binding on drug distribution and elimination.

Regarding drug distribution, plasma and tissue protein binding have opposing effects. Since plasma proteins are largely retained within the plasma compartment, drugs which are highly bound to plasma proteins tend to have low apparent volumes of distribution (V_d). Warfarin, the best-known example of a drug which is highly bound to plasma albumin (>99%), has a low V_d (0.1 l/kg), indicating poor diffusion into the tissues (Tillement et al. 1978). Drugs which are highly bound to tissues tend to have a high V_d. This, of course, will depend on the type of tissue to which the drug is bound and its abundance in the body. For drugs which are highly bound to both plasma and tissues, the V_d will depend on the relative binding in both sites. With tricyclic antidepressants and phenothaizines, for example, almost all of the drug in plasma is bound to albumin; however, due to extensive tissue binding of these agents, the circulating drug in plasma represents only a small fraction of the total drug in the body (Koch-Weser and Sellers 1976). The equilibrium which exists between tissue and plasma binding is represented in Fig. 2. As mentioned earlier, only unbound drug is available to diffuse to receptor sites and exert therapeutic effects.

The next phase in the pharmacokinetic process is drug elimination, involving primarily the liver and kidney. It is obvious that if a drug is highly bound to tissues, it will be protected against elimination since much of the drug will be outside the plasma compartment and will therefore not be "delivered" to the kidney or liver for elimination. In addition, the extent of plasma protein binding can have marked effects on drug clearance by these organs. Consider first drug metabolism. For most drugs, plasma protein binding is protective in that the affinity for plasma binding sites exceeds that for metabolising systems. It is therefore only free (unbound) drug that is metabolised during passage through the liver (or other metabolic site). Such drugs would be considered to have a low hepatic extraction ratio (low tendency to be removed from the blood during its passage through the liver). On the other hand, if a drug has a high extraction ratio, the attraction to metabolic enzyme systems in the liver will overcome the binding affinity for plasma proteins and bound drug will be stripped from the binding sites to undergo metabolic change (McElnay and D'Arcy 1983). This will often involve active uptake of the drug into liver cells. In the case of high extraction drugs, plasma protein binding will facilitate drug metabolism by maintaining high total drug concentrations in the plasma (via decreased drug distribution) and will essentially act as a carrier system, taking the drug to its site of destruction. The metabolism of highly extracted drugs, e.g., propranolol, are dependent upon liver blood-flow (Rowland 1980).

A similar situation pertains to the kidney. Glomerular filtration, being a passive process, allows only free (unbound) drug to pass through the glomerulus in the normal healthy kidney, since protein molecules are too large to be filtered. Plasma protein binding therefore protects a drug against elimination in the kidney by glomerular filtration. As the filtrate passes down the renal tubules, back diffusion of drug occurs when the concentration of free drug in the renal tubule (due to water reabsorption) exceeds the free

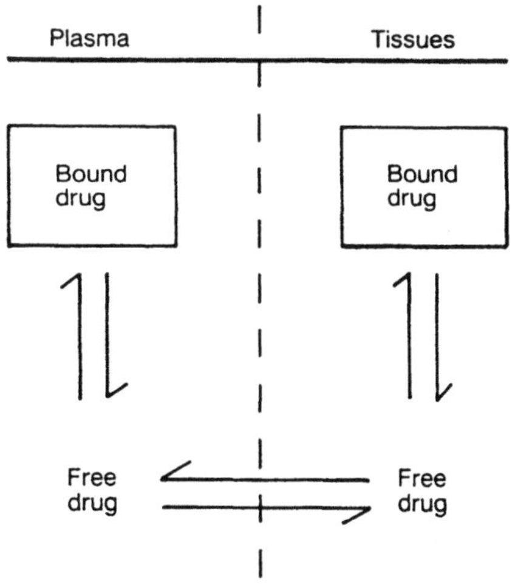

Fig. 2. Distribution of drug between plasma and tissue compartments. Only free (pharmacologically active) drug can diffuse out of the plasma; the equilibrium would therefore move towards the right after displacement from plasma binding and towards the left after displacement from tissue binding sites. Free drug concentrations in the plasma and tissue compartments are equal. Quantities of drug bound will depend on the amount of binding materials present in both plasma and tissues and their respective binding capacities. (After TILLEMENT 1980)

drug concentration in plasma. High binding in plasma helps maintain a high concentration gradient for the back diffusion of drugs which are lipid soluble at the pH of tubular filtrate, again helping to prevent drug elimination. Although glomerular filtration is the main mechanism of drug elimination via the kidneys, a number of drugs, e.g., penicillins, are actively secreted into tubular urine. As in the case of high extraction in the liver, the affinity for these active processes overcomes the binding affinity in plasma. Therefore plasma protein binding tends to facilitate elimination by limiting drug distribution, thereby allowing more drug to be presented to the kidney for elimination.

D. Displacement of Drugs from Binding Sites

In the case of plasma protein binding, most displacement of drugs from plasma binding sites is via a direct competition of the two drugs involved for the same binding sites. Drugs with a higher association constant will displace drugs with a lower association constant at common binding sites. Probably the best-known example of such a displacement interaction is between warfarin and phenylbutazone, with phenylbutazone successfully displacing warfarin from its binding. In this type of interaction the phenylbutazone is termed the "displacer" drug and warfarin the "displaced" drug, the latter term being usually reserved for the drug with the higher potential to give rise to adverse effects (Koch-Weser and Sellers 1976). Since the displacement is competitive, the binding of both drugs will be decreased. For substantial displacement to take place, the displacer must occupy the majority of the available binding sites, thereby substantially lowering the binding sites available to the primary drug (Rowland and Tozer 1995). For this to take place the molar concentration of drug in plasma must exceed the molar concentration of albumin (approximately 150 μg/ml for a drug with a molecular weight of 250). As a result of these prerequisites relatively few drugs are commonly referred to as displacers, e.g., salicylic acid, valproic acid and phenylbutazone. Since the molar concentration of α_1-acid glycoprotein is 9–23 μM in plasma, lower molar concentrations of basic drugs will be required to fulfil the latter displacer criterion (Rowland and Tozer 1995). In certain cases the parent drug may not itself be involved in a displacement interaction, but rather a metabolite may act as the displacer, e.g., trichloracetic acid displaces warfarin while the parent drug chloral hydrate does not (Sellers and Koch-Weser 1970).

As well as the competitive displacement interactions, the binding of one drug to, e.g., albumin, can give rise to changes in the conformation of the protein and thereby change the shape of the binding sites for a second drug. This normally gives rise to a decreased binding of the second drug although occasionally with allosteric effects the binding affinity of the second drug is increased. Aspirin influences the binding of certain drugs by this non-competitive mechanism, e.g., flufenamic acid and phenylbutazone, via permanently acetylating lysine residues of albumin (Pinkard et al. 1973). The alkylating agents carmustine and mechlorethamine give rise to decreased binding of penbutolol to α_1-acid glycoprotein. Dialysis against saline for 24 h does not restore the free fraction of penbutolol. The authors who conducted this latter research concluded that treatment of cancer patients with alkylating agents could alter the serum proteins and modify their binding capacity and that this should be taken into account when co-administering other basic drugs, e.g., methadone (Aguirre et al. 1994).

The study of the binding of drugs to tissue components had been much more limitied that those involving plasma; however, it is likely that both competitive and non-competitive binding displacement also takes place at tissue binding sites.

E. Therapeutic Consequences of Plasma Binding Displacement Drug-Drug Interactions

The interest in plasma protein binding displacement interactions was first aroused by the well-known interaction between warfarin and phenylbutazone (Aggeler et al. 1967), which results in a marked, and clinically significant, increased prothrombin time. It was correctly demonstrated that phenylbutazone displaced warfarin from its plasma binding sites (albumin); however, it was incorrectly assumed that the plasma binding displacement was the underlying mechanism of the interaction. Similarly, the increased pharmacological effect of tolbutamide in the presence of sulphonamides (CHRISTENSEN et al. 1963) was attributed to plasma binding displacement of tolbutamide by the sulphonamide. Again, if one examines the binding of these latter two drugs in vitro, there is undoubted displacement of the tolbutamide. If, for example, one places both sets of drugs together with plasma in an equilibrium dialysis cell, the free concentration and free fraction (free drug/total drug) of warfarin and tobutamide will be elevated, with corresponding decreases in the percentage of drug bound to the albumin component in plasma. In these early days plasma binding displacement was heralded as an important interaction mechanism, and the relative ease with which binding characteristics of drugs both singly and in combination could be determined led to many papers being published on drug-drug displacement interactions.

The problem with these data, however, was that the work was carried out in vitro and whereas displaced drug cannot distribute further than the confines of a dialysis cell, in the in vivo situation, displaced drug can distribute out of the plasma compartment into the relatively large "sink" of tissues. This moves the equilibrium situation presented in Fig. 2 in an anticlockwise direction. In addition, immediately after displacement has taken place, increased amounts of free drug will be presented to sites of elimination (liver and kidney). These compensatory effects of redistribution and increased elimination of free drug (assuming the drug has a low extraction ratio) lead to only a transient increase in free (pharmacologically active) drug in the plasma, whilst at equilibrium (steady-state) the free concentration of the drug will equate to pre-interaction levels. In addition, since distribution of drug from plasma to its receptor is often not instantaneous, but via a first-order process with comparable rate to general drug distribution, this will also reduce the extent of the transient increase in effect of the displaced drug. If the displacer drug is given orally, rapid compensatory effects may prevent even the transient increases in free drug concentration from taking a place. The displacement does, however, lead to important changes in the displaced drug's apparent volume of distribution (this will be increased), the total concentration of drug in plasma (decreased; Fig. 3) and the free fraction of drug in plasma (increased). This means that the same pharmacological effect will be achieved from a reduced total (free and bound) serum concentration of the drug. This type of effect has been clearly demonstrated in research which examined the effects of sodium salicylate on the elimination of indomethacin in dogs. Indomethacin was administered intravenously (bolus followed by an infusion) to yield steady-state plasma levels in the dogs. This was followed by administration of sodium salicylate either intravenously or via a duodenal fistual. Sodium salicylate administration gave rise to an immediate fall in indomethacin plasma concentrations by some 60%–70%. Total systemic clearance and biliary excretion of indomethacin

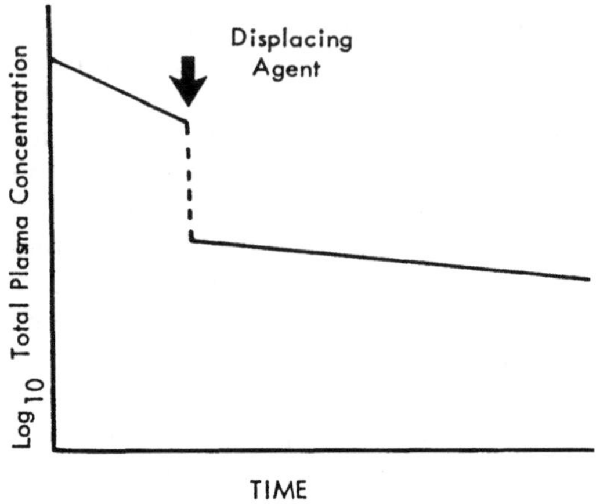

Fig. 3. Typical plasma concentration-time profile for a drug interaction involving plasma protein binding placement. An almost immediate drop in total plasma concentration (due to an increased V_d) is seen. Assuming clearance is unchanged, elimination half-life in this case is shown to be increased (decreased slope of elimination curve). (After D'ARCY and McELNAY 1982)

were increased as was the V_d, while plasma protein binding was decreased (BECK et al. 1990).

If follows therefore that interactions which have in the past been attributed solely to plasma binding displacement must have a different underlying mechanism. The best example of this is the now familiar warfarin–phenylbutazone interaction. As long ago as 1974, LEWIS et al. discovered a metabolic mechanism for this interaction. Phenylbutazone was shown to stereo-selectively inhibit the metabolism of S-warfarin (warfarin is available commercially as a racemic mixture and S-warfarin is five to six times more potent than R-warfarin; O'REILLY 1971). There was a corresponding increase in the elimination of R-warfarin. The increased proportion of the more potent S-warfarin, however, gave rise to the increased anticoagulation observed. Azapropazone is another example of a non-steroidal anti-inflammatory agent which gives rise to markedly increased prothrombin times when administered together with warfarin (POWELL-JACKSON 1977). Although azapropazone is a potent displacer of warfarin from plasma binding sites (McELNAY and D'ARCY 1980), this displacement cannot explain the increased pharmacological effect seen, and, although it has never been investigated, it would appear that a metabolic effect similar to that with phenylbutazone takes place (azapropazone has structural similarities to phenylbutazone). (Although many other non-steroidal anti-inflammatory agents may not increase prothrombin time, it should be borne in mind that they can cause gastrointestinal bleeding, which can be exacerabated in anticoagulated patients.)

Other interactions which were initially described as being due to plasma binding displacement, and which have now been demonstrated to have other mechanisms responsible for the interaction, include: warfarin/sulphamethoxazole and sulphinpyrazone – inhibition of metabolism (O'REILLY 1980; TOON et al. 1986); methotrexate/salicylate – inhibition of renal clearance (LIEGLER et al. 1970; STEWART et al. 1991); phenytoin/valproate – inhibition of metabolism (PERUCCA et al. 1980) and tolbutamide/sulphonamides – inhibition of metabolism (POND et al. 1977).

An interesting example of decreased plasma protein binding and decreased clearance was described by FERRAZZINI et al. (1990). This involved the interaction between trimethoprim-sulphamethoxazole and methotrexate in children with leukaemia. The free fraction of methotrexate was increased in all patients (mean 37.4%–52.2%) with concomitant administration of trimethoprim-sulphamethoxazle. Although plasma clearance did not change significantly, renal clearance of free methotrexate decreased significantly. The changes in protein binding and renal clearance of methotrexate resulted in a mean 66% increase in the systemic exposure to the drug. The authors suggested that this may exaplain the myelotoxicity often observed when the drugs are administered together. Work in the rabbit suggests that a similar interaction involving both decreased protein binding and renal clearance when ketoprofen is coadministered with methotrexate (PERRIN et al. 1990). Although decreased elimination is undoubedly the underlying cause of the interactions, plasma protein binding displacement tends to complicate the interpretation of the findings with regard to measured serum drug concentrations.

Although in most cases plasma binding displacement interactions per se do not give rise to clinically significant interactions, there are three main exceptions to this general rule:

I. Rapid Intravenous Infusion of Displacing Agent

If the displacing agent is given rapidly as a bolus injection, due to rapid displacement of the displaced drug, the free concentration could increase dramatically, and, since the compensatory mechanisms may take a short time to reduce the free drug concentration to pre-interaction levels, the displaced drug could reach receptors or enter compartments that it would not normally reach. The best-known example of such a displacement interaction involves the intravenous administration of sulphonamides to neonates. This results in the displacement of plasma-bound bilirubin, which in turn enters the CSF, resulting in kernicterus. Regarding drug-drug interactions, although conceivable, such interactions are unlikely to arise because the displaced drug would rapidly move down the newly created unbound concentration gradient into the large tissue water space and because intravenous injections are not normally administered rapidly in practice due to the adverse reactions often experienced (ROWLAND and TOZER 1995). A case report of an interaction in which this mechanism may have been involved has, however, recently been published. KISHORE et al. (1993) described a case of acute verapamil toxicity in a patient with chronic verapamil toxicity, possibly precipitated by the intravenous administration of the highly protein bound drugs ceftriaxone and clindamycin. Administration of the ceftriaxone and clindamycin led to complete heart block, requiring cardiopulmonary resuscitation and insertion of a temporary pacemaker. The patient reverted to spontaneous sinus rhythm some 16h later. The authors suggested that ceftriaxone, clindamycin or both agents may have precipitated the toxicity via displacing verapamil from its protein binding sites.

II. Parenteral Administration of Displaced Drug Having High Extraction Ratio

Consider the case of a drug with a high extraction ratio which is administered parenterally. Following an interaction that leads to displacement from plasma protein binding sites, the drug will distribute into the tissues and other body fluids, resulting in an initial decrease in plasma total drug concentrations. However, clearance of the displaced drug will remain unchanged and unless the dose of the drug is decreased the drug will begin to accumulate in the body so that the total (free + bound) drug concentration in plasma will return to the same value as prior to the interaction. (Steady-state plasma concentration of total (free + bound drug) equals rate of drug administration divided by clearance.) This will result in increased free drug concentrations in the plasma and therefore an enhanced therapeutic effect. Examples of drug interactions involving this latter mechanism are difficult to find – one drug which fulfils the criteria of having a high extraction ratio and being administered repeatedly by intravenous injection is lignocaine (ROLAN 1994). If the drug has a high extraction ratio in the liver, and is administered orally, due to decreased oral bioavailability as a result of a higher first-pass effect, the drug would not be expected to accumulate in the body and any effect of displacement on free drug concentrations at steady state is likely to be small (ROWLAND and TOZER 1995). However, if the high extraction ratio pertains to drug elimination in the kidney, a build-up of free drug concentration in the plasma after displacement could be anticipated even after oral administration (McELNAY and D'ARCY 1983).

III. Therapeutic Drug Monitoring and Drug Displacement from Plasma Binding Sites

For drugs with a narrow therapeutic range (narrow concentration range between therapeutic and toxic serum concentrations), to assist in patient management, therapeutic drug monitoring (TDM) is often performed. TDM involves the measurement of drug concentration in plasma or whole blood and adjustment of dosage based on measured drug concentrations. Total concentration of drug is the normal measurement made; it is clear from Fig. 3 that this concentration will fall after drug displacement and redistribution has taken place. The normal response to a decreased plasma drug concentration during therapeutic drug monitoring would be to increase the dose of drug or decrease the dosing interval. This would obviously be inappropriate since the free drug concentration remains at pre-interaction levels. Increased dosage would lead to drug accumulation and increased steady-state free drug concentration with attendant toxicity. Clinicians therefore must take displacement interactions into account when performing therapeutic drug monitoring. A possible solution is to monitor free drug concentrations in the plasma rather than total drug concentrations. Alternatively saliva drug concentrations of the drug of interest can be measured since these correlate with free drug concentrations (LINDUP 1975; MCAULIFFE et al. 1977). An example of this approach has also been suggested by TRNAVSKA *et al.* (1991), who examined the kinetics of lamotrigine using both plasma and saliva monitoring in patients undergoing long term antiepileptic therapy. The authors suggested that salivary monitoring represented a non-invasive alternative in therapeutic drug monitoring. These approaches are, however, seldom used clinically.

An example of a drug interaction involving drugs which require therapeutic drug monitoring is that between phenytoin and sodium valproate. Much attention has been focused on this interaction. FRIEL et al. (1979), for example, found that the mean total phenytoin plasma concentrations declined from 19.7 to 15.3 μg/ml when valproic acid was added to the regimen of 12 outpatients. In 20 patients receiving both drugs routinely, the median phenytoin free fraction was 15.8%, which contrasted with a much lower free fraction value (9.1%) in 40 patients who were receiving either phenytoin alone or phenytoin and phenobarbitone. The authors suggested that the fall in total concentrations of phenytoin could lead to inappropriate adjustment of phenytoin dosage. Theoretical consideration of the interaction by KOBER et al. (1978) led these authors to conclude that valproic acid, with its high therapeutic plasma concentrations, would increase the free fraction of phenytoin by 100%. PISANI and DIPERRI (1981) found a significant fall in plasma and a concurrent significant increase in salivary phenytoin when valproate (200 mg intravenously) was administered to ten epileptic patients treated long term with phenytoin. KNOTT et al. (1982) in a much larger study involving measurement of salivary and plasma phenytoin in 42 epileptic patients, indicated that saliva phenytoin concentrations bore the same close relationship to unbound therapeutically active phenytoin in patients receiving both drugs as it did in patients receiving phenytoin alone, whereas total plasma phenytoin concentrations did not. In a more recent study, MAY and RAMBECK (1990) examined the serum concentrations of phenytoin in 28 epileptic patients who were treated with phenytoin and valproate and in 15 patients who were treated with phenytoin in the absence of valproate. The results indicated that there were greater than normal fluctuations in serum phenytoin concentrations when valproate was coadministered. It was, however, noted that serum fluctuations in the concentrations of free (unbound) phenytoin did not differ between the two treatment groups. Samples were taken at 0800, 1100, 1400, 1700, and in part at 2000 hours.

F. Therapeutic Consequences of Tissue Binding Displacement Interactions

Displacement of drugs from tissue binding sites gives rise to the opposite effect to that of plasma binding displacement, i.e. the equilibrium depicted in Fig. 2 moves in a clockwise direction, giving rise to redistribution of drug from the tissues back into the plasma compartment. The plasma compartment, being relatively small in comparison to the tissue compartment, may be unable to buffer the impact of increasing amounts of free drug and therefore adverse sequelae may be more likely. Since increased free drug is forced into the plasma compartment, the total (free + bound) drug concentration will initially be increased (Fig. 4) and there will be a reduction in the apparent volume of distribution of the displaced drug. Elimination mechanisms will of course come into play and the rate of elimination of the displaced drug in, for example, the liver and the kidney will increase due to increased concentrations of free drug. Assuming binding within the plasma remains unaffected and drug clearance remains the same, both the mean total and the mean free drug concentrations will re-equilibrate to their original pre-interaction values. This re-equilibration, unlike the case with plasma binding displacement in which redistribution into the "sink" of the tissues takes place very quickly, will require some four to five half-lives. Therefore for some drugs with a long half-life, mean plasma concentrations of free drug may remain elevated for prolonged periods. Although clearance will not be effected, the reduction in V_d will result in a decrease in the elimination half-life, which in turn will lead to greater than normal fluctuations in plasma levels of the displaced drug during a dosing interval. A major drug interaction can take place if the displacing drug also inhibits the elimination of the displaced agent as occurs in some of the known interactions which occur by this mechanism. The best-known interactions that involve tissue displacement are with the cardiac glycoside, digoxin. The most extensively studied interaction involves digoxin and quinidine (Fig. 5). Administration of quinidine to a patient stabilised on digoxin leads, as expected, to an increased digoxin concentration in plasma. In one clinical trial, serum digoxin concentrations increased in 25 of 27 subjects during co-administration of quinidine from a mean of 1.4 ng/ml to 3.2 ng/ml. Signs of digoxin toxicity (anorexia, nausea and vomiting) were noted in 16 patients (59%) during concomitant quinidine therapy (REIFFEL et al. 1979). The data presented in Fig. 5 indicate clearly that the volume of distribution of digoxin is decreased during concomitant quinidine treatment – mean

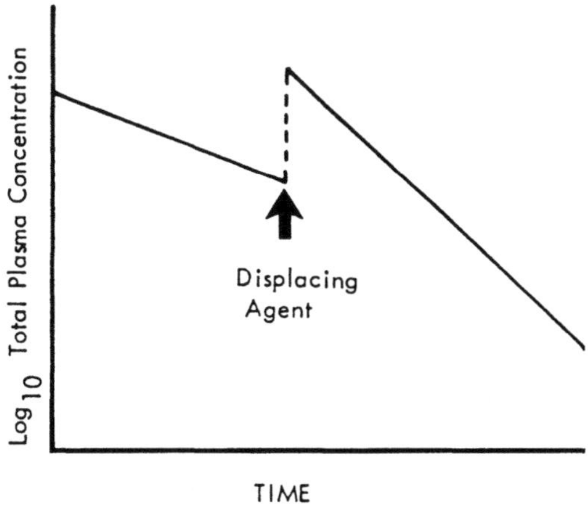

Fig. 4. Typical plasma concentration time profile for a drug interaction involving tissue binding displacement. An almost immediate increase in total plasma concentration (due to a decreased V_d) is seen. Elimination half-life is shown to be increased. This is because of increased amounts of both free and bound drug being presented at sites of elimination. (After D'ARCY and MCELNAY 1982)

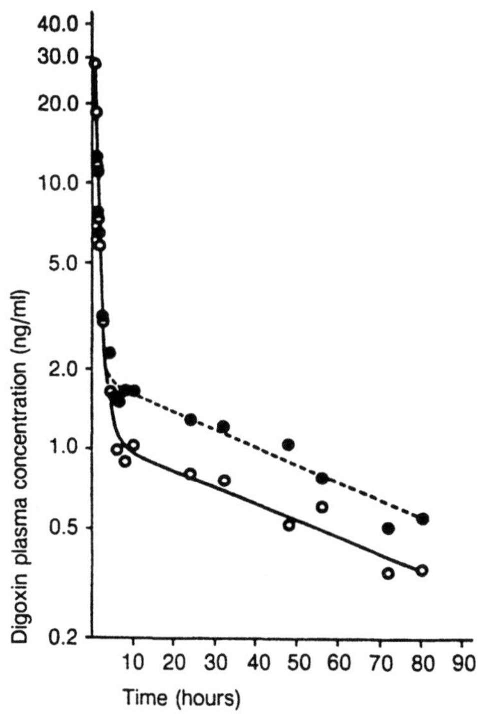

Fig. 5. Semilogarithmic plot of plasma concentration of digoxin as a function of time after the intravenous administration of digoxin before quinidine treatment (○—○) and during quinidine treatment (●—●) in one subject. (After HAGER et al. 1979)

values for V_d were 10.87 l/kg for digoxin alone and 7.35 l/kg for digoxin administration to six patients receiving quinidine (Hager et al. 1979). These authors suggested that the results could be explained by the displacement of digoxin from tissue binding sites by quinidine. Renal clearance of digoxin is also decreased by quinidine, probably due to inhibition of renal tubular secretion, and therefore this compensatory mechanism is compromised, a situation which will further exacerbate the outcome of the interaction. It is obviously necessary to carefully monitor serum digoxin concentrations during concomitant quinidine treatment and, indeed, Doering (1979) has suggested that digoxin dosage should be halved during quinidine therapy, as on average digoxin concentration doubles during the interaction. There is, however, no substitute to concentration monitoring since there is great variability in the magnitude of the quinidine-induced effects, e.g., one patient may have no induced increase in digoxin serum concentration while another may have a fivefold increase (Bigger 1979). Since discovery of the interaction involving quinidine, several other structurally related drugs have been shown to be involved in a similar interaction with digoxin. These drugs include the other antiarrhythmics amiodarone and propafenane. Disopyramide is free from interaction with digoxin and could be used to replace quinidine in patients receiving digoxin, since it has similar properties and indications (Koch-Weser 1979). Choroquine, an antimalarial drug, which is also used as a disease modifying agent in the treatment of rheumatoid arthritis, has been shown to increase digoxin plasma levels in dogs, presumably via a tissue binding displacement interaction (McElnay et al. 1985).

An interaction involving both displacement at plasma and tissue binding sites will have an even greater potential for adverse effects since free drug concentrations could be elevated to an even more marked extent. Sulphaethidole, for example, may displace dicloxacillin at both sites (De Sante et al. 1980).

G. Displacement of Drugs from Binding Sites by Endogenous Materials

The best-known example of a drug-drug interaction, which also involves endogenous materials, involves the parenteral anticoagulant heparin (Desmond et al. 1980). These authors have shown that heparin may cause a 150%–250% increase in the measured free fraction of diazepam, chlordiazepoxide and oxazepam, but no change in that of lorazepam, in normal non-fasting subjects. The changes were noted within 90 s of administration of 100 units of heparin; binding returned to baseline in 30–45 min. The authors reported a twofold increase in free fatty acid concentrations post administration of heparin and that these endogenous materials were responsible for the decreased binding of the affected benzodiazepines. Since sampling canulae are commonly flushed with heparin to keep them patent during clinical trials, such an effect could give rise to erroneous data. Heparin, for example, has also been shown to alter the plasma binding of phenytoin (CRAIG et al. 1976), quinidine (KESSLER et al. 1979), propranolol (WOOD et al. 1979a,b) and warfarin (CHAKRABARTI 1978).

Although initially the displacement via endogenous fatty acids appeared to be an in vivo effect, a careful study by GIACOMINI et al. (1980) indicated that the displacement takes place after sample collection. Via in vitro hydrolysis of plasma triglycerides by lipases, which are released and activated after in vivo administration of heparin, free fatty acid concentrations in blood samples are increased. The problem can largely be overcome by the use of a lipase inhibition in the sample collection vial and it is recommended that this precaution be taken when performing protein binding analyses with drugs which can be displaced by free fatty acids.

H. Disease States and Altered Plasma Protein Binding

Endogenous materials, which can act as displacing agents, can also be elevated in a number of disease states. This phenomenon will give rise to essentially the same pharmacokinetic picture as is seen in the presence of exogenously administered displacing drugs. Free fatty acid concentrations are elevated, for example, in stress or hyperlipidaemia, bilirubin in jaundice (BARRE et al. 1983) and 2-hydroxybenzoylglyine in uraemia (FISET et al. 1986). In insulin-dependent diabetes mellitus (IDDM) the increased serum concentrations of free fatty acids gives rise to decreased binding of valproic acid (Fig. 6; GRAINGER-ROUSSEAU et al. 1989). This is not surprising since valproic acid itself is a fatty acid and is therefore involved in competition with endogenous fatty acids for albumin binding sites.

Other factors which can give rise to changed binding in disease, and which can obviously complicate drug-drug binding interactions, include alteration in serum albumin concentrations (as in renal and hepatic disease), alteration in albumin distribution due to an increased unidirectional permeability of capillaries and impaired return of albumin to the plasma compartment due to impaired drainage of lymph vessels (BARRE et al. 1983). Conformational changes in binding protein molecules resulting in changes in drug binding affinity can also take place, e.g., via glycosylation of albumin (SHAKLAI et al. 1984). An increased concentration of α_1-acid glycoprotein induced by body stress, e.g., myocardial infarction or inflammation (PIAFSKY et al. 1978; PIAFSKY 1980), can lead to increased binding of basic drugs. Increased plasma concentrations of this latter protein are also seen in pregnancy. Marked decreases in albumin concentrations can be found in a range of debilitating disease states including burns, cancer, cardiac failure, cystic fibrosis, renal failure, nutritional deficiency, liver diseases and post surgery (TILLEMENT et al.

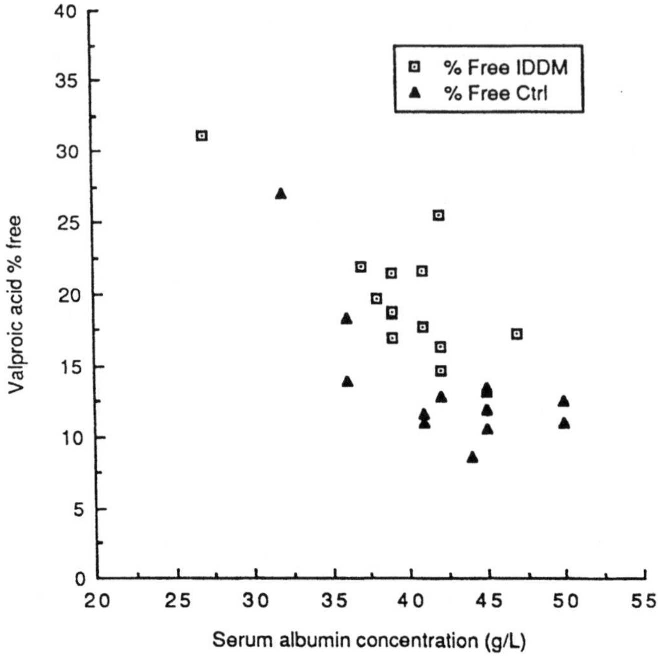

Fig. 6. Scattergraph of valproic acid % free vs serum albumin concentration in serum from insulin-dependent diabetes mellitus patients and age-matched controls (total serum concentration 90 μg/ml). (After Grainger-Rousseau et al. 1989)

1978; Øie 1986). All these conditions will complicate drug-drug interactions at plasma binding sites. Dasgupta et al. (1994) found that endogenous compounds present in uraemic serum led to a decreased binding of valproate, but that the displacement of valproate by salicylate was less marked in uraemic serum when compared with normal serum. A review of the influence of disease on plasma protein binding has been presented by Grainger-Rousseau et al. (1989) and a summary table from that review is presented as Table 1. In the present context, this table serves two purposes; first it draws attention to specific drug-disease interactions involving plasma protein binding and secondly data on the free fraction of a range of drugs which are bound to plasma proteins are presented. The latter information is important bearing in mind that clinically significant drug-drug displacement interactions in plasma usually involve drugs which are greater than 90% bound to plasma proteins at normal therapeutic concentrations. Although pregnancy is not a disease, it has been included in Table 1 since binding of some drugs will change during pregnancy. It has recently been suggested, for example, that during pregnancy and the 1st month after delivery, both total and free valproate serum concentrations should be closely monitored to determine the lowest effective dose (Johannessen 1992).

Table 1. Influence of diseases on drug plasma protein binding

Drug	Disease	% Free or free fraction (control vs. patient)	Significant difference	Clinical significance/ comment	References
Carbamazepine	Alcoholic liver disease, renal failure, rheumatoid arthritis, ulcerative colitis	22.7% vs. 19.5%	Y	Not clinically significant. Caution in interpreting total drug concentrations. Correlation with AGP shown	BARRUZI et al. 1986
Diazepam	Early pregnancy Mid pregnancy Late pregnancy	1.8% vs. 1.9% 1.8% vs. 2.1% 1.8% vs. 2.6%	N Y Y	Large V_d and therapeutic index means change is not clinically important in chronic use. If used in status epilepticus may have potentiated action in late pregnancy due to increased free fraction	PERUCCA et al. 1981
	Uraemia	2.9% vs. 4.5%	–	Not indicated. Site II not affected by carbamylation. Endogenous binding inhibitors implicated	CALVO et al. 1982
	Uraemia Kidney transplant Nephrotic syndrome	1.64% vs. 3.23% 1.50% vs. 2.11% 1.60% vs. 3.55%	Y Y Y	Not indicated, but an important difference implicated when interpreting kinetic data with regard to type of renal renal disease and the protein involved in binding	GROSSMAN et al. 1982
	Chronic cardiac failure	≈1% vs. 1.5%	N	Does not alter diazepam binding site affinity	FITCHL et al. 1983
	Acute uraemia Chronic uraemia	2% vs. 6% 2% vs. 4%	Y Y	Large therapeutic index means that drug effect can be monitored clinically as clinical end point can be determined safely	TIULA and NEUVONEN 1986
	Age-related decrease in renal function	1.8% vs. 3.0%	Y	Pharmacokinetic changes but no clinical implication	TIULA and ELFVING 1987
Digitoxin	Chronic cardiac failure	6% vs. 5.9%	N	Not significant	FITCHL et al. 1983
	End-stage renal disease (haemodialysis)	2.0% vs. 2.5%	N	Difference too small to have any therapeutic consequence	LOHMAN and MERKUS 1978
Disopyramide	MI	0.25%, vs. 0.15% (at 2 mg/l) 0.53% vs. 0.32% (at 5 mg/l)	Y	Binding varies with drug and protein concentration. May be clinically significant as AGP concentration decreases post MI	DAVID et al. 1983
	MI	0.2% vs. 0.13%	Y	Not indicated. Pharmacokinetic change suggested. Interpretation of total drug levels caution	CAPLIN et al. 1985
Flecainide	MI	39% vs. 47%	Y	Not indicated. Endogenous compounds may lead to displacement and decreased binding	CAPLIN et al. 1985
Imipramine	Chronic cardiac failure	19% vs. 18%	N	Drug binding not affected	FITCHL et al. 1983
Lignocaine	Uraemia	30.7% vs. 20.8%	Y	Interpreting kinetic data	GROSSMAN

Table 1. *Continued*

Drug	Disease	% Free or free fraction (control vs. patient)	Significant difference	Clinical significance/ comment	References
	Kidney transplant	33.7% vs. 24.6%	Y	with regard to the type of renal disease and the	et al. 1982
	Nephrotic syndrome	30.4% vs. 34.2%	N	protein involved in binding may be important	
	NIDDM	32% vs. 30%	N	Not indicated (but no change expected)	O'Byrne et al. 1988
Lorcainide	Cardiac arrhythmia	26.03% vs. 24.07%	N	Not significant	Somani et al. 1984
	Renal disease	26.03% vs. 29.04%	N	Not significant	
Metoclopramide	Renal disease	0.6% vs. 0.59%	N	Not significant	Webb et al. 1986
Phenylbutazone	Alcoholic hepatitis Alcoholic cirrhosis	6% vs. 13% 6% vs. 19%	Y Y	Not known. Bilirubin and hypoalbuminaemia implicated in binding defect	Brodie and Boobis 1978
Phenytoin	Early pregnancy[a] Mid pregnancy Late pregnancy	9.7% vs. 10.6% 9.7% vs. 10.9% 9.7% vs. 12.6%	Y Y Y	Important in drug monitoring interpretation. Decreased albumin concentration in pregnancy implicated	Perucca et al. 1981
	Chronic cardiac failure	≈ 15% vs. 15%	N	Not significant	Fitchl et al. 1983
	Acute uraemia Chronic uraemia	10% vs. 25% 10% vs. 24%	Y Y	Important clinically in drug monitoring, although free concentration remains the same due to pharmacokinetic compensation	Tiula and Neuvonen 1986
	Age-related decrease in renal function	10% vs. 13.5%	Y	Important in drug monitoring	Tiula and Elfving 1987
Prednisolone	Portosystemic shunt	17.7% vs. 28.6%	Y	Not significant as elimination and V_d altered; therefore, normalising free concentration	Bergrem et al.1983
	Nephrotic syndrome	2.26×10^3 vs. 4.20×10^3 M^{-1} (albumin K_a) 2.12×10^7 vs. 3.44×10^7 M^{-1} (transcortin K_a)	Y N	Not known. Altered pharmacokinetic disposition, due to change in binding	Frey and Frey 1984
Propranolol	Chronic cardiac failure	14% vs. 14%	N	AGP binding not affected	Fitchl et al. 1983
	Cancer, MI and IHD, infection, heart failure, COPD, CVA, miscellaneous	10.8% vs. 5.5%	Y	Not clinically significant but may be important when interpreting drug levels	Paxton and Briant 1984
	Acute uraemia Chronic ureaemia	10% vs. 9% 10% vs. 8.9%	N N	Not significant Not indicated	Tiula and Neuvonen 1986
Salicylate	Alcoholic hepatitis Alcoholic cirrhosis	27% vs. 34% 27% vs. 41%	N Y	Not known. Bilirubin and hypoalbuminaemia implicated in binding defect	Brodie and Boobis 1978

Table 1. *Continued*

Drug	Disease	% Free or free fraction (control vs. patient)	Significant difference	Clinical significance/ comment	References
Sulphadiazine	Alcoholic hepatitis Alcoholic cirrhosis	46% vs. 58% 46% vs. 51%	Y N	Not known. Bilirubin and hypoalbuminaemia implicated in binding defect	Brodie and Boobis 1978
Sulphisoxazole	Uraemia	5.2% vs. 21.8%		Not indicated. Carbamylation of drug binding site I implicating in defect; site II not affected	Calvo et al. 1982
Theophylline	Acute illness in COPD	54.6% vs. 69.7% 7.4% vs. 8.1 (μg/ml)	Y N	Not significant as free concentration is not is not changed significantly	Zarowitz et al. 1985
Tolfenamic acid	Renal disease Liver disease	0.08% vs. 0.17% 0.08% vs. 0.29%	Y Y	Not clear. High affinity for red blood cells means that these may act as reserve binding sites when free fraction increases	Laznicek and Senius 1986
Valproic acid	Renal disease	8.4% vs. 20.3%	Y	Not clear. May lead to increased incidence of toxicity. Important in drug monitoring interpretation	Gugler and Mueller 1978
	Early pregnancy[a] Mid pregnancy Late pregnancy	9.4% vs. 11.5% 9.4% vs. 12.1% 9.4% vs. 14.6%	Y Y Y	Important in drug monitoring. Decreased albumin serum concentration implicated in binding defect	Perucca et al. 1981
	IDDM	6.2% vs. 7.6%	Y	Not significant if diabetes is well controlled. If poorly controlled increased free concentration may be greater	Gatti et al. 1987
Verapamil	Arrhythmia	–	–	Not indicated. Pharmacokinetic change may be implicated	McGowan et al. 1982
	Liver disease	0.099 vs. 0.16	Y	Clinical end point effective to titrate dose so change not significant in therapy. Change may be relevant when interpreting kinetic data and total drug concentrations	Giacomini et al. 1984
Warfarin	Chronic cardiac failure	1% vs. 1%	N	No change expected	Fitchl et al. 1983
Warfarin	NIDDM	1.1% vs. 1%	N	Not indicated (but no change expected)	O'Byrne et al. 1988
Zomepirac	Uraemia	1.4% vs. 4%	Y	Decrease due to endogenous inhibitors Not clinically significant	Pritchard et al. 1983

Y, yes; N, no; MI, myocardial infarction; IHD, ischaemic heart disease; COPD, chronic obstructive pulmonary disease; CVA, cerebrovascular accident; AGP, α_1-acid glycoprotein; K_a, drug-protein association constant; V_d, apparent volume of distribution; NIDDM, non-insulin-dependent diabetes mellitus; IDDM, insulin-dependent diabetes mellitus.

[a] Although pregnancy is not a disease, it can give rise to changed binding; therefore it has been included in this table.

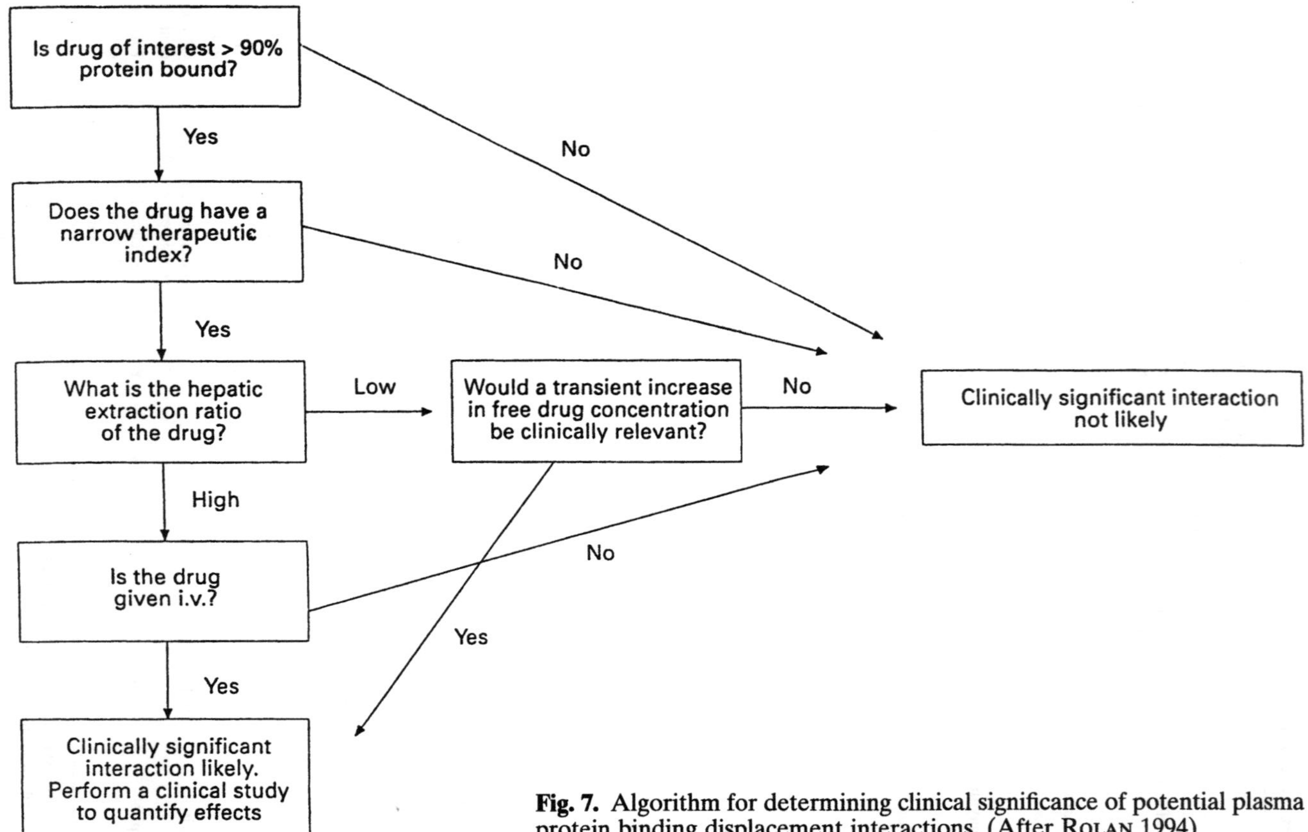

Fig. 7. Algorithm for determining clinical significance of potential plasma protein binding displacement interactions. (After Rolan 1994)

I. Conclusions

Plasma protein binding displacement has been overestimated and overstated as a mechanism of drug-drug interactions since the effects are of a transient nature; it is only considered to be problematic with regard to adverse clinical outcomes under specific conditions. Drug interactions, which in the past have been attributed to plasma binding displacement, have another underlying mechanism, usually involving decreased elimination of the drug in the liver or kidney. ROLAN (1994) has drawn up an algorithm to help assist in determining whether displacement interactions are likely to be clinically significant. A copy of this algorithm is presented in Fig. 7. If a plasma protein binding interaction is suspected, and the drug has a narrow therapeutic range, it is advisable to monitor free (unbound) concentrations of the drug as a guide to dosage adjustment. This is also the case in patients who have diseases which give rise to changed binding. Tissue binding displacement interactions, since re-equilibration of free drug concentrations post an interaction is dependent on the drug's half-life, rather than on redistribution, can be more problematic and this type of interaction can be complicated by decreased elimination of the displaced drug, as is seen in the case of the classical interaction between digoxin and quinidine.

References

Aggeler PM, O'Reilly RA, Leong L, Kowitz PE (1967) Potentiation of anticoagulant effect of warfarin by phenylbutazone. N Engl J Med 276:496–501

Aguirre C, Rodriguez-Sasiain JM, Calvo R (1994) Decrease in penbutolol protein binding as a consequence of treatment with some alkylating agents. Cancer Chemother Pharmacol 34:86–88

Allen LM, Creaven PJ (1974) Binding of a new antitumour agent, thalicorpine to DNA. J Pharm Sci 63:74–475

Barre J, Houin G, Brunner RF, Bree F, Tillement JP (1983) Disease induced modifications of drug pharmacokinetics. Int J Clin Pharmacol Res 3:215–226

Baruzzi A, Contin M, Perucca E, Albani F, Riva R (1986) Altered serum protein binding of carbamazepine in disease states associated with an increased α_1-acid glycoprotein concentration. Eur J Clin Pharmacol 31:85–89

Beck WS, Dietzel K, Geisslinger G, Engler H, Vergin H, Brune K (1990) Effects of sodium salicylate on elimination kinetics of indomethacin and bile production in dogs. Drug Metab Dispos 18(6):962–967

Belpaire FM, Wynant P, Van Trappen P, Dhont M, Verstraete A, Bogaert MG (1995) Protein binding of propranolol and verapamil enantiomers in maternal and foetal serum. Br J Clin Pharmacol 39:190–193

Bergrem H, Ritland S, Opendal I, Bergran A (1983) Prednisolone pharmacokinetics and protein binding in patients with portosystemic shunt. Scand J Gastroenterol 18:273–276

Bigger JT (1979) The quinidine-digoxin interaction. What do we know about it? N Engl J Med 301:779–781

Bree F, Urien S, Nguyen P, Riant P, Albengres E, Tillement JP (1990) A re-evaluation of the HSA-piroxicam interaction. Eur J Drug Metab Pharmacol 15(4):303–307

Brodie LMJ, Boobis D (1978) The effect of chronic alcoholic ingestion and alcoholic liver disease on binding of drugs to serum proteins. Eur J Clin Pharmacol 13:435–438

Calvo R, Carlos R, Erill S (1982) Effects of carbamylation of plasma proteins and competitive displacers on drug binding in uraemia. Pharmacology 24:248–252
Caplin JL, Johnston A, Hamer J, Camm AJ (1985) The acute changes in serum binding of disopyramide and flecainide after myocardial infarction. Eur J Clin Pharmacol 28:253–255
Chakrabarti SK (1978) Cooperativity of warfarin binding with human serum albumin induced by free fatty acid anion. Biochem Pharmacol 27:739–743
Christensen LK, Hansen JM, Kristensen M (1963) Sulphaphenazole-induced hypoglycaemic attacks in tolbutamide-treated diabetics. Lancet II:1298–1301
Craig WA, Evenson MA, Ramgopal V (1976) The effect of uraemia, cardiopulmonary bypass and bacterial infection on serum protein binding. In: Benet (ed) The effect of diseases states on drug pharmacokinetics. Am Pharm Assoc, Washington, pp 125–136
Crooks MJ, Brown KF (1974) The binding of sulphonylureas to serum albumin. J Pharm Pharmacol 26:304–311
D'Arcy PF, McElnay JC (1982) Drug interactions involving the displacement of drugs from plasma protein and tissue binding sites. Pharmacol Ther 17:211–220
Dasgupta A, Jaques M (1994) Reduced in vitro displacement of valproic acid from protein binding by salicylate in uremic sera compared with normal sera. Role of uremic compounds. Am J Clin Path 101:349–353
David BM, Ilett KF, Whitford EG, Stenhouse NS (1983) Prolonged variability in plasma protein binding of disopyramide after acute myocardial infarction. Br J Clin Pharmacol 15:435–441
Davis DR, Yeary RA (1975) Bilirubin binding to hepatic Y and Z protein (Ligandin): tissue bilirubin concentration in phenobarbital treated Gunn rat. Proc Soc Exp Biol Med 148:9–13
De Sante KA, Dittert LW, Stavchansky S, Doluisio JT (1980) Influence of sulfaethidole on the human pharmacokinetics of dicloxacillin. J Clin Pharmacol 20:535–542
Desmond PV, Roberts RK, Wood AJJ, Dunn GD, Wilkinson GR, Schenker S (1980) Effect of heparin administration on plasma binding of benzodiazepines. Br J Clin Pharmacol 9:171–175
Doering W (1979) Quinidine-digoxin interaction: pharmacokinetics, underlying mechanism and clinical implications. N Engl J Med 301:400–404
Doherty JE, Perkins WH, Flanigan WJ (1967) The distribution and concentration of tritiated digoxin in human tissues. Ann Int Med 66:116–124
Ferrazzini G, Klein J, Sulh H, Chung D, Griesbrecht E, Koren G (1990) Interaction between trimethoprim-sulfamethoxazole and methotrexate in children with leukemia. J Pediatr 117(5):823–826
Fiset C, Valee F, LeBel M, Gergeron MG (1986) Protein binding of ceftriaxone: comparison of three techniques of determination and the effect of 2-hydroxybenzoylglycine, a drug binding inhibitor in uraemia. Ther Drug Monit 8:483–489
Fitchl B, Meister W, Schmied R (1983) Serum protein binding of drugs is not altered in patients with severe chronic cardiac failure. Int J Clin Pharmacol Ther Toxicol 21:241–244
Frey FJ, Frey BM (1984) Altered plasma protein binding of prednisolone in patients with the nephrotic syndrome. Am J Kidney Dis 3:339–348
Friel PN, Leal KW, Wilensky AJ (1979) Valproic acid-phenytoin interaction. Ther Drug Monit 1:243–248
Gatti G, Grema F, Attardo-Parrinello G, Frantion P, Aguzzi F, Perucca E (1987) Serum protein binding of phenytoin and valproic acid in insulin-dependent diabetes mellitus. Ther Drug Monit 9:389–391
Giacomini KM, Giacomini JC, Blaschke TF (1980) Absence of effect of heparin on the binding of prazosin and phenytoin to plasma proteins. Biochem Pharmacol 29:3337
Giacomini KM, Massoud N, Wong FM, Giacomini JC (1984) Decreased binding of verapamil to plasma proteins in patients with liver disease. J Cardiovasc Pharmacol 6:924–928

Grainger-Rousseau TJ, McElnay JC, Collier PS (1989) The influence of disease on plasma protein binding of drugs. Int J Pharm 54:1–13
Grossman SH, Davis D, Kitchell BB, Shand DG, Routledge PA (1982) Diazepam and lidocaine plasma protein binding in renal disease. Clin Pharmacol Ther 31:350–357
Gugler R, Mueller G (1978) Plasma protein binding of valproic acid in healthy subjects and in patients with renal disease. Br J Clin Pharmacol 5:441–446
Hager WD, Fenster P, Mayersohn M, Perrier D, Graes P, Marcus FI, Goldman S (1979) Digoxin-quinidine interaction. N Engl J Med 300:1238–1241
Hasegawa T, Hara K, Hata S (1994) Binding of dorzolamide and its metabolite, N-deethylated dorzolamide, to human erythrocytes in vitro. Drug Metab Disposit 22:377–382
Johannessen SI (1992) Pharmacokinetics of valproate in pregnancy: mother-foetus-newborn. Pharm Weekbl (Sci Ed) 14:114–117
Jusko WJ, Gretch M (1976) Plasma and tissue binding of drugs in pharmacokinetics. Drug Metab Rev 5:43–140
Kessler KM, Leech RC, Spann JF (1979) Blood collection techniques, heparin and quinidine protein binding. Clin Pharmacol Ther 25:204–210
Kishore K, Raina A, Misra V, Jonas E (1993) Acute verapamil toxicity in a patient with chronic toxicity: possible interaction with ceftriaxone and clindamycin. Ann Pharmacother 27:877–880
Knott C, Hamshaw-Thomas A, Reynolds F (1982) Phenytoin-valproate interaction: importance of saliva monitoring in epilepsy. Br Med J 284:13–16
Kober A, Ekman B, Sjoholm I (1978) Direct and indirect determination of binding constants of drug-protein complexes with microparticles. J Pharm Sci 67:107–109
Koch-Weser J (1979) Disopyramide. N Engl J Med 300:957–962
Koch-Weser J, Sellers EM (1976) Binding of drugs to serum albumin (first of two parts). N Engl J Med 294:311–316
Laznicek M, Senius KEO (1986) Protein binding of tolfenamic acid in the plasma from patients with renal and hepatic disease. Eur J Clin Pharmacol 30:591–596
Lewis RJ, Trager WF, Chan KK, Breckenridge AK, Orme M, Rowland M, Schary W (1974) Warfarin: stereochemical aspects of its metabolism and the interaction with phenylbutazone. J Clin Invest 53:1607–1617
Liegler DG, Henderson ES, Hahn MA, Oliverio VT (1970) The effect of organic acids on renal clearance of methotrexate in man. Clin Pharmacol Ther 10:849–857
Lindup WE (1975) Drug-albumin binding. Biochem Soc Trans 3:635–640
Lohman JJHM, Merkus FWHM (1987) Plasma protein binding of digitoxin and some other drugs in renal disease. Pharm Weekbl (Sci Ed) 9:75–78
May T, Rambeck B (1990) Fluctuations of unbound and total phenytoin concentrations during the day in epileptic patients on valproic acid comedication. Ther Drug Monit 12:124–128
McAuliffe JJ, Sherwin AL, Leppik IE, Fayle SA, Pippenger CE (1977) Salivary levels of anticonvulsants: a practical approach to drug monitoring. Neurology (Minneap) 27:409–413
McElnay JC, D'Arcy PF (1980) Displacement of albumin-bound warfarin by anti-inflammatory agents in vitro. J Pharm Pharmacol 32:709–711
McElnay JC, D'Arcy PF (1983) Protein binding displacement interactions and their clinical importance. Drugs 25:495–513
McElnay JC, Sidahmed AM, D'Arcy PF, McQuade RD (1985) Chloroquine-digoxin interaction. Int J Pharm 26:267–274
McGowan FX, Reiter MJ, Pritchett EI, Shand DG (1982) Verapamil plasma binding: relationship to α_1-acid glycoprotein and drug efficacy. Clin Pharmacol Ther 33:485–490
O'Byrne PO, O'Connor P, Feely J (1988) Plasma protein binding of lignocaine and warfarin in non-insulin dependent diabetes mellitus. Br J Clin Pharmacol 26: 648
Øie S (1986) Drug distribution and binding. J Clin Pharmacol 26:583–586

O'Reilly RA (1971) Interaction of several coumarin compounds with human and canine plasma albumin. Mol Pharmacol 7:209–218
O'Reilly RA (1980) Stereoselective interaction of trimethoprim-sulfamethoxazole with the separated enantiomorphs of racemic warfarin in man. N Engl J Med 303:33–36
Paxton JW, Briant RH (1984) a_1-acid glycoprotein concentrations and propranolol binding in elderly pateints with acute illness. Br J Clin Pharmacol 18:806–810
Perrin A, Milano G, Thyss A, Cambon P, Schneider M (1990) Biochemical and pharmacological consequences of the interaction between methotrexate and ketoprofen in the rabbit. Br J Cancer 62:736–741
Perucca E, Hebdige S, Frigo GM, Gatti G, Lecchini S, Crema A (1980) Interaction between phenytoin and valproic acid: plasma protein binding and metabolic effects. Clin Pharmacol Ther 28:779–789
Perucca E, Ruprah M, Richens A (1981) Altered drug binding to serum proteins in pregnant women: therapeutic relevance. J R Soc Med 74:422–426
Piafsky KM (1980) Disease induced changes in the plasma binding of basic drugs. Clin Pharmacokinet 5:246–262
Piafsky KM, Borga O, Odar-Cederlof L, Johansson C, Sjovqist F (1978) Increased plasma protein binding of propanolol and chlorpromazine mediated by disease induced elevations of plasma a_1-acid glycoprotein. N Engl J Med 299:1435–1439
Pinkard RN, Hawkins D, Farr RS (1973) The infleunce of acetylsalicylic acid on the binding of acetrizote to human albumin. Ann New York Acad Sci 226:341–354
Pisani FD, Di Perri RG (1981) Intravenous valproate: effects on plasma and saliva phenytoin levels. Neurology 31:467–470
Plumbridge TW, Aarons LJ, Brown JR (1978) Problems associated with analysis and interpretation of small molecule/macromolecule binding data. J Pharm Pharmacol 30:69–74
Pond SM, Birkett DJ, Wade DN (1977) Mechanisms of inhibition of tolbutamide metabolism: phenylbutzone, oxphenbutazone, sulphaphenazole. Clin Pharmacol Ther 22:573–579
Powell-Jackson PR (1977) Interaction between azapropazone and warfarin. Br Med J 1:1193–1194
Pritchard JF, O'Neill PJ, Affrime MB, Lowenthal DT (1983) Influence of uraemia, haemodialysis and nonesterified fatty acids on zomepirac plasma protein binding. Clin Pharmacol Ther 34:681–688
Reiffel JA, Leahey EB, Drusin RE, Heissenbuttel RH, Lovejoy W, Bigger JT (1979) A previously unrecognised drug interaction between quinidine and digoxin. Clin Cardiol 2:40–42
Rolan PE (1994) Plasma protein binding displacement interactions – why are they still regarded as clinically important? Br J Clin Pharmacol 37:125–128
Rothschild MA, Oratz M, Schreiber SS (1973) Albumin metabolism. Gastroenterol 64:324–337
Rowland M (1980) Plasma protein binding and therapeutic drug monitoring. Ther Drug Monit 2:29–37
Rowland M, Tozer TN (1995) In: Clinical pharmacokinetics, concepts and applications. Williams and Wilkins, Philadelphia, pp 270–289
Salazar-Bookaman MM (1994) Relevance of drug-melanin interactions to ocular pharmacology and toxicology. J Ocul Pharmacol 10:217–239
Scatchard G (1949) The attractions of protein for small molecules and ions. Ann NY Acad Sci 51:660–672
Sellers EM, Koch-Weser J (1970) Potentiation of warfarin induced hypoprothrombinemia by chloral hydrate. N Engl J Med 283:827–831
Shaklai N, Garlick RL, Bunn HF (1984) Nonenzymatic glycosylation of human serum albumin alters its conformation and function. J Biol Chem 259:3812–3817
Somani P, Simon V, Gupta RK, King P, Shapiro RS, Stockard H (1984) Lorainide kinetics and protein binding in patients with end-stage renal disease. Int J Clin Pharmacol Ther Toxicol 223:121–125

Stewart CF, Fleming RA, Germain BF, Seleznick MJ, Evans WE (1991) Aspirin alters methotrexate disposition in rheumatoid arthritis patients. Arthritis Rheum 34:1514–1520
Tillement JP (1980) Plasma binding of drugs. Pharm Int 1:64–65
Tillement JP, Lhoste F, Gidicelli JF (1978) Disease and drug protein binding. Clin Pharmacokinet 3:144–154
Tiula E, Elfving S (1987) Serum protein binding of phenytoin, diazepam and propranolol in age-related decrease in renal function. Ann Clin Res 129:163–169
Tiula E, Neuvonen PJ (1986) Effect of total drug concentration on the free fraction in uraemic sera. Ther Drug Monit 8:27–31
Toon S, Low L, Gibaldi M, Trager WF, O'Reilly RA, Motley ChH (1986) The warfarin-sulphinpyrazone interaction: stereochemical considerations. Clin Pharmacol Ther 39:15–24
Trnavska Z, Krejocova H, Tkaczykovam, Salcmanova Z, Elis J (1991) Pharmacokinetics of lamotrigine (Lamictal) in plasma and saliva. Eur J Drug Metab Pharmacokin 3:211–215
Webb D, Buss DC, Fifield R, Bateman DN, Routledge PA (1986) The plasma protein binding of metoclopramide in health and renal disease. Br J Clin Pharmacol 21:334–336
Wood M, Shand DG, Wood AJ (1979a) Altered drug binding due to use of indwelling heparinized cannulas (heparin lock) for sampling. Clin Pharmacol Ther 25:103–107
Wood M, Shand DG, Wood AJ (1979b) Propranolol binding in plasma during cardiopulmonary bypass. Anesthesiol. 51:512–516
Zarowitz B, Shlom J, Eichenhorn MS, Popovich J (1985) Alternations in theophylline protein binding in acutely ill patients and COPD. Chest 87:766–769

CHAPTER 5

Drug Interactions and Drug-Metabolising Enzymes

P.F. D'ARCY

A. General Introduction

Induction and inhibition of cytochrome P450 enzymes are mechanisms that underly some of the more serious drug-drug interactions; it is therefore pertinent that the structure, distribution and essential roles of cytochromes P450 are reviewed.

Cytochrome P450 is not a single species of protein; the system is actually a collection of isoenzymes, all of which possess an iron atom in a porphyrin complex. They catalyse different types of oxidation reactions and under certain circumstances may catalyse other types of reaction such as reduction (TIMBRELL 1993).

Since the early experiments of CONNEY et al. (1956; CONNEY 1967) it has often been observed that the rate of oxidative metabolism of various substrates can differ markedly depending on the age, sex, species or the extent of exposure of the animal to different inducing agents. A number of different investigators have directed studies to evaluate the number and types of cytochrome P450s that exist in a single organ. In particular they questioned whether specific types of reactions catalysed by the microsomal electron transport system required specific cytochrome P450, or whether many cytochrome P450s have a rather broad substrate specificity differing only in the rate of catalysis of each reaction.

SCHENKMAN (1993) has well documented the beginning of this interest and study into the cytochrome P450 monooxygenase. It commenced with the compilation of knowledge of the metabolism of xenobiotics by WILLIAMS (1959), who described the many different metabolites produced from xenobiotics in vivo by animals. From the late 1940s into the 1960s knowledge was advanced by BRODIE et al. (1955), who was one of the first to start in vitro biochemical studies on the metabolism of xenobiotics by oxidative enzymes. Work from MILLER's laboratory augmented these studies (MÜLLER and MILLER 1953) and it soon became clear that liver microsomes were the source of NADPH-dependent, oxidative enzymes capable of metabolising a number of xenobiotics.

Since these early studies in the 1950s more than 800 different xenobiotics, many of which are therapeutic drugs, have been shown to be substrates for liver microsomal oxidative enzymes. The major types of oxidation reaction catalysed by the cytochrome P450 system can be subdivided into: oxidation or hydroxylation (e.g., many drugs including paroxetine), deamination (e.g., amphetamine), dealkylation (e.g., morphine), sulphoxidation (e.g., chlorpromazine, paroxetine), desulphuration (e.g., thiopentone), dehalogenation (e.g., halogenated anaesthetics) and glucuronidation (e.g., paroxetine).

Certain oxidative reactions of xenobiotics are also catalysed by enzymes other than the cytochrome P450 monooxygenase system, for example the microsomal FAD-containing monooxygenase (ZEIGLER 1984, 1985; DAMANI 1988). This is responsible for the N-oxidation of tertiary amines such as dimethylaniline and trimethylamine. The enzyme requires NADPH and oxygen; the substrate specificity also includes secondary amines and sulphides, thioethers and thiocarbamates, and organophosphates (TIMBRELL 1993). Alcohols may be oxidised by alcohol dehydrogenase; xanthine oxidase catalyses the oxidation of nitrogen heterocyclics such as the purine hypoxanthine. Some amines such as tyramine are substrates for monoamine oxidases; diamines such as putrescine are metabolised by the soluble enzyme diamine oxidase. The peroxidases are also involved in the oxidation of xenobiotics, the most important being prostaglandin synthase, which is known to catalyse the oxidation of *p*-phenetidine, a metabolite of phenacetin, a process that may be involved in the nephrotoxicity of the drug (TIMBRELL 1993).

Extensive work by many investigators including AXELROD (1955), POSNER et al. (1961), SLADEK and MANNERING (1966), ALVARES et al. (1967) and LU and COON (1968) isolated and purified a number of cytochrome P450 genes from animals and at the current time some 154 cytochrome P450 genes have been characterised from humans, animals, insects and plant species.

During the last 10 years or so, a vast amount of work has been done on the identification and characterisation of human xenobiotic-metabolising cytochrome P450s. Most studies have been done on hepatic and pulmonary enzymes due to the availability of tissues from tissue donor programmes.

BLACK (1993) has reviewed the cytochrome P450 structure and function and GONZALEZ (1993) cytochrome P450 in humans. Mammalian P450 enzymes are tightly bound in both microsomes (endoplasmic reticulum) and mitochondria and have been purified by means of detergent solubilisation. Purified detergent-free preparations exhibit a highly amphiphilic character and exist as micellar aggregates of approximately six protomers (DEAN and GRAY 1982). The enzymes are discrete gene products of about 57000 molecular weight and contain one equivalent of b-type heme per polypeptide.

The most abundantly expressed cytochrome P450s in human liver are the CYP3A subfamily. A brief description on this subfamily will serve as an example of the whole genus. Other subfamilies in human tissues comprise CYP1A, CYP2B, CYP2C, CYP2D, CYP2E, and CYP2F; the locus, function and activities of these are shown in Table 1.

Table 1. Cytochromes P450 in human. (Based on GONZALEZ 1993)[a]

Subfamily	Members	Chromosome (locus)	Established functions/activities	
			Metabolises	Induced by
CYP1A1	CYP1A1 CYP1A2	15q22-qter	Aflatoxin B_1 Arylamine and its promutagens Acetaminophen Benzo[α]pyrene Caffeine Food mutagens Phenacetin Warfarin	Ketocoazole Omeprazole Polycyclic aromatic hydrocarbons Tobacco smoke
CYP2A	CYP2A6 CYP2A7	19q12-13.2	Aflatoxin B_1 *N*-Nitrosodiethylamine Warfarin	
CYP2B	CY2B6 CY2B7	19	Warfarin	Phenobarbitone
CYP2C	CY2C9 CYP2C10 CYP2C17 CYP2C18 CYP2C19	10q24.1-24.3	Mephentoin Tolbutamide Warfarin	
CYP2D	CYP2D6 CYP2D7P CYP2D8P	22q11-2qter	Debrisoquine Sparteine Nitrosamine Carcinogen (NNK)	Tobacco smoke
CYP2E	CYP2E1	10	Acetone Caffeine Chlorzoxazone Paracetamol Solvents *N*-Nitrosodimethylamine	Ethanol
CYP2F	CYP2F1	19	Naphthylamine	
CYP3A	CYP3A3 CYP3A4 CYP3A5 CYP3A7	7q21-22.1	Aflatoxins Benzo[α]pyrrene Cyclosporin Erythromycin 17-Ethynyloestradiol Lidocaine Midazolam Nifedipine Quinidine Warfarin	Dexamethasone Steroids

[a] Other cytochrome P450s have been detected in animals; multiple gene conversions have occurred among some animal CYP genes, rendering them quite dissimilar to their counterparts in humans.

CYP3A cytochrome P450s account for up to 60% of total cytochrome P450 in some liver specimens (Shimada and Guengerich 1989; Guengerich and Kim 1990). A CYP3A protein has been isolated from human livers on the basis of its ability to oxidise the calcium channel blocking drug nifedipine (Guengerich et al. 1986). Interestingly, a cytochrome P450, designated CYP3A7, that is absent in adults has been found in foetal human liver. Although this is the most abundantly expressed cytochrome P450 in the foetus, its function is unknown.

There is a large degree of interindividual variability in the expression of CYP3A proteins, this could be due to the presence or absence of inducers or due to a genetic polymorphism. CYP3A proteins are also abundantly expressed in extrahepatic tissues, especially the gut (Watkins et al. 1987; DeWaziers et al. 1990). This is believed to have important implications for the metabolism of orally administered erythromycin, and other drugs that serve as a substrate for this enzyme. Immunocorrelation studies have established that CYP3A4 and probably CYP3A3 metabolise numerous drugs and toxins including aflatoxins, benzo[α]pyrene, cyclosporin, erythromycin, lidocaine, 17-ethynyloestradiol, midazolam, quinidine and warfarin.

This subfamily is therefore important not only for drug clearance but also for carcinogen activation, properties shared by many of the other subfamilies of cytochrome P450s. The structure of human CYP3A genes has not yet been determined but the locus has been found on chromosome 7 (7q21-q22.1) (Brooks et al. 1988).

B. Extrahepatic Microsomal Forms of Cytochrome P450

Cytochrome P450 has been shown to exist in many extrahepatic tissues in human and animals, including the gastrointestinal tract, lung and brain. Hydroxylation of benzo[α]pyrene has been induced in the oesophagus, forestomach, duodenum, small intestine, caecum and colon of animals treated with 1,2-benzanthrene, a well-known inducer of cytochrome P450 (Wattenberg et al. 1962; Wattenberg and Leon 1962). The small intestine has been demonstrated to catalyse the dealkylation of phenacetin (Pantuck et al. 1975) as well as 7-ethoxycoumarin. The presence of an active cytochrome P450-dependent mixed function oxidase system in tumour cell microsomes illustrates the potential for both drug metabolism and carcinogen activation (Strobel et al. 1993).

Cytochrome P450 content and activity are quite low in the lung when compared with those of the liver. Much of the investigative work has been done using rabbit lung microsomes (see, for example, Philpot et al. 1975; Williams et al. 1984; Matsubara et al. 1987) since this species has higher cytochrome P450 content and activities than other animal species and man (see, for example, Philpot et al. 1975; McManus et al. 1980). The lungs contain several cytochrome P450 isoenzymes with different substrate specificities (see review by Philpot and Wolf 1981).

The presence of cytochrome P450s in the lung is of obvious importance since this tissue is the portal of entry of various airborne contaminants and potential carcinogens and, in so far as therapy is concerned, it is a site for absorption of a variety of aerosolised inhalants and a site for drugs reaching the lung via the circulatory system (ARINÇ 1993).

Microsomal and mitochondrial forms of cytochrome P450 have been found in the brain although in low concentrations (see review by WARNER and GUSTAFSSON 1993). Most of the cytochrome P450 in the brain remains uncharacterised. The brain is not very responsive to well-known inducers of hepatic P450 such as barbiturates and polycyclic aromatic hydrocarbons (SRIVASTAVA and SETH 1983). A notable exception to this resistance to induction is the long-term administration of phenytoin (VOLK et al. 1988).

Brain cytochrome P450 is hormonally regulated and there is a five to tenfold increase of P450 content in the hypothalamic preoptic area and the olfactory lobes during pregnancy, lactation and in response to dihydrotestosterone treatment during pregnancy and lactation. Induction occurs in both the microsomal and mitochondrial fractions (WARNER et al. 1989).

Novel physiological functions of brain cytochrome P450 are thought to include:

1. Regulation of the intracellular levels of cholesterol (GUPTA et al. 1986; RENNERT et al. 1990).
2. Signal transduction – metabolities of arachidonic acid, formed through action of cytochrome P450, are thought to be the intracellular signals involved in the release of peptide hormones from the hypothalamus and pituitary (SNYDER et al. 1983; CASHMAN et al. 1987).
3. Regulation of vascular tone via P450-dependent metabolites of arachidonic acid (SACERDOTI et al. 1988; MURPHY et al. 1988).
4. Activation or inactivation of neurosteroids, e.g., dehydroepiandrosterone synthesised in the brain through the activity of cytochrome P450scc is believed to have a role in neuronal function (LE GOASCOGNE et al. 1987).

The role of brain cytochrome P450 in central nervous system toxicity through the activation of procarcinogens, the formation of cytotoxic metabolites from industrial solvents and the generation of reactive oxygen radicals during the catalytic cycle needs to be clarified. Aromatic hydrocarbons, *N*-nitroso compounds, triazines, hydrazines and common industrial solvents can produce gliomas in animals (MALTONY et al. 1982). Some of the more common solvents in paints are substrates for cytochrome P450 and exposure to industrial solvents is known to produce neurotoxicity in man (MALLOV 1976; ELOFSSON et al. 1980; MAIZLISH et al. 1985). It is also a distinct possibility that there is in situ formation of carcinogens from procarcinogens and that some of the toxicity of these solvents is due to P450-driven metabolism within the CNS (WARNER and GUSTAFSSON 1993).

C. Genetic Polymorphism

There are wide variations in cytochrome P450 activity in both humans and animals. DALY and IDLE (1993) have suggested, in their comprehensive review on animal and human cytochrome P450 polymorphisms, that this variation may be due to either a genetic polymorphism which results in either lack of protein synthesis or synthesis of a protein with reduced enzyme activity, or to a variation in the level of cytochrom P450 gene expression due to regulation by factors such as hormones or xenobiotics.

The inability of up to 12% of individuals from a variety of racial groups to metabolise the antihypertensive agent debrisoquine to 4-hydroxy-debrisoquine has been known since the 1970s (MAHGOUB et al. 1977). Pedigree studies showed that the poor metaboliser phenotype was inherited in an autosomal recessive manner (EVANS et al. 1980). Similar findings have been reported for the antiarrhythmic and oxytocic agent sparteine (EICHELBAUM et al. 1979), and other compounds including phenformin (OATES et al. 1982) and bufuralol (DAYER et al. 1982). Further studies indicated that a single enzyme was responsible for the metabolism of all these drugs (BERTILSSON et al. 1980). At least 25 other compounds including beta-blockers, antihypertensives, antiarrhythmics and antidepressants are known to be metabolised by debrisoquine hydroxylase (see review by BROSEN and GRAM 1989).

Cloning and sequencing of the CYP2D genes has assisted identification of the mutations which give rise to the debrisoquine polymorphism in humans, and a number of variant alleles associated with the poor phenotype have been identified. Debrisoquine phenotype may have important consequences with regard to both response to drug therapy and exposure to xenobiotics. For example, the use of the tricyclic antidepressant nortriptyline, the neuroleptic perphenzazine, or the antiarrhythmics propafenone and flecainaide may cause problems in both rapid extensive metabolisers (low and inadequate serum concentrations) and in poor metabolisers (toxic serum concentrations) (BROSEN and GRAM 1989). Fortunately, however, the majority of patients exhibit a normal metabolism of these agents and gain maximum therapeutic benefit with minimal toxicity.

Some interesting studies on a possible association between susceptibility to lung cancer and debrisoquine metabolic phenotype have been carried out by AYESH et al. (1984). They found that only 1.6% of a group of bronchial carcinoma patients were poor metabolisers of debrisoquine compared with 9.0% of a group of smoking controls. The molecular basis of this possible association between the poor metaboliser phenotype and decreased lung cancer susceptibility has been suggested by CRESPI et al. (1991) to relate to a reduced ability to activate tobacco smoke-derived procarcinogens.

An association between the poor metaboliser phenotype and susceptibility to early-onset Parkinson's disease has been suggested by BARBEAU et al.

(1985) as possibly due to environmental exposure to pesticides similar to MPP^+ which can produce a parkinsonian syndrome in man (DAVIS et al. 1979). However, subsequent studies have failed to confirm these suggestions.

Debrisoquine was an early example used to demonstrate polymorphism; other drugs including the anticonvulsant mephenytoin, the hypoglycaemic tolbutamide, the antihypetensive nifedipine, and antipyrine and caffeine have all been shown to exhibit metabolic polymorphism in humans. Apparently tolbutamide polymorphism is distinct from the debrisoquine polymorphism but is allied to that of phenytoin. Nifedipine metabolic polymorphism is currently somewhat debatable; the cytochrome P450 responsible for the majority of nifedipine metabolism is the glucocorticoid-inducible CYP3A4, although 10%–20% of human livers contain an additional nifedipine-metabolising cytochrome P450, CYP3A5. Thus variability in nifedipine metabolism may be the result of several factors including glucocorticoid levels and CYP3A5 expression.

Similarly, there is a possible polymorphism in the metabolism of coumarins. Cytochrome P450 CYP2A6 shows coumarin-7-hydroxylase activity and a cDNA variant of this enzyme CYP2A6v, which lacks enzyme activity, may give rise to polymorphism in the metabolism of coumarin. CYP1A2 is the enzyme responsible for the oxidative metabolism of caffeine and is polymorphic, with approximately 40% of an American population showing rapid metabolism of this agent [see review by DALY and IDLE (1993) and references therein].

D. Age and Disease and Cytochrome P450

Age-related changes in pharmacokinetic behaviour, specifically of drug metabolism, have often been attributed to changes in the cytochrome P450 based enzymes. However, studies in humans have suggested that the changes observed in clearance of oxidatively metabolised drugs are more likely due to physiological changes in the elderly rather than to changes in their cytochrome P450 systems. There are few, if any, detectable changes related to ageing in cytochrome P450 protein, mRNA or activities associated with ageing (SCHMUCKER et al. 1990).

There have been relatively few studies on the influence of disease states on metabolism of xenobiotics. As might be expected, different diseases of the liver have been investigated but have been found to influence metabolism differently (HOYUMPA and SCHENKER 1982). Indeed, some metabolic pathways do not seem to be influenced by liver damage at all, for example the glucuronidation of paracetamol, morphine and oxazepam was not affected by liver cirrhosis in man although the oxidation of barbiturates, antipyrine, and methadone and the conjugation of salicylates with glycine were all depressed by cirrhosis (TIMBRELL 1993). Clinical influenza is known to affect drug-metabolising enzymes as also is influenza vaccine and possibly other immunisation procedures. It is likely that the latter are in response to the production of interferon (see D'ARCY 1984a; GRIFFIN et al. 1988).

Study of these aspects of drug metabolism in the elderly is important since about 20% of the population of Western-developed countries is over the age of 65 years and the elderly are one of the fastest-growing segments of the community. This has importance in the pharmacotherapy since this segment is also the heaviest consumer of medicines. It is a proven fact that the extensive use of medication increases the possibility of adverse reactions to therapy and it seems likely that this predisposition in the elderly is related to pharmacokinetic or pharmacodynamic changes and not to senescent changes in the inducibility or nature of their hepatic or extrahepatic P450-driven drug metabolism (see review by BIRNBAUM 1993).

E. Clinical Importance of Enzyme Induction or Inhibition

Many of the major interactions between drugs are due to hepatic P450 enzymes being affected by the previous administration of other drugs. Some drugs act as potent enzyme inducers whilst others inhibit the enzyme systems.

I. Enzyme Inducers

Enzyme induction has been defined by GELEHRTER (1979) as an adaptive increase in the number of molecules of a specific enzyme, secondary either to an increase in its rate of synthesis or to a decrease in its rate of degradation. It is believed that high substrate concentration, and not hormonal regulation, is responsible for increased enzyme synthesis. If the demands on an enzyme system exceed its maximal functional capacity, adaptation should entail a parallel increase in enzymes and their containing structure.

Cytochrome P450 and its associated electron transport carriers are membrane-bound proteins in mammalian tissue; thus the induction of hepatic microsomal enzymes by phenobarbitone and other drugs has been shown conclusively to stimulate an increase in the smooth endoplastic reticulum, synthesis of enzyme protein, phospholipid and haem, DNA-dependent RNA synthesis, and the synthesis of specific mRNAs (REMMER and MERKER 1963; PARKE 1975; LINDELL et al. 1977). A further study of the effect of phenobarbitone on rat liver also showed that there were no significant alterations of the activities of any of the DNA-dependent RNA polymerases, which suggested that the primary stimulus of enzyme induction was more likely to be on post-transcriptional events such as enhanced stabilisation or decreased turnover of hepatic RNA, than on the transcription of new RNA (LINDELL et al. 1977).

Increased protein synthesis also appears to occur in the induction of the hepatic microsomal mixed-function oxidases following treatment of animals with the carcinogenic polycyclic hydrocarbons and, as with phenobarbitone, microsomal ribonuclease is inhibited. There are, however, a number of differences between these two inducing agents and these include an absence of marked hypertrophy, no marked increase in phospholipid synthesis, an increase of cytochrome P448 and not P450 and a change in substrate specificity (PARKE 1979).

High demands for protein and haem synthesis and the synthesis of choline for phosphatidylcholine make corresponding demands on dietary folate. The prolonged administration of the enzyme inducer phenobarbitone to rats on a low-folate diet has led to inhibition of the expected P450 microsomal induction and to increased folate deficiency. Rats treated daily with phenytoin plus phenobarbitone show evidence of folate deficiency.

The rate of metabolism of alcohol in alcoholics who have recently been drinking can be more than twice that in abstinent subjects. This aspect of induced metabolism is non-specific and also extends to many drugs. For example, the serum level of the anticonvulsant drug phenytoin falls more rapidly in alcoholics than it does in abstinent controls. This reduces the duration of anticonvulsant action and there is a very real danger of precipitating seizures if large amounts of alcohol are taken. Likewise the oral hypoglycaemic agent tolbutamide and the anticoagulant warfarin have a shorter half-life in the blood of alcoholics than in that of abstinent controls due to alcohol having an inductive effect of liver microsomal enzymes.

Alcohol is often the "silent" cause of drugs failing to achieve their full therapeutic effect. Interactions between alcohol and medication have been reviewed by D'ARCY and MERKUS (1981).

Exposure to tobacco smoke, in both active and passive smoking, will induce liver enzymes. A single injection of 3,4-benzpyrene, a common component of tobacco smoke, can induce rats to synthesise an enzyme system in the liver microsomes and non-hepatic tissues that is fully capable of hydroxylating 3,4-benzpyrene to non-carcinogenic products (CONNEY et al. 1957). Plasma levels and clinical effects of the analgesics pentazocine and propoxyphene are reduced in smokers. Interactions between tobacco smoking and drugs have been reviewed by D'ARCY (1984b).

Environmental pollutants may also affect P450 activity; thus chlorinated hydrocarbon pesticides are potent enzyme inducers and can exert long-lasting effects due to their storage in body fat; for example, KOLMODIN et al. (1969) showed accelerated metabolism of phenazone in farm workers exposed to DDT and gamma benzene hexachloride (lindane).

Theoretically, the same enzyme systems may be used to reduce the serum level of pesticides; the use of phenytoin and phenobarbitone anticonvulsants has been suggested as a means of lowering body pesticide levels when these become excessive. Certainly it has been shown that the DDT level in milk from dairy cows was reduced when they were treated with phenobarbitone.

Some antibiotics are enzyme inducers and, if used at constant dosage for long periods of time, will stimulate their own hepatic metabolism, which may then become the prime cause of their own inadequate levels in the blood. This problem has particular relevance in, for example, the long-term treatment of tuberculosis. ACOCELLA et al. (1972) have shown that rifampicin administered constantly at 600 mg/day for 7 days gave significantly lower blood levels on the 7th day than it did on the 1st day of treatment. Development of tolerance to drug regimens is certainly caused, at least in part, by enzyme induction.

Because so many therapeutic drugs have been shown to be implicated in cytochrome P450 induction, it is obviously impossible to describe or discuss the effects of their interactions even concisely in a chapter of this type. As a compromise therefore some examples of common P450-inducing agents are given in Table 2, while some recent reports of serious drug-drug interactions due to enzyme induction are shown as examples in Table 3.

In surveying the literature it soon becomes evident that reports of interactions due to enzyme inhibition are far more numerous than those caused by enzyme induction. This is not really surprising since the former are more readily observed in the clinic and they evoke therapeutic overdosage or toxicity, while the latter cause a reduced therapeutic effect of medication which may not immediately be noticed or reported.

II. Enzyme Inhibitors

There are many examples of drugs slowing or inhibiting the metabolic degradation of other drugs and thus causing an increase in both their duration and intensity of pharmacological actions. Drugs causing P450 enzyme inhibition include chloramphenicol, the H_2-receptor blocker, cimetidine, monoamine oxidase inhibitors (MAOIs) (e.g., iproniazid, nialamide), *p*-aminosalicylic acid

Table 2. Some examples of compounds and drugs acting as enzyme inducers

Alcohol
Chlorinated hydrocarbon pesticides
Polycyclic hydrocarbons (e.g., dioxin) and other environmental pollutants
Tobacco smoking
Many therapeutic drugs, including:
Barbiturates
Carbamazepine
Carbenoxolone
Ciprofloxacin
Clofibrate
Glucagon
Glucocorticoids (e.g., dexamethasone)
Isoniazid
Phenybutazone
Phenytoin
Primidone
Rifampicin
Theophylline

Table 3. Examples of serious drug-drug interactions due to enzyme induction (illustrative only, not comprehensive)

Drug combination	Sequelae
Ciprofloxacin/phenytoin	Lowered phenytoin plasma levles; danger of breakthrough seizures in epileptics (DILLARD et al. 1992)
Combined oral contraceptive/phenprocoumon	Increased clearance of the anticoagulant in seven patients reported; bleeding may occur if oral contraceptives are withdrawn (MÖNIG et al. 1990)
Phenytoin/theophylline	Both drugs show a reduced plasma concentration; poor seizure control and poor control of asthma may result (ADEBAYO 1988)
Phenytoin/verapamil	Subnormal and subtherapeutic plasma concentrations of verapamil (WOODCOCK et al. 1991)
Rifampicin/cyclosporin	Reduced cyclosporin plasma concentrations; kidney graft rejected (LANGHOFF and MADSEN 1983)
Rifampicin/dapsone	Subtherapeutic plasma concentrations of dapsone in HIV-infected patients suffering from *Pneumocystis carinii* pneumonia (HOROWITZ et al. 1992)
Rifampicin/disopyramide	Effective plasma concentrations of the antiarrhythmic were difficult to achieve in spite of dosage increase; on stoppage of rifampicin it took 3–5 days for inductive effect to disappear (STAUM 1990)
Rifampicin/oral contraceptives	Reduction of oral contraceptive efficacy; breakthrough bleeding and danger of unplanned pregnancy (D'ARCY 1986; ANGLE et al. 1991)
Rifampicin/propafenone	Reduced propafenone plasma concentrations and re-emergence of ventricular arrhythmias in a previously stabilised patient (CASTEL et al. 1990)

(PAS), pheniprazine, the two narcotics pethidine and morphine, and the antibacterial 4-quinolones. Some other agents, for example, chlorcyclizine, glutethimide and phenylbutazone, can inhibit or induce drug metabolism by enzymes depending upon whether they are administered acutely or chronically. Inhibition generally requires only a single dose of a compound rather than repeated doses as does induction; the environmental impact of inhibition is probably less than that of induction although inhibition may also be relevant in the work place, for example, workers exposed to the solvent dimethylformamide are less likely to suffer alcohol-induced flushes than those not exposed, possibly due to the inhibition of alcohol metabolism (TIMBRELL 1993).

Enzyme inhibition is especially important in relation to the therapeutic use of MAOIs and tricyclic depressants, the coumarin anticoagulants, the oral hypoglycaemics, the xanthine oxidase inhibitors (e.g., azathioprine, mercaptopurine) and the immunosuppressants (e.g., cyclosporin). Many of these drug-drug interactions can have serious, even life-threatening, sequelae. Table 4 presents information on known inhibitors of P450 enzymes while Table 5 gives examples of serious drug-drug interactions caused by this mechanism.

Because of the clinical importance of many drug-drug interactions involving P450 inhibition, it is important that the mechanisms of enzymatic inhibition by drugs be understood. Unfortunately, the nature of the procedures involving inhibition of P450 enzymes is not as well understood as the mechanisms of enzyme induction.

TIMBRELL (1993) in his comprehensive review of the biotransformation of xenobiotics, has explained that there are many different types of inhibitor of the microsomal monooxygenase system. There are reversible inhibitors which appear to bind as substrates and are competitive inhibitors, such as

Table 4. Some examples of compounds and drugs acting as enzyme inhibitors

Cobalt chloride
Industrial solvents (e.g., carbon tetrachloride, dimethylformamide)
Organophosphorus compounds
Many therapeutic drugs including:
Allopurinol
Amiodarone
Chloramphenicol
Cimetidine
Corticosteroids
Cyclophosphamide
Danazol
Dextropropoxyphene
Dicoumarol
Erythromycin
Felbamate
Fluconazole
Glibenclamide
Influenza vaccine
Iproniazid
Itraconazole
Ketoconazole
MAOI antidepressants
Moricizine
Nicoumalone
4-Quinolone and fluoroquinolone antibacterials (e.g., ciprofloxacin, enoxacin, lomefloxacin, norfloxacin, ofloxacin, perfloxacin)
Selegiline
Serotonin re-uptake inhibitor antidepressants (SRIs)
Tamioxifen
Triacetyloleandomycin

Table 5. Serious drug-drug interactions due to enzyme inhibition (illustrative only, not comprehensive)

Drug combination	Sequelae
Allopurinol/cyclosporin	Cyclosporin plasma levels grossly increased; danger of nephrotoxicity (GORRIE et al. 1994)
Amiodarone/theophylline	Theophylline serum concentrations increased; theophylline intoxication reported (SOTO et al. 1991)
Cyclosporin/nifedipine	Competition for metabolism by cytochrome P450pcn led to severe flushing and other signs of nifepidine toxicity in psoriatic patients (MCFADDEN et al. 1989)
Danazol/warfarin	Anticoagulant effect potentiated; bleeding complications possible; other mechanisms may be involved (MEEKS et al. 1992; BOOTH 1993)
Erythromycin/tacrolimus	Dramatic rise in plasma concentrations of the immunosuppressant, tacrolimus; danger of CNS toxicity or nephrotoxicity (SHAEFFER et al. 1994)
Felbamate/warfarin	Anticoagulation enhanced; 50% reduction of warfarin dosage required (TISDEL et al. 1994)
Fluconazole/nortriptyline	Steady-state serum concentration of nortriptyline increased when fluconazole was initiated and toxicity was evident (GANNON and ANDERSON 1992)
Fluconazole/phenytoin	Blood phenytoin concentration raised; danger of serious phenytoin toxicity (CADLE et al. 1994)
Fluconazole/warfarin	Over anticoagulation in a stabilised patient; bleeding episodes reported (SEATON et al. 1991)
Fluoroquinolones/warfarin	Anticoagulant effects enhanced by fluoroquinolones; danger of bleeding (CONSUMERS' ASSOCIATION 1993)
Glibenclamide/warfarin	Anticoagulation potentiated; bruising and soft tissue bleeding with large haematomas reported (JASSAL 1991)
Ketoconazole or itraconazol/astemizole or terfenadine	Plasma concentrations of the antihistamine astemizole increased greatly; serious cardiac arrhythmias reported (AHMAD 1992; NOTICEBOARD 1992)
Moricizine/warfarin	Anticoagulant effects enhanced; danger of bleeding. Other mechanisms may also apply (SERPA et al. 1992)
Nicardipine and other Ca^{+}-channel blockers/ cyclosporin	Increased cyclosporin blood concentrations; danger of nephrotoxicity; other mechanisms may be involved (TJIA et al. 1989)
Proguanil/warfarin	Anticoagulant effects enhanced; severe haematuria (ARMSTRONG et al. 1991)
Selegiline/pethidine	Enhanced toxicity of pethidine; life-threatening interaction reported (ZORNBERG et al. 1991)
Tamoxifen/warfarin	Grossly elevated anticoagulation; bleeding disorders reported; mechanisms other than enzyme inhibition may also apply (LODWICK et al. 1987; COMMITTEE ON SAFETY OF MEDICINES 1989; TENNI et al. 1989)

dichlorobiphenyl, which inhibits the O-demethylation of *p*-nitroanisole. There are inhibitors which are metabolised to compounds which bind strongly to the active site of the enzyme, such as piperonyl butoxide, which probably acts by forming an inactive complex with cytochrome P450, which becomes irreversibly inhibited. Other inhibitors, such as carbon tetrachloride, cyclophosphamide, carbon disulphide and allylisopropylacetamide, destroy cytochrome P450. Finally, there are those which interfere with the synthesis of the enzyme such as cobalt chloride, which interferes with the synthesis of the haem portion of the enzyme.

Other enzymes which are also involved in xenobiotic metabolism may also be inhibited, for example, monoamine oxidase inhibitor antidepressants cause a decreased metabolism of dietary amines, such as tyramine, and this has caused serious and sometimes fatal interactions when cheese is eaten (see GRIFFIN et al. 1988).

Cimetidine is known to decrease the clearance of other drugs that are largely eliminated by hepatic metabolism. It does this mainly by binding to the hepatic microsomal cytochrome P450 mixed-function oxidase system, thus inhibiting the metabolism of other drugs. However, with cimetidine other mechanisms may be involved since it also reduces liver blood flow and this reduced perfusion may potentiate the action of drugs (e.g., propranolol) whose systemic clearance is highly dependent upon liver blood flow.

WOLFF et al. (1993) have emphasised that more knowledge about structural features of compounds inhibitory for cytochrome P450 enzymes is required to evaluate whether one drug may be inhibitory for the metabolism of one or more drugs. They have suggested and described two approaches to elucidate the specificity of cytochrome P450 enzymes on a molecular level: firstly, to determine the three-dimensional structure of the active site by X-ray crystallography and use of this information for the development of substrate and inhibitor models and, secondly, the design of pharmacophor models of substrates and inhibitors on the basis of quantitative structure-activity relationships or by using molecular modelling methods.

F. Conclusions

It may be apparent from these brief accounts of interactions involving P450 enzyme induction or inhibition that many of the involved drugs are those on which patients are carefully stabilised often for long periods of time (e.g., anticoagulants, antidepressants, hypoglycaemics, hypotensives, oral contraceptives). Past experience has shown that it is these drug-stabilised patients who are at special risk from any interaction that will influence the potency or availability of their medication. This is especially so for the elderly patients, whose medication needs are generally greater than younger persons and who have a substantially greater risk than the younger patient of experiencing adverse reactions to medication.

In discussing the role of cytochrome P450 enzymes in the mechanisms of these interactions some caveats must be introduced about the information that has been used to establish their veracity. Much of the current knowledge of the cytochrome P450 enzyme systems has been gained from studies in animals. However, animal species differ from each other and from humans in respect to their pattern of cytochrome P450 enzymes involved in detoxification or bioactivation of xenobiotics, and cytotoxic and genotoxic chemicals. It is important to know the specificity of these enzymes and their distribution among human and animal species if animals studies are to be useful in understanding actual interactions or predicting likely interactions between therapeutic drugs.

It is also tempting to focus all attention on the role of the cytochrome P450 enzymes; however, this enzyme system is but one of the many that may exist in the cell for the metabolism of xenobiotics. Cytochrome P450 has been the easiest to study because it is a coloured protein that can readily be evaluated spectrophotometrically, but it is not all of drug metabolism and other enzyme systems do exist which should not be neglected if a full understanding is to be gained of how drugs function as enzyme inducers (ESTABROOK 1979).

Although the majority of oxidation reactions of xenobiotics are catalysed by the cytochrome P450 monooxygenase system (NEBERT and GONZALEZ 1985), certain oxidative reactions of xenobiotics are also catalysed by enzymes other than the cytochrome P450 monooxygenase system, for example the microsomal FAD-containing monooxygenase (ZEIGLER 1984, 1985; DAMANI 1988). This is responsible for the N-oxidation of tertiary amines such as dimethylaniline and trimethylamine. The enzyme requires NADPH and oxygen; the substrate specificity also includes secondary amines and sulphides, thioethers and thiocarbamates, and organophosphates (TIMBRELL 1993). Alcohols may be oxidised by alcohol dehydrogenase; xanthine oxidase catalyses the oxidation of nitrogen heterocyclics such as the purine hypoxanthine. Some amines such as tyramine are substrates for monoamine oxidases; diamines such as putrescine are metabolised by the soluble enzyme diamine oxidase. The peroxidases are also involved in the oxidation of xenobiotics, the most important being prostaglandin synthase, which is known to catalyse the oxidation of *p*-phenetidine, a metabolite of phenacetin, a process that may be involved in the nephrotoxicity of the drug (TIMBRELL 1993).

The role of enzyme systems in the metabolism of therapeutic agents is indeed a fascinating study; the knowledge base is ever growing and the genetic basis of enzyme source and action is gradually being revealed. Patient therapy can only benefit from these disclosures, and safety in therapy is one of the prime contributions that such knowledge may provide. Much of the empirical approach to therapy is being replaced by judgements based upon proven fact and one might well look forward to the future when synergistic drug-drug interactions are not hazardous but only beneficial.

References

Acocella G, Bonollo L, Garimoldi M, Mainardi M, Tenconi LT, Nicolis FB (1972) Kinetics of rifampicin and isoniazid administered alone and in combination to normal subjects and patients with liver disease. Gut 13:47–53

Adebayo GI (1988) Interaction between phenytoin and theophylline in healthy volunteers. Clin Exp Pharmacol Physiol 15:883–887

Ahmad SR (1992) USA antihistamine alert. Lancet 340:542

Alvares AP, Schilling G, Levin W, Kuntzman R (1967) Studies on the induction of CO-binding pigments in liver microsomes by phenobarbital and 3-methylcholanthrene. Biochem Biophys Res Commun 29:521–526

Angle M, Huff PS, Lea JW (1991) Interactions between oral contraceptives and therapeutic drugs. Outlook 9:1–6

Arinç E (1993) Extrahepatic microsomal forms: lung microsomal cytochrome P450 isozymes. In: Schenkman JB, Greim H (eds) Cytochrome P450. Springer, Berlin Heidelberg New York, pp 373–386

Armstrong G, Beg MF, Scahill S (1991) Warfarin potentiated by proguanil. Br Med J 3034:789

Axelrod J (1955) The enzymatic demethylation of ephedrine. J Pharmacol Exp Ther 114:430–438

Ayesh R, Idle JR, Ritchie JC, Crothers MJ, Hetzel MR (1984) Metabolic oxidation phenotypes as markers for susceptibility to lung cancer. Nature 311:169–170

Barbeau A, Cloutier T, Roy M, Plasse L, Paris S, Poirier J (1985) Ecogenetics of Parkinson's disease: 4-hydroxylation of debrisoquine. Lancet II:1213–1216

Bertilsson L, Dengler HJ, Eichelbaum M, Schultz HU (1980) Pharmacogenetic covariation of defective N-oxidation of sparteine and 4-hydroxylation of debrisoquine. Eur J Clin Pharmacol 17:153–155

Birnbaum LS (1993) Changes in cytochrome P450 in senescence. In: Schenkman JB, Greim H (eds) Cytochrome P450. Springer, Berlin Heidelberg New York, pp 477–492

Black SD (1993) Cytochrome P450 structure and function. In: Schenkam JB, Greim H (eds) Cytochrome P450. Springer, Berlin Heidelberg New York, pp 156–168

Booth CD (1993) A drug interaction between danazol and warfarin. Pharm J 250:439–440

Brodie BB, Axelrod J, Cooper JR, Gaudette LE, La Duc BN, Mitowa C, Udenfriend S (1955) Detoxification of drugs and other foreign compounds by liver microsomes. Science 121:603–604

Brooks BA, McBride OW, Dolphin CT, Farrall M, Scambler PJ, Gonzalez FJ, Idle JR (1988) The gene CYP3 encoding P45PCN1 (nifedipine oxidase) is tightly linked to the gene COLIA2 encoding collagen type I alpha on 7q21-p22.1. Am J Hum Genet 43:280–284

Brosen K, Gram LF (1989) Clinical significance of the sparteine/debrisoquine oxidation polymorphism. Eur J Clin Pharmacol 36:537–547

Cadle RM, Zenon GJ III, Rodriguez-Barradas MC, Hamill RJ (1994) Fluconazole induced symptomatic phenytoin toxicity. Ann Pharmacother 28:191–195

Cashman JR, Hanks D, Weiner RI (1987) Epoxy derivatives of arachidonic acid are potent stimulators of prolactin secretion. Neuroendocrinology 46:245–251

Castel JM, Cappiello E, Leopaldi D, Latini R (1990) Rifampicin lowers plasma concentrations of propafenone and its antiarrhythmic effect. Br J Clin Pharmacol 30:155

Committee on Safety of Medicines (1989) Serious interaction between tamoxifen and warfarin. Committee on Safety of Medicines, London (Current problems, no 26)

Conney AH (1967) Pharmacological implications of microsomal enzyme induction. Pharmacol Rev 19:317–366

Conney AH, Miller EC, Miller JA (1956) The metabolism of methylated aminoazo dyes. V. Evidence for induction of enzyme synthesis in the rat by 3-methylcholanthrene. Cancer Res 16:450–459

Conney AH, Miller EC, Miller JA (1957) Substrate-induced synthesis and other properties of benzpyrenehydroxylase in rat liver. J Biol Chem 228:753–766
Consumers' Association (1993) Fluoroquinolones reviewed. Drug Ther Bull 31(18):69–72
Crespi CL, Penman BW, Gelboin HV, Gonzalez FJ (1991) A tobacco smoke-derived nitrosamine 4-(methylnitrosamino)-1-(3-pyridyl)-1-butanone is activated by multiple human cytochrome P450s including the polymorphic human cytochrome P4502D6. Carcinogenesis 12:1197–1201
Daly AK, Idle JR (1993) Genetics: animal and human cytochrome P450 polymorphisms. In: Schenkman JB, Grein H (eds) Cytochrome P450. Springer, Berlin Heidelberg New York, pp 433–446
Damani LA (1988) The flavin-containing monooxygenase as an amine oxidase. In: Gorrod JW, Oelschlager H, Caldwell J (eds) Metabolism of xenobiotics. Taylor and Francis, London, pp 59–70
D'Arcy PF (1984a) Tobacco smoking and drugs: a clinically important interaction? Drug Intell Clin Pharm 18:302–307
D'Arcy PF (1984b) Vaccine-drug interactions. Drug Intell Clin Pharm 18:697–700
D'Arcy PF (1986) Drug interactions with oral contraceptives. Drug Intell Clin Pharm 20:353–362
D'Arcy PF, Merkus FWHM (1981) Alcohol and drug interactions. Pharm Int 2:273–280
Davis GC, Williams AC, Markey SP, Ebert MH, Caine ED, Reichert CM, Kopin IJ (1979) Chronic parkinsonism secondary to intravenous injection of meperidine analogues. Psychiatry Res 1:249–254
Dayer P, Balant L, Courvoisier F, Kupfer A, Kubli A, Georgia A, Fabre J (1982) The genetic control of bufuralol metabolism in man. Eur J Drug Metab Pharmacokinet 7:73–77
Dean WL, Gray RD (1982) Relationship between state of aggregation and catalytic activity for cytochrome P-450 LM2 and NADPH-cytochrome P-450 reductase. J Biol Chem 257:14679–14685
DeWaziers I, Cugnenc P, Yang CS, Leroux JP, Beaune P (1990) Cytochrome P-450 isozymes, epoxide hydrolase and glutathione transferase in rat and human hepatic and extrahepatic tissues. J Pharmacol Exp Ther 253:387–394
Dillard ML, Fink RM, Parkerson R (1992) Ciprofloxacin-phenytoin interaction. Ann Pharmacother 26:263
Eichelbaum M, Spannbrucker N, Steincke B, Dengler HJ (1979) Defective N-oxidation of sparteine in man: a new pharmacogenetic defect. Eur J Clin Pharmacol 17:153–155
Elofsson S-A, Gamberale F, Hindmarsh T, Iregren A, Isaksson A, Johnsson I, Knave B, Lydahl E, Mindus P, Persson HE, Philipson B, Steby M, Struwe G, Söderman E, Wennberg A, Widen L (1980) Exposure to organic solvents. A cross-sectional epidemiological investigation on occupationally-exposed car and industry spray painters with special reference to the nervous system. Scand J Work Environ Health 6:239–273
Estabrook RW (1979) Concluding remarks. In: Estabrook RW, Lindenlaub E (eds) The induction of drug metabolism. Schattauer, Stuttgart, pp 643–645 (Symposia Medica Hoechst, vol 14)
Evans DAP, Mahgoub A, Sloan TP, Idle JR, Smith RL (1980) A family and population study of the genetic polymorphism of debrisoquine oxidation in a British white population. J Med Genet 17:102–105
Gannon RH, Anderson ML (1992) Fluconazole-nortriptyline drug interaction. Ann Pharmacother 26:1456–1457
Gelehrter TD (1979) Enzyme induction in mammals – an overview. In: Estabrook RW, Lindenlaub E (eds) The induction of drug metabolism. Schattauer, Stuttgart, pp 7–24
Gonzalez FJ (1993) Cytochrome P450 in humans. In: Schenkman JB, Greim H (eds) Cytochrome P450. Springer, Berlin Heidelberg New York, pp 239–257

Gorrie M, Beaman M, Nicholls A, Blackwell P (1994) Allopurinol interaction with cyclosporin. Br Med J 308:113
Griffin JP, D'Arcy PF, Speirs CJ (1988) A manual of adverse drug interactions, 4th edn. Wright, London
Guengerich FP, Kim DH (1990) In vitro inhibition of dihydropyridine oxidation and aflatoxin B_1 activation in human liver microsomes by naringin and other flavonoids. Carcinogenesis 11:2275–2279
Guengerich FP, Marin MV, Beaune PH, Kremers P, Wolff T, Waxman DJ (1986) Characterization of rat and human liver microsomal cytochrome P-450 forms involved in nifedipine oxidation, a prototype for genetic polymorphism in oxidative drug metabolism. J Biol Chem 261:5051–5060
Gupta A, Sexton RC, Rudney H (1986) Modulation of regulatory oxysterol formation and low density lipoprotein suppression of 3-hydroxy-3-methyl glutaryl coenzyme A (HMG-Coa) reductase activity by ketoconazole. A role for cytochrome P450 in the regulation of HMG-Coa reductase in rat intestinal epithelia cells. J Biol Chem 261:8348–8356
Horowitz HW, Jorde UP, Wormser GP (1992) Drug interactions in use of dapsone for Pneumocystis carinii prophylaxis. Lancet 339:747
Hoyumpa AM Jr, Schenker S (1982) Major drug interactions: effect of liver disease, alcohol and nutrition. Ann Rev Med 33:113–149
Jassal SV (1991) Warfarin potentiation by glibenclamide. Br Med J 303:789
Kolmodin B, Azarnoff DL, Sjöqvist F (1969) Effect of environmental factors on drug metabolism: decreased plasma half-life of aminopyrine in workers exposed to chlorinated hydrocarbon insecticides. Clin Pharmacol Ther 10:638–642
Langhoff E, Madsen S (1983) Rapid metabolism of cyclosporin and prednisone in kidney transplant patients receiving tuberculostatic treatment. Lancet II:1031
Le Goascogne C, Robel P, Gouezow M, Snanes N, Baulieu E-E, Waterman M (1987) Neurosteroids: cytochrome P450 in the rat brain. Science 237:1212–1214
Lindell TJ, Ellinger R, Warren JT, Sundheimer D, O'Malley AF (1977) The effect of acute and chronic phenobarbital treatment on the activity of rat liver deoxyribonuclease acid-dependent ribonucleic acid polymerases. Mol Pharmacol 13:426–434
Lodwick R, McConkey B, Brown AM (1987) Life threatening interaction between tamoxifen and warfarin. Br Med J 295:1141
Lu AYH, Coon MJ (1968) Role of hemoprotein P450 in fatty acid ω-hydroxylation in a soluble enzyme from liver microsomes. J Biol Chem 243:1331–1332
Mahgoub A, Idle JR, Dring LG, Lancaster R, Smith RL (1977) Polymorphic hydroxylation of debrisoquine in man. Lancet II:584–586
Maizlish NA, Langolf GD, Whitehead LW, Fine LJ, Alberts JW, Goldberg J, Smith P (1985) Behavioral evaluation of workers exposed to mixtures of organic solvents. Br J Ind Med 43:257–262
Mallov JS (1976) MBK neuropathy among spray painters. JAMA 235:1455–1457
Maltony C, Ciliberti A, Caretti D (1982) Experimental contributions in identifying brain potential carcinogens in the petrochemical industry. Ann NY Acad Sci 381:216–249
Matsubara S, Yamamoto S, Sogawa K, Yokotani N, Fujii-Kuriyama Y, Haniu M, Shively JE, Gotoh O, Kusunose E, Kusunose M (1987) cDNA cloning and inducible expression during pregnancy of the mRNA for rabbit pulmonary prostaglandin ω-hydroxylase (cytochrome P-450P2). J Biol Chem 262:13366–13371
McFadden JP, Pontin JE, Powles AV, Fry L, Idle JR (1989) Cyclosporin decreased nifedipine metabolism. Br Med J 299:1224
McManus ME, Boobis AR, Pacifici GM, Frempang RY, Brodie MJ, Kahn GC, Whyte C, Davies DS (1980) Xenobiotic metabolism in the human lung. Life Sci 26:481–487
Meeks ML, Mahaffey KW, Katz MD (1992) Danazol increases the anticoagulant effects of warfarin. Ann Pharmacother 26:641–642

Mönig H, Baese C, Heidemann HT, Ohnhaus EE, Schulte HM (1990) Effect of oral contraceptive steroids on the pharmacokinetics of phenprocoumon. Br J Clin Pharmacol 30:115–116
Müller GC, Miller JA (1953) The metabolism of methylated aminoazo dyes. II. Oxidative demethylation by rat liver homogenates. J Biol Chem 202:579–587
Murphy RC, Falk JR, Lumni S, Yadagari P, Zirrolli JA, Balazy M, Masferrer JL, Abraham NG, Schwartzman ML (1988) 12(R)-Hydroxyeicosatrienoic acid: a vasodilator cytochrome P450-dependent arachidonate metabolite from bovine corneal epithelium. J Biol Chem 263:17197–17202
Nebert DW, Gonzalez EJ (1985) Cytochrome P-450 gene expression and regulation. Trends Pharmacol Sci 6:160–164
Noticeboard (1992) Warning about astemizole. Lancet 340:1155
Oates S, Shah RR, Idle JR, Smith RL (1982) Genetic polymorphism of phenformin 4-hydroxylation. Clin Pharmacol Ther 32:81–89
Pantuck EJ, Hsiao KC, Kuntzman R, Conney AH (1975) Intestinal metabolism of phenacetin in the rat: effect of charcoal-broiled beef and rat chow. Science 187:744–745
Parke DV (1975) Induction of the drug-metabolising enzymes. In: Parke DV (ed) Enzyme-induction. Plenum, London, p 207
Parke DV (1979) The responsiveness of cells to various drug inducers. In: Estabrook RW, Lindenlaub E (eds) The induction of drug metabolism. Schattauer, Stuttgart, pp 105–111
Philpot RM, Wolf CR (1981) The properties and distribution of the enzymes of pulmonary cytochrome P450 dependent monooxygenase systems. In: Hodgson E, Bend JR, Philpot RM (eds) Reviews in biochemical toxicology, vol 3. Elsevier, Amsterdam, pp 51–76
Philpot RM, Arinç E, Fouts JR (1975) Reconstitution of the rabbit pulmonary microsomal mixed-function oxidase system from solubilized components. Drug Metab Dispos 3:118–126
Posner HS, Mitoma C, Udenfriend S (1961) Enzymatic hydroxylation of aromatic compounds. II. Further studies of the properties of the microsomal hydroxylating system. Arch Biochem Biophys 94:269–279
Remmer H, Merker HJ (1963) Drug induced changes in the liver endoplasmic reticulum: association with drug metabolizing enzymes. Science 14:1657–1658
Rennert H, Fischer RT, Alkvarez JG, Trzaskos JM, Strauss JM (1990) Generation of regulatory oxysterols: 26-hydroxylation of cholesterol by ovarian mitochondria. Endocrinology 127:738–746
Sacerdoti D, Abraham NG, McGiff JC, Schwartzman ML (1988) Renal cytochrome P450-dependent metabolism of arachidonic acid in spontaneously hypertensive rats. Biochem Pharmacol 37:521–527
Schenkman JB (1993) Historical background and description of the cytochrome P450 monooxygenase system. In: Schenkman JB, Greim H (eds) Cytochrome P450. Springer, Berlin Heidelberg New York, pp 3–13
Schmucker DL, Woodhouse KW, Wang RK, Wynne H, James OF, McManus M, Kremers P (1990) Effects of age and gender on in vitro properties of human liver microsomal monooxygenases. Clin Pharmacol Ther 48:365–374
Seaton TL, Ceklum CL, Back DJ (1991) Possible potentiation of warfarin by fluconazole. Ann Pharmacother 24:1177–1178
Serpa MD, Cossolias J, McCreevy MJ (1992) Moricizine-warfarin: a possible drug interaction. Ann Pharmacother 26:127
Shaeffer MS, Collier D, Sorrell MF (1994) Interaction between FK 506 and erythromycin. Ann Pharmacother 28:280–281
Shimada T, Guengerich FP (1989) Evidence for cytochrome P-450_{NF}, the nifedipine oxidase, being the principal enzyme involved in the bioactivation of aflatoxins in human liver. Proc Natl Acad Sci USA 86:462–465

Sladek NE, Mannering GJ (1966) Evidence for a new P-450 hemoprotein in hepatic microsomes from methylcholanthrene treated rats. Biochem Biophys Res Commun 24:668–674

Snyder GD, Capdevila J, Chacos N, Manna S, Flack JR (1983) Action of luteinizing hormone-releasing hormone: involvement of novel arachidonic acid metabolites. Proc Natl Acad Sci USA 80:3504–3507

Soto J, Sacrisán JA, Arellano F, Mazas J (1991) Possible theophylline-amiodarone interaction. Ann Pharmacother 24:1115

Srivastava SP, Seth PK (1983) 7-Ethoxycoumarin O-deethylase activity in rat brain microsomes. Biochem Pharmacol 32:3657–3660

Staum JM (1990) Enzyme induction: rifampicin-disopyramide interaction. Ann Pharmacother 24:701–703

Strobel HW, Stralka DJ, Hammond DK, White T (1993) Extrahepatic microsomal forms: gastrointestinal cytochromes P450, assessment and evaluation. In: Schenkman JB, Greim H (eds) Cytochrome P450. Springer, Berlin Heidelberg New York, pp 363–371

Tenni P, Lalich DL, Byrne MJ (1989) Life threatening interaction between tamoxifen and warfarin. Br Med J 298:93

Timbrell JA (1993) Biotransformation of xenobiotics. In: Ballantyne B, Marrs T, Turner P (eds) General and applied toxicology, vol 1. Macmillan, Basingstoke, pp 89–119

Tisdel KA, Israel D SA, Kolb KW (1994) Warfarin-felbamate interaction: first report. Ann Pharmacother 28:805

Tjia JF, Back DJ, Breckenridge AM (1989) Calcium channel antagonists and cyclosporin metabolism: in vitro studies with human liver microsomes. Br J Clin Pharmacol 28:362–365

Volk B, Amelized Z, Anagnostopoulos J, Knoth R, Oesch F (1988) First evidence of cytochrome P450 induction in the mouse brain by phenytoin. Neurosci Lett 84:219–224

Warner A, Gustafsson J-A (1993) Extrahepatic microsomal forms: brain cytochrome P450. In: Schenkman JB, Greim H (eds) Cytochrome P450. Springer, Berlin Heidelberg New York, pp 387–397

Warner M, Tollet P, Carlström K, Gustafsson J-Å (1989) Endocrine regulation of cytochrome P450 in the rat brain and pituitary gland. J Endocrinol 122:341–349

Watkins PB, Wrighton PB, Schuetz EG, Molowa DT, Guzelian PS (1987) Identification of glucocorticoid-inducible cytochromes P-450 in the intestinal mucosa of rats and man. J Clin Invest 80:1029–1036

Wattenberg LW, Leon JL (1962) Histochemical demonstration of reduced pyridine nucleotide dependent polycyclic hydrocarbon metabolizing systems. J Histochem Cytochem 10:412–420

Wattenberg LW, Leong JL, Strand PJ (1962) Benzpyrene hydroxylase activity in the gastrointestinal tract. Cancer Res 22:1120–1125

Williams DE, Hale SE, Okita RT, Masters BSS (1984) A progtaglandin ω-hydroxylase cytochrome P-450 (P-450PGω) purified from lungs of pregnant rabbits. J Biol Chem 259:14600–14608

Williams RT (1959) Detoxification mechanisms. Cunningham, London

Wolff T, Strobl G, Greim H (1993) Structural models for substrates and inhibitors of cytochrome P450 enzymes. In: Schenkman JB, Greim H (eds) Cytochrome P450. Springer, Berlin Heidelberg New York, pp 195–207

Woodcock BG, Kirsten R, Nelson K, Rietbrock S, Hopf R, Kaltenbach M (1991) A reduction in verapamil concentrations with phenytoin. N Engl J Med 325: 1179

Zeigler DM (1984) Metabolic oxygenation of organic nitrogen and sulphur compounds. In: Mitchell JR, Horning MG (eds) Drug metabolism and drug toxicity. Raven, New York, pp 33–53

Zeigler DM (1985) Molecular basis for N-oxygenation of sec- and tert-amines. In: Gorrod JW, Damani LA (eds) Biological oxidation of nitrogen in organic molecules. Ellis Horwood, Chichester, pp 45–52
Zornberg GL, Bodkin JA, Cohen BA (1991) Severe adverse interactions between pethidine and selegiline. Lancet 337:246

CHAPTER 6

Drug Interactions Involving Renal Excretory Mechanisms

A. SOMOGYI

A. Introduction

The kidney represents the final elimination organ for virtually all foreign substances irrespective of whether they are cleared unchanged by the kidney or whether as metabolites formed in the body (predominantly the liver). It had long been regarded that filtration at the glomerulus was the sole excretory mechanism in the kidney for foreign substances. However, in 1874, Heidenhain demonstrated the phenomenon of secretion by canine tubular cells using the dye indigo carmine, and Marshall and colleagues in the 1920s conclusively showed secretion of organic anions by the proximal tubules. Thus the concept of filtration and secretion by the kidney came into existence. In 1929 Møller and co-workers coined the term clearance, in which they mathematically described the excretion of substances by the kidney. It is somewhat surprising that the concepts developed by these early pioneers of renal physiology are still highly applicable in pharmacokinetics today. The active tubular secretion of organic cations was shown for the first time by Rennick in the dog using tetraethylammonium and by Sperber using a variety of organic cations in the chicken. Further details of the history of renal physiology/pharmacology can be found in several reviews (PETERS 1960; WEINER and MUDGE 1964; RENNICK 1972).

B. Mechanisms of Renal Excretory Clearance

I. Anatomy and Physiology of the Kidney

The major function of the kidney is to remove waste products from the blood and to regulate fluid and electrolyte content in the body. The functional unit of the kidney is the nephron, which filters, secretes and reabsorbs molecules. Blood supply to the kidneys is via the left and right renal arteries at the rate of approximately 1200 ml/min or 20%–25% of cardiac output.

II. Glomerular Filtration

Filtration involves the forcing of fluid containing dissolved substances through the endothelial-capsular membrane. This membrane, which acts as a filter, allows some materials to pass through whilst restricting the passage of others. The resulting filtrate is free of protein and substances with a molecular weight of greater than 40000 and about 20 Å in diameter. The rate of glomerular filtration is about 180 l/day or 120 ml/min.

The pharmacokinetic consequence of glomerular filtration is that the clearance rate of drugs at the glomerulus [CL(GFR)] can be approximated to the unbound fraction (fu) in plasma of the drug multiplied by the glomerular filtration rate (GFR).

$$CL(GFR) = fu \times GFR$$

Hence for drugs which are highly bound ($fu < 0.1$) in plasma, clearance at the glomerulus is substantially less than the glomerular filtration rate, and the maximum clearance of any drug at the glomerulus will be equal to the glomerular filtration rate.

The mechanisms by which drugs can interact at the glomerulus are: (a) displacement from protein-binding sites. The pharmacokinetic consequence will be that fu will increase and clearance of the drug by the glomerulus will also increase in direct proportion to the increase in fu. There are very few examples in the literature of drug-drug interactions occurring exclusively at the glomerulus. (b) Where the glomerulus is physically damaged through the nephrotoxic action of a xenobiotic. In this case, the maximum clearance of a drug will be equal to the GFR, irrespective of the value of the drug's fu. However, this type of interaction is poorly documented and would rarely occur on its own.

III. Tubular Secretion

The blood flow to the kidney then passes to the tubular network, where active transport systems (requiring energy) are capable of removing drugs and endogenous substances from blood and secreting them into the lumen of the kidney. This occurs at the site of the proximal tubule, where there are a number of transporters for organic anions and organic cations. There are specific transporters for organic cations and organic anions at both the luminal (brush border) and contraluminal membranes of the epithelial cells lining the proximal tubule.

1. Physiological Considerations

The blood flow to the proximal tubule via the renal arteries is approximately 85% of total renal blood flow. Thus the maximum clearance of any drug by tubular secretion at this site is equivalent to the total renal blood flow of 1.2 l/min.

2. Cellular Mechanisms

Our increasing understanding of the cellular mechanisms of transport of drugs across the epithelial membranes of the proximal tubule has been accomplished by the use of a number of techniques such as isolated perfused and non-perfused tubules, isolated perfused kidneys, plasma membrane vesicles and primary or continuous kidney cell lines. The results of transport studies using these techniques have allowed prediction of drug interactions in humans (SOMOGYI et al. 1989).

a) Transport Processes for Organic Anions

This has been reviewed by several groups (MØLLER and SHEIKH 1983; ULLRICH 1994). Although not fully evaluated in humans, the renal transporter of organic anions (*p*-aminohippuric acid being the most commonly used probe substrate) is located at the contraluminal cell membrane (Fig. 1). The cellular uptake of organic anions has been termed a tertiary active transport process, since the primary transporter is the ATP sodium/potassium transport step, which activates a sodium-dicarboxylate transporter, creating an α-ketoglutarate gradient which is exchanged for the organic anion. At the luminal membrane, the transport mechanism(s) has not been identified. The specificity of the contraluminal transporter has been summarized by ULLRICH (1994); the substrate must have a hydrophobic domain of greater than 4 Å; the strength of the ionic charge on the substrate is a determinant of the interaction with the transporter; the transporter also accepts hydrophobic substrates and, finally, electron-attracting side groups (such as Cl, Br) augment the interaction of the substrate with the transporter.

b) Transport Processes for Organic Cations

This has been reviewed on a number of occasions (PETERS 1960; RENNICK 1981). Secretion of organic cations across the mammalian kidney occurs along

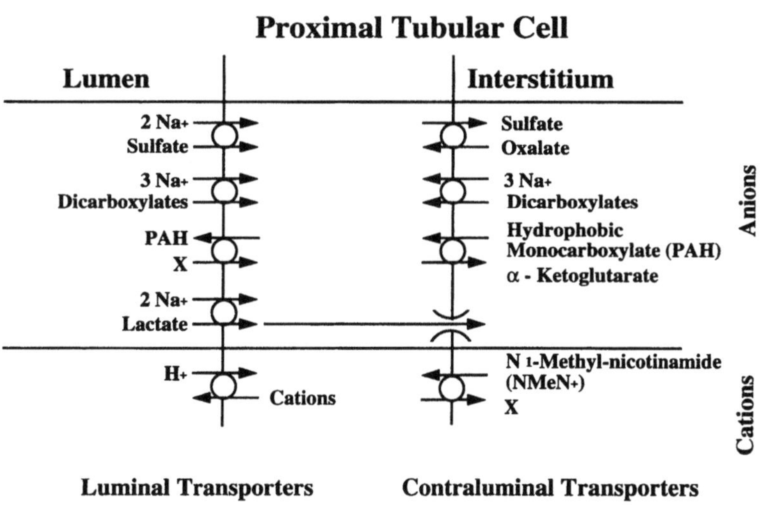

Fig. 1. Location of transporters for organic anions and cations in the renal proximal tubule (from ULLRICH 1994). Note: PAH is the organic anion *p*-aminohippuric acid

the entire length of the proximal tubule. At the contraluminal membrane there is evidence for one common transport system. At the luminal membrane, the transport of organic cations occurs by an electroneutral H^+/organic cation system. The driving force at this luminal interface is the sodium/hydrogen transporter, which transports sodium ions into the cell and hydrogen ions out (Fig. 1). More than one transport system exists for organic cations at the luminal membrane, since choline has a high affinity for a transporter which is not H^+ driven. The key substrate requirements are similar to those of the contraluminal organic anion transporter (hydrophobicity, ionic charge strength and hydrogen bond formation). Substrates invariably contain a nitrogen atom (ULLRICH 1994). The cloning and sequencing of such organic cation transporters is only in its infancy but will provide further valuable information on substrate specificity.

3. Pharmacokinetic Evidence for Tubular Secretion

To determine whether a drug in vivo undergoes net active tubular secretion, knowledge of the value of its renal excretory clearance is required. This is usually obtained from pharmacokinetic studies in which plasma concentrations and the amount of drug recovered in urine as the unchanged form of the drug are quantified. Renal clearance (CL_R) is calculated as the amount excreted in urine divided by the area under the plasma concentration-time curve. When this value is greater than clearance by the glomerulus [CL(GFR); see equation above], active secretion is inferred. That is, if $CL_R >$ CL(GFR) then the drug is secreted by the proximal tubules. For example, cimetidine has a total body clearance of 600 ml/min and 70% of the dose is recovered unchanged in urine. Hence its renal clearance is $0.7 \times 600 = 420$ ml/min. This is greater than clearance at the glomerulus [$0.8(fu) \times 120 = 96$ ml/min], allowing one to conclude that cimetidine also undergoes net renal tubular secretion.

This knowledge allows one to examine the sites in the kidney for the renal elimination of drugs. In terms of renal tubular secretion, when renal clearance indicates tubular secretion then the chemical nature of the drug needs to be ascertained. If the drug is an organic anion then one can assume that the drug is secreted by the organic anion transporter and if an organic cation then it is secreted by the organic cation transporter. A key feature of a drug undergoing tubular secretion is that it will compete with another drug for secretion. In all studies to date there is little evidence that competition for transport is non-competitive but rather competitive. Thus the drug with the higher affinity for the transporter will be a more potent inhibitor of the transport of a similar chemical substance for the same transporter. Hence organic anions will compete with one another for transport (e.g., probenecid and penicillin) and organic cations will compete with one another (e.g., cimetidine and procainamide). This is the most common type of drug-drug interaction involving renal excretory mechanisms and will form the basis of this review.

IV. Tubular Reabsorption

As drug passes down the tubular lumen into the distal tubule and collecting duct, passive diffusion (driven by the concentration gradient) of drug across the cellular membrane back into the blood can occur. This process of passive reabsorption is applicable to the unionized lipid-soluble form of the drug. Thus the pK_a of the drug and the pH of the tubular fluid (urine) become important, as only the unionized form is sufficiently lipid soluble to diffuse through the epithelial membranes of the tubule cells. Hence any drug which alters urine pH or urine flow rate (which alters the concentration gradient) has the potential for altering the tubular reabsorption of other drugs. The significance of any interaction at this site is often of minor importance since the basic requirement is the degree of lipid solubility of the drug and in general lipid-soluble drugs are extensively metabolized by the liver rather than being extensively cleared by the kidney. The few exceptions include the ephedrine series of drugs, salicylate and chlorpropamide.

C. Drug Interactions Involving Tubular Secretion

Of the mechanisms for renal drug excretion, drug interactions predominantly involve tubular secretion. Many in vivo studies in humans have been conducted characterizing interactions involving drugs which are either organic anions or organic cations.

I. Organic Anions

The most commonly investigated interacting drug has been probenecid with many published studies showing that it reduces the renal clearance of drugs which are organic anions.

1. Probenecid

It was the discovery of penicillin by Lord Florey which ultimately led to the discovery of probenecid. Penicillin although effective had a short duration of action due to its very high renal clearance and short half-life. A number of strategies were investigated to overcome this therapeutic problem. One strategy was to try and block the renal excretion of penicillin. The development of probenecid (Beyer et al. 1951) heralded the clinical practice of elevating blood concentrations of organic anion drugs by blocking their renal tubular secretion. Since then, probenecid has been used as a prototypic in vitro and in vivo probe to assess whether an organic anion drug is secreted by the proximal tubules. This has been documented for a number of drugs but, in particular, penicillin and cephalosporin antibiotics. In this review, the symbol ± always refers to the standard deviation.

a) Allopurinol

ELION et al. (1968) showed in three patients with gout that the administration of 1 g probenecid increased the renal clearance of oxypurinol (the active metabolite of allopurinol) from about 12 ml/min to 36 ml/min, without any change in inulin clearance (a measure of GFR). The mechanism for this increase in renal clearance involves inhibition by probenecid of the active luminal re-uptake of oxypurinol, via the uric acid transport system.

b) Enprofylline and Dyphylline

In six normal healthy subjects, 1 g probenecid reduced the renal clearance of enprofylline, administered as a 1-mg/kg intravenous infusion, from 17 ± 5 l/h to 8 ± 3 l/h ($P < 0.01$) (BORGÅ et al. 1986), with a reduction in total body clearance and increase in half-life. Similarly, in 12 healthy subjects, the total body clearance of dyphylline (a dihydroxypropyl derivative of theophylline) was reduced ($P < 0.001$) by almost 50% from 178 ± 8 to 101 ± 7 ml/h per kilogram by 1 g probenecid (MAY and JARBOE 1981). Enprofylline and dyphylline are organic acid methylated xanthines and have renal clearance values greater than the glomerular filtration rate, indicating net tubular secretion.

c) Frusemide

Frusemide is an organic acid, highly protein bound (unbound fraction is less than 0.1), loop diuretic which produces its pharmacological effect(s) at the luminal rather than the peritubular site of the renal tubule. Access to the luminal site is mainly by active secretion by the organic anion transport system. It was hypothesized that administration of probenecid could blunt the diuretic and natriuretic effect of frusemide by blocking access to its site of action. Four healthy normal subjects received a single intravenous injection of 1 mg/kg frusemide with and without pretreatment of 0.5 g probenecid 8 and 2 h before the frusemide dose. Probenecid reduced the renal clearance of frusemide from 134 ± 23 to 63 ± 6 ml/min ($P < 0.05$). However, there was no significant difference in the 6-h urine volume or sodium excretion (HONARI et al. 1977). When frusemide was administered as an intravenous infusion in four subjects, the reduction in frusemide renal clearance was accompanied by a rise in plasma frusemide concentrations, but urine flow rate (29 ± 4 ml/min frusemide alone and 15 ± 4 ml/min with probenecid) and excreted fraction of filtered sodium (19% ± 7% alone and 14% ± 9% with probenecid) declined significantly ($P < 0.05$) (HONARI et al. 1977). In six normal, healthy subjects, the effect of probenecid (0.5 g every 6 h for 3 days) on the disposition and effect of frusemide 40 mg intravenously was evaluated (HOMEIDA et al. 1977). Probenecid significantly ($P < 0.01$) reduced the renal clearance of frusemide from 90 ± 24 to 20 ± 12 ml/min. In the first 30 min after frusemide administration, there was a significant ($P < 0.01$) reduction in sodium excretion rate from 3.1 ± 0.6 to 2.2 ± 0.7 mmol/min. However, there was no difference in the ratio between the urinary frusemide concentration and urinary sodium concentration. In nine subjects who received an intravenous infusion of frusemide and inulin for over 2h, the administration of probenecid 500mg twice daily reduced the frusemide renal clearance to inulin clearance ratio from 0.92 ± 0.21 to 0.44 ± 0.18 ($P < 0.05$) (ANDREASEN et al. 1980). It was calculated that, when frusemide was administered alone, 2% of the urinary concentration of frusemide was derived from filtration at the glomerulus with the remaining 98% being secreted by the organic anion transport system. When probenecid was co-administered, the values were 8% and 92%, respectively. Probenecid only slightly reduced urinary sodium excretion from 24.7% of the filtered load to 21.0% ($P < 0.05$). These studies have supported the hypothesis that frusemide's pharmacodynamic effects are more associated with the rate of urinary frusemide excretion than with plasma concentrations, and have highlighted the usefulness of a competitive inhibitor of secretion to evaluate the mechanisms of the natriuretic effect of this class of drug.

d) Indomethacin

BABER et al. (1978), in a study in 17 patients with rheumatoid arthritis, administered 25mg indomethacin three times a day and demonstrated that probenecid 0.5g twice a day significantly increased plasma concentrations of indomethacin by an average of 63%, but there was no difference ($P > 0.05$) in the renal clearance of indomethacin (3.0 ± 2.5ml/h per kilogram indomethacin alone versus 2.7 ± 2.5ml/h per kilogram indomethacin plus probenecid). However, probenecid significantly ($P < 0.01$) reduced the apparent renal clearance of the glucuronide conjugate of indomethacin from 271 ± 158ml/min to 126 ± 188ml/min. This study highlighted a new type of renal drug interaction, in which the renal clearance of a conjugated metabolite rather than the parent drug was inhibited by another drug via competition for active tubular secretion. However, it is not known whether the glucuronidation of indomethacin occurs in the kidney (proximal tubule cells) and whether probenecid blocks access of indomethacin to the site of renal glucuronidation or inhibits renal glucuronidation.

e) Clofibrate

In four healthy male subjects who received clofibrate (500mg orally every 12h) with and without 6 hourly probenecid, steady-state concentrations of unbound clofibric acid were increased almost fourfold (VEENENDAAL et al. 1981). The renal clearance of unbound clofibric acid was significantly ($P <$ 0.05) reduced by probenecid from 14 ± 5 to 3 ± 1ml/min, which is consistent with inhibition of tubular secretion.

f) Sulphinpyrazone

In four patients with gout, the administration of probenecid reduced the renal clearance of sulphinpyrazone from 21 ± 11 to 8 ± 5ml/min, and when the degree of protein binding (fu <0.02) was considered, the renal tubular secretion of this organic anion was inhibited by a mean of 75% (as inulin clearance remained unchanged) (PEREL et al. 1969).

g) *Zidovudine*

This antiviral agent undergoes extensive glucuronidation to form the ether glucuronide GAZT (3′-azido-3′-deoxy-5′-β-D-glucopyranuronosylthymidine) (DE MIRANDA et al. 1989). In seven patients with AIDS or AIDS-related complex, the administration of probenecid 500 mg every 4 h substantially altered the disposition of oral zidovudine (2 mg/kg 8 hourly). Probenecid significantly ($P < 0.01$) increased the area underneath the plasma concentration-time profile of zidovudine from 0.97 ± 0.04 to 2.00 ± 0.77 mg/l × h. In addition, the area underneath the plasma concentration-time curve of the glucuronide metabolite was also significantly increased by probenecid from 6.9 ± 2.0 to 14.0 ± 5.7 mg/l × h ($P < 0.01$). In four of the subjects from whom urine could be collected, the renal clearance of zidovudine was not altered by probenecid (188 ± 65 vs. 141 ± 31 ml/min per 1.73 m^2). However, the apparent renal clearance of the glucuronide metabolite was substantially and significantly reduced ($P < 0.01$) by probenecid from 293 ± 46 to 114 ± 50 ml/min per 1.73 m^2 (Fig. 2). In a pilot study in two healthy subjects (HEDAYA et al. 1990), probenecid 500 mg every 6 h for 3 days reduced the renal clearance of zidovudine from 4.38 and 5.13 ml/min per kilogram to 2.68 and 3.27 ml/min per kilogram, respectively. For the glucuronide conjugate, probenecid substantially reduced the renal clearance from 11.6 and 11.0 to 2.82 and 2.44 ml/min per kilogram, respectively. Both groups attributed their findings to inhibition

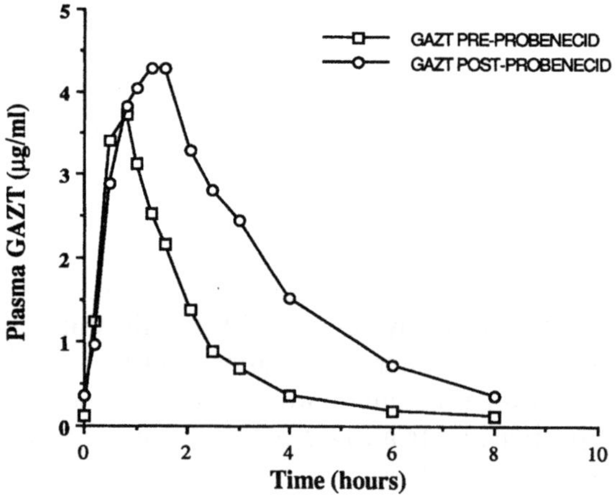

Fig. 2. Plasma concentration-time profile of the glucuronide conjugate of zidovudine (GAZT) following administration of zidovudine alone (pre-probenecid) and after concomitant administration of probenecid (post-probenecid) in seven patients. Probenecid significantly reduced the apparent renal clearance of GAZT. (From DE MIRANDA et al. 1989)

of the renal tubular secretion of this acidic glucuronide metabolite by probenecid, in a similar manner to indomethacin (see Sect. C.I.1.d). However, like the interaction with indomethacin, it is not known whether zidovudine glucuronidation occurs in the kidney in humans and whether probenecid prevents access of zidovudine to the proximal tubular cells (via competition for transport into the cells) or whether it inhibits renal glucuronidation.

h) Acyclovir

In three patients, LASKIN et al. (1982) showed that probenecid 1 g orally 1 h before acyclovir (administered as an intravenous infusion of 5 mg/kg) significantly ($P < 0.05$) reduced the renal clearance of acyclovir from 248 ± 80 to 168 ± 48 ml/min per 1.73 m^2, with no alteration in creatinine clearance (a measure of GFR). Acyclovir is both a weak acid (pK_a 2.3) and weak base (pK_a 9.3) and it is likely, given the inhibition shown with probenecid, that the tubular secretion of acyclovir is predominantly via the organic anion transporter.

i) Penicillins

Since the discovery that probenecid could increase serum penicillin concentrations in humans (BURNELL and KIRBY 1951), many studies (too numerous to cite) have shown that the renal clearance and hence tubular secretion of those penicillins which are secreted can be reduced by probenecid, resulting in elevated plasma concentrations. In a comprehensive study into the renal tubular secretion of benzylpenicillin and its inhibition by probenecid (OVERBOSCH et al. 1988), four healthy volunteers underwent three different studies. In each study, benzylpenicillin was infused to achieve three different steady-state plasma concentrations. In the second and third studies, probenecid was administered by continuous infusion at low (31.3 mg/h in one subject, 21.4 mg/h in the other three subjects) and high (between 93.8 and 200.3 mg/h) doses resulting in plasma probenecid concentrations of 6.7 and 35.1 mg/l, respectively. Probenecid, in a concentration-dependent manner, increased the unbound plasma concentrations of benzylpenicillin with an estimated probenecid EC_{50} of 52.3 mg/l. When benzylpenicillin was administered alone at the three different infusion rates, the data were able to be fitted to a non-linear equation in which the maximum tubular excretion rate for benzylpenicillin (range 2493–4131 mg/h) and EC_{50} (plasma benzylpenicillin concentration at which half the maximum excretion rate was achieved) of 22.6 to 60.5 mg/l could be derived. When probenecid was administered in studies 2 and 3, the ED_{50} (probenecid dose at which 50% of the tubular transport is inhibited) was estimated to be between 13.2 and 108.5 mg/h. These investigators extrapolated their data to the clinical setting and suggested that the commonly used daily dose of 2 g/day of probenecid is likely to be close to the maximum effective dose for inhibition of the tubular secretion of benzylpenicillin.

Probenecid has also been shown to reduce the renal clearance of nafcillin from 2.0 ± 0.5 ml/min per kilogram to 0.56 ± 0.17 ml/min per kilogram ($P <$

0.05) (Waller et al. 1982) in five healthy subjects, and increase ticarcillin plasma concentrations (Corvaia et al. 1992); penicillins which have a renal clearance greater than clearance due to glomerular filtration alone.

Of all the drugs interacting with probenecid, inhibition of the renal tubular secretion of penicillins is considered to be the prototypic interaction for drugs involving the kidney.

j) Quinolone Antibiotics

Several of these antibiotics are predominantly cleared by the kidney with renal clearance values in excess of the glomerular filtration rate, indicating net tubular secretion. In eight healthy male subjects, probenecid significantly ($P < 0.01$) reduced the renal clearance of cinoxacin to creatinine clearance ratio from 1.32 ± 0.48 to 0.32 ± 0.17 (Rodriguez et al. 1979). Since cinoxacin is only 40% unbound in plasma, these data indicate that there is little tubular reabsorption. Similar results were reported by Israel et al. (1978), in which probenecid reduced the renal clearance of cinoxacin from 153 ± 60 ml/min per 1.73 m^2 to 66 ± 9 ml/min per 1.73 m^2 ($P < 0.001$) in four healthy subjects. For ciprofloxacin, with a renal clearance of over 200 ml/min, probenecid has been shown to reduce its renal clearance by approximately 50% in healthy subjects (Wingender et al. 1985); however, for fleroxacin, which has a renal clearance of unbound drug of about 140 ml/min, probenecid had no effect on its renal clearance in five healthy subjects (Weidekamm et al. 1987). These results again support the hypothesis that for this class of antibiotic, if renal clearance is greater than clearance by filtration, then tubular secretion is by, at least in part, the organic anion transport system.

k) Cephalosporin Antibiotics

Interactions between cephalosporin antibiotics and probenecid have been comprehensively reviewed by Brown (1993). A large number of studies have been published investigating the effect of probenecid on the disposition of the cephalosporins. Some studies have shown a significant interaction to occur in which plasma concentrations are increased, whilst others have not. The two criteria for an interaction to occur in the kidney are, firstly, the kidney must eliminate a large proportion of the drug. For the cephalosporins this can range from 40% for ceftriaxone to 95% for cefuroxime (Brown 1993). The second criterion is that the renal clearance of the drug must exceed the clearance attributed to filtration at the glomerulus. For some cephalosporin antibiotics such as moxalactam, renal clearance does not exceed the glomerular filtration clearance and hence probenecid has no effect on their renal clearance (Table 1). Two studies will be cited to illustrate the effects of probenecid on the renal disposition of cephalosporins. In eight healthy male subjects, the effect of probenecid 500 mg twice daily for 2 days prior to and on the morning of the study was investigated to determine its effect on the disposition of cephradine given as a 500-mg intravenous dose (Roberts et al. 1981). For cephradine,

Table 1. Association between evidence of net renal tubular secretion of cephalosporin antibiotics and inhibition of total body clearance by probenecid (Data obtained from BROWN 1993; GILMAN et al. 1990). Note: the list of cephalosporins is not exhaustive

Drug name	Secreted[a]	Secretion inhibited by probenecid[b]
Cefaclor	Yes	Yes
Cefadroxil	Yes	Yes
Cephalexin	Yes	Yes
Cephalothin	Yes	Yes
Cefamandole	Yes	Yes
Cefazedone	Yes	Yes
Cefazolin	Yes	Yes
Cefmetazole	Yes	Yes
Cefmenoxime	Yes	Yes
Cefonicid	Yes	Yes
Ceforanide	No	No
Cefotaxime	No	No
Cefoxitin	Yes	Yes
Cefradine	Yes	Yes
Ceftazidine	No	No
Ceftizoxime	No	No
Ceftriaxone	No	No
Cefuroxime	No	No
Moxalactam	No	No

[a] Results of human pharmacokinetic studies indicate the renal clearance of unbound drug is greater than GFR.
[b] Total body clearance reduced by probenecid.

90% of the dose is excreted unchanged in urine, and in normal healthy subjects renal clearance is approximately 400 ml/min and the degree of plasma binding is low. Hence this drug's major clearance is by tubular secretion. Probenecid significantly ($P < 0.01$) increased the area under the serum cephradine concentration-time curve from 24 ± 9 to 57 ± 11 mg/l×h. This was entirely due to a significant ($P < 0.01$) reduction in renal clearance from a mean of 22 l/h to 12 l/h. Interpretation of the data is consistent with inhibition of the renal tubular secretion of cephradine by probenecid via the organic anion transport system. As a second example, the disposition of cefonicid (500 mg intramuscularly) was investigated when probenecid (1 g orally) was administered to eight healthy subjects (PITKIN et al. 1981). Approximately 90% of the dose of cefonicid is excreted unchanged in urine, and its renal clearance is only approximately 30 ml/min. However, this antibiotic is highly bound in plasma, with an unbound fraction of 0.05. Hence although the renal clearance is low it is mainly via tubular secretion. Probenecid significantly ($P < 0.001$) increased the area under the cefonicid plasma concentration-time curve from 346 ± 34 mg/l × h to 724 ± 99 mg/l × h. The terminal half-life of cefonicid was prolonged from 4.9 ± 0.8 h to 7.5 ± 1.4 h and the renal clearance of cefonicid was substantially reduced in the 4- to 6-h time period from 30.2 ± 7.1 ml/min to 7.8 ± 2.0 ml/min ($P < 0.001$) (Pitkin et al. 1981). The investigators calculated that the secretion rate of cefonicid was reduced by probenecid from a mean of 744 to 246 μg/min. These data are again entirely consistent with inhibition of renal tubular secretion. Table 1 is a summary of data on the effect of probenecid on the renal disposition of the cephalosporin antibiotics. In all cases, cephalosporin antibiotics whose disposition (hence renal clearance) is altered by probenecid have renal clearance values indicating net tubular secretion.

l) Famotidine

This organic cation histamine H_2 antagonist has a renal clearance in humans of almost three times the glomerular filtration rate, indicating net renal tubular secretion presumably by the organic cation transport system. In eight young healthy male volunteers (Inotsume et al. 1990) the administration of probenecid 1000 mg 2 h before, 250 mg 1 h before and 250 mg with a 20-mg oral dose of famotidine resulted in the area underneath the plasma famotidine concentration-time curve being significantly increased ($P < 0.01$) from 424 ± 54 ng/ml × h to 768 ± 110 ng/ml × h. The renal clearance of famotidine was reduced from 297 ± 54 ml/min to 107 ± 14 ml/min ($P < 0.01$). Using the famotidine unbound fraction in plasma and creatinine clearance values, these authors calculated that the clearance of famotidine due to tubular secretion was significantly ($P < 0.01$) reduced by 89% (from 196 ± 61 ml/min to 22 ± 12 ml/min). The mechanism of this interaction is unclear as famotidine is an organic cation, whereas probenecid's inhibitory effect is considered to be specific for organic anions. It may well be that probenecid is a polysubstrate which may interact with the organic anion transport system and suitable substrates of the organic cation transport system. This has been investigated in vitro with similar substances (Ullrich et al. 1994).

2. Methotrexate

The antineoplastic and antirheumatoid drug methotrexate is approximately 50% excreted unchanged in urine (depending on the dose). Its renal clearance is approximately 75 ml/min and the unbound fraction in plasma is 0.5. It has been shown that methotrexate, being an organic anion undergoes renal tubular secretion by the organic anion transport system (Liegler et al. 1969) and there is also some evidence for active tubular reabsorption (Hendel and Nyfors 1984). Although its renal clearance increases with urine pH, suggestive of passive reabsorption, this was attributed to greater solubility of the drug in urine (Sand and Jacobsen 1981). Several drug interactions have been reported involving the renal clearance of methotrexate. Since the drug has a narrow therapeutic index, these interactions have sometimes been manifested in overt methotrexate toxicity. Drugs which interact renally with methotrexate are reviewed below.

a) Probenecid

In four patients administered intravenous methotrexate 200 mg/m², the co-administration of probenecid (500–1000 mg) resulted in serum concentrations at 24 h which were four times higher (mean = 0.4 mg/l) than with four other patients not receiving probenecid (mean = 0.09 mg/l) (Aherne et al. 1978). The renal clearance of methotrexate was 104 ± 34 ml/min, which was significantly ($P < 0.05$) reduced to 46 ± 15 ml/min in those patients also taking probenecid. In a more comprehensive study, three patients receiving high-dose methotrexate (3.0 g/m²) also received probenecid (dose not stated, but serum concentrations were between 21 and 116 mg/l) (Howell et al. 1979). There was a significant increase in serum methotrexate concentrations, but, as only three subjects were studied, the overall disposition of methotrexate was not statistically significantly altered. Nevertheless, at 24 h after the dose of methotrexate, probenecid resulted in a threefold increase in plasma methotrexate concentrations. In addition, in CSF at both 24 and 72 h, methotrexate concentrations were increased by a factor of 2.8- and 4.2-fold, respectively. The half-life of methotrexate in CSF was not altered by probenecid. This is a clinically important interaction as methotrexate toxicity (bone marrow depression) has been reported in some patients receiving probenecid. In the monkey, probenecid has been shown to completely inhibit the renal tubular transport of methotrexate, resulting in renal clearance values being equivalent to filtration at the glomerulus (Bourke et al. 1975).

b) Penicillins

There have been no detailed pharmacokinetic studies investigating the interaction between methotrexate and penicillins. However, a number of case reports involving different penicillins have shown marked (penicillin by 35%, piperacillin by 66%, ticarcillin by 40%, dicloxacillin by 93%) reductions in the clearance of methotrexate (Bloom et al. 1986). The clinical importance of this interaction was highlighted in patients receiving low-dose methotrexate for psoriasis (Mayall et al. 1991). Five patients taking between 7.5 and 12.5 mg methotrexate daily developed neutropenia and thrombocytopenia when also receiving amoxycillin, flucloxacillin, benzylpenicillin or piperacillin. Three of these patients died. The clinical seriousness of this interaction highlights the importance of drug interactions involving renal tubular secretion mechanisms.

c) Urinary Alkalinizers

Methotrexate can also undergo passive renal tubular reabsorption depending on urine pH and the degree of ionization (Sand and Jacobsen 1981). It has been shown that alkalinization of urine (pH >7) results in a decrease in serum methotrexate concentrations of between 40% and 77% (Nirenberg et al. 1977). Alkalinization of urine increases the degree of ionization of this weak acid, preventing passive reabsorption and hence increasing its renal clearance. However, Sand and Jacobsen (1981) interpreted their findings in terms of increasing the solubility of methotrexate in urine, although this was not directly assessed.

d) Non-steroidal Anti-inflammatory Drugs

There have been a number of pharmacokinetic studies and case reports highlighting a clinically and mechanistically significant interaction between methotrexate and this class of drug.

α) Aspirin and Salicylates

Eight patients with rheumatoid arthritis and normal renal function were investigated to evaluate the effect of aspirin on the disposition of methotrexate (STEWART et al. 1990a). On the 1st day of the study they were given 10mg methotrexate by intravenous bolus and after 7 days of aspirin 975mg orally four times daily, another intravenous bolus dose of methotrexate was given. Aspirin significantly ($P < 0.05$) increased the area underneath the plasma methotrexate concentration-time curve from 2.7 ± 0.5 to $3.4 \pm 0.9\,\mu$mol/l × h. The total body clearance of methotrexate was reduced slightly from 141 ± 27 to 112 ± 23 ml/min ($P < 0.05$) and the renal clearance of methotrexate was reduced from 94 ± 25 to 77 ± 48 ml/min; however, this was not statistically significant. Due to the complex mechanism(s) (secretion and reabsorption) in the renal clearance of methotrexate, collection of the 24-h urine in this study may have been inappropriate to detect time- and concentration-dependent changes in renal clearance. In nine patients receiving between 7.5 and 15mg methotrexate orally, the addition of choline magnesium trisalicylate 2.25–4.5 g/day decreased the total body clearance of methotrexate from 168 ± 50 to 128 ± 57 ml/min ($P < 0.05$) and reduced the renal clearance of methotrexate from 117 ± 35 to 84 ± 27 ml/min (TRACY et al. 1992). In 12 rheumatoid arthritis patients receiving low-dose (5–10 mg/m^2 once a week) methotrexate, the effect of aspirin (45–50 mg/kg per day for 16 days) on the intravenous pharmacokinetics of 10 mg/m^2 methotrexate was investigated (FURST et al. 1990). Aspirin had no effect on the total body clearance of methotrexate (urine was not collected), or on the disposition of the metabolite 7-hydroxymethotrexate. Finally, MAIER et al. (1986) proposed that the report by ST. CLAIR et al. (1985) on methotrexate toxicity in three patients could have been attributed to the high doses (3.9–5.0 g/day) of aspirin received by those patients.

β) Indomethacin

There has been to systematic pharmacokinetic study investigating the interaction with this nonsteroidal anti-inflammatory drug. However, there have been case reports of acute renal failure and death, particularly in elderly patients taking indomethacin and methotrexate (ELLISON and SERVI 1985). It is unclear whether the mechanism involves competition between indomethacin and methotrexate for active renal tubular secretion or, more likely, in the elderly patient who requires the vasodilatory prostaglandins to maintain renal perfusion, addition of indomethacin, which blocks the synthesis of these prostaglandins, can result in acute renal failure resulting in reduced clearance of methotrexate with its associated toxicities.

γ) Ketoprofen

Although described as a pharmacokinetic study, details were lacking in a retrospective study in 36 patients who were on a number of cycles of high-dose (mean 3200 mg/m²) methotrexate. Four of the nine patients who developed severe toxicity were also taking ketoprofen and three of these patients died. There was a substantial increase in the serum methotrexate concentration in those patients with toxicity (Thyss et al. 1986).

δ) Naproxen

In a comprehensive study, 15 patients aged between 30 and 78 years with rheumatoid arthritis received oral and intravenous methotrexate 15 mg with and without naproxen at a dose of 1000 mg/day for 26 days (Stewart et al. 1990b). Twelve of the patients completed all four arms of the study and, for the three who did not complete the study, withdrawal was not associated with drug treatment. There was no significant difference in the total body clearance of methotrexate with or without naproxen (103 ± 35 alone and 113 ± 48 ml/min with naproxen). In addition, the renal clearance of methotrexate from either routes (oral and intravenous) was not altered by naproxen. It should be noted that these patients had normal creatinine clearance values (80 ml/min per 1.73 m²). Tracy et al. (1992) showed significant ($P < 0.05$) but small decreases in the total body clearance of methotrexate (168 ± 50–131 ± 42 ml/min) in nine patients taking naproxen 750–1250 mg/day, with no statistically significant difference in renal clearance (117 ± 35 ml/min methotrexate alone and 96 ± 38 ml/min with naproxen). Ahern et al. (1988) also found no difference in plasma methotrexate concentrations in three patients with and without concomitant naproxen dosing. There has, however, been a report of serious toxicity in patients receiving methotrexate and naproxen (Singh et al. 1986).

ε) Ibuprofen

In nine patients with rheumatoid arthritis, the addition of ibuprofen 2.4–3.6 g/day significantly ($P < 0.05$) reduced the total and renal clearances of methotrexate from 168 ± 50 to 101 ± 37 ml/min and from 117 ± 35 to 70 ± 27 ml/min, respectively (Tracy et al. 1992).

Other non-steroidal anti-inflammatory drugs such as diclofenac, diflunisal, flurbiprofen, sulindac and azapropazone have been investigated, albeit usually with small numbers ($n = 1$–3) of patients. Hence, data interpretation has not allowed conclusions to be deduced, although methotrexate toxicity occurred in some patients. The interaction between methotrexate and non-steroidal anti-inflammatory drugs is important as there have been several deaths related to this interaction. The life-threatening nature of the interaction suggests that non-steroidal anti-inflammatory drugs ought to be co-administered with extreme caution to patients on methotrexate.

The mechanism for the interaction is likely to be one or a combination of: (a) competition for active tubular secretion between the non-steroidal anti-inflammatory drug and methotrexate as both are organic anions. Since methotrexate renal clearance is concentration dependent and active secretion appears to occur at low doses (HENDEL and NYFORS 1984), the interaction could be more significant at low methotrexate doses; and (b) The non-steroidal anti-inflammatory drugs can decrease renal perfusion via their inhibitory action on cyclooxygenase, resulting in a reduction in renal vasodilatory prostaglandins, and a decrease in filtered and secreted methotrexate, causing a reduction in renal clearance and increase in plasma concentrations. This second mechanism for the interaction would be particularly relevant for the elderly patient whose renal function can be easily compromised by non-steroidal anti-inflammatory drugs.

II. Organic Cations

The most commonly investigated *interacting* drugs have been cimetidine, other H_2 receptor antagonists and trimethoprim.

1. Cimetidine

Cimetidine has been shown to reduce the renal clearance of the following organic cations: procainamide and its active metabolite *n*-acetylprocainamide, triamterene, ranitidine, amiloride, metformin, pindolol, nicotine and quinidine.

*a) Procainamide/*n*-Acetylprocainamide*

Several studies have been conducted in healthy subjects and patients and case reports have highlighted the clinical importance of this interaction.

In the original study (SOMOGYI et al. 1983), six healthy subjects took 1 g procainamide orally with and without 400 mg cimetidine (1 h before the procainamide dose and 200 mg 4-hourly for 12 h thereafter) with blood and urine samples collected over 12 and 48 h, respectively. Cimetidine significantly ($P < 0.025$) increased the area under the plasma concentration-time curve of procainamide by an average of 44% and prolonged the half-life from an harmonic mean of 2.92 to 3.68 h (Fig. 3). This was due to a marked reduction ($P < 0.025$) in the renal clearance of procainamide from 347 ± 46 to 196 ± 11 ml/min (Fig. 4). In the same study, the area under the plasma concentration-time curve for *n*-acetylprocainamide was significantly ($P < 0.025$) increased by an average of 25% with a commensurate reduction ($P < 0.025$) in renal clearance from 258 ± 60 to 197 ± 49 ml/min. Similar data were obtained by CHRISTIAN et

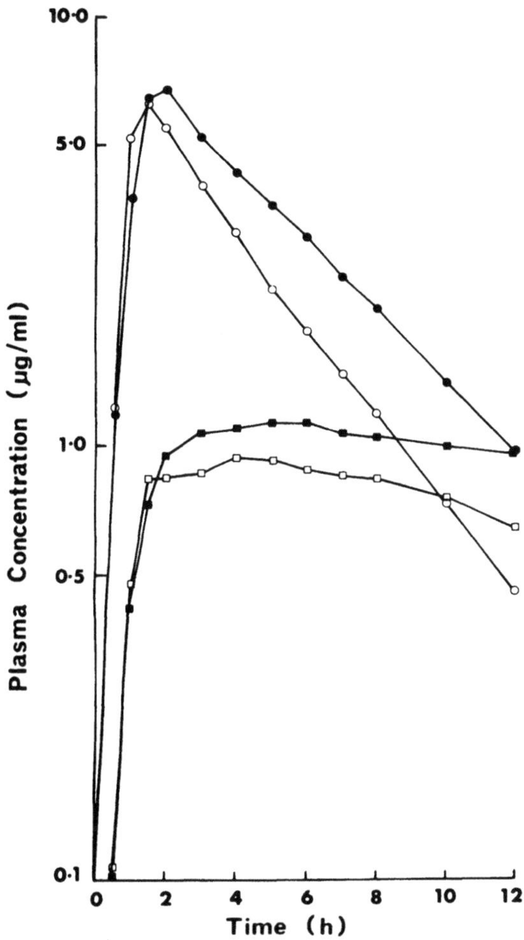

Fig. 3. Plasma concentration-time profiles of procainamide (*round symbols*) and *n*-acetylprocainamide (*square symbols*) in six subjects who received 1g oral procainamide alone (*open symbols*) and when co-administered with cimetidine (*closed symbols*). (From Somogyi et al. 1983)

al. (1984) in nine healthy subjects in which cimetidine increased the area under the plasma concentration-time curve of procainamide by 40% ($P < 0.0005$), prolonged its half-life by 24% and reduced its renal clearance by 36% ($P < 0.0005$). In addition, the area under the plasma concentration-time curve of *n*-acetylprocainamide increased ($P < 0.03$) by 26% but there was no change in renal clearance. In seven healthy Chinese subjects administered separate single oral doses of 200mg and 400mg cimetidine (Lai et al. 1988), the area under the plasma concentration-time curves of procainamide (500mg oral dose) was increased by 24% ($P < 0.01$) and 38% ($P < 0.05$), respectively, with significant ($P < 0.01$) reductions in renal clearance of 31% and 41%, respectively. These data reinforced the mechanism of competitive inhibition of tubu-

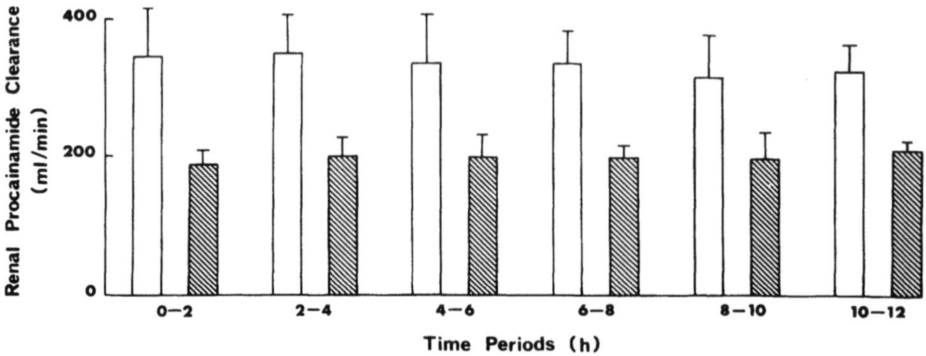

Fig. 4. Mean ± SD 2 hourly renal clearance values of procainamide when given alone as 1 g procainamide to six healthy subjects (*open bars*) and when co-administered with cimetidine (*hatched bars*). The reduction in renal clearance by cimetidine was significant ($P < 0.02$). (From Somogyi et al. 1983)

lar secretion of procainamide by cimetidine. Only the higher dose (400 mg) of cimetidine significantly increased (mean 45%) the area under the plasma concentration-time curve and reduced the renal clearance (16%) of *n*-acetylprocainamide. In a steady-state pharmacokinetic study in six healthy young male subjects who received 500 mg sustained release procainamide every 6 h for 13 doses, cimetidine 1200 mg/day for 4 days significantly ($P < 0.01$) increased the procainamide area under the plasma concentration-time curve by a mean of 43% and significantly ($P < 0.01$) decreased renal clearance by a mean of 34% (Rodvold et al. 1987). Steady-state serum *n*-acetylprocainamide concentrations were significantly increased ($P < 0.001$) by a mean of 42%. The results of these four studies in healthy subjects are similar and all groups proposed the mechanism to involve inhibition of the active tubular secretions of procainamide and *n*-acetylprocainamide by cimetidine via competitive inhibition for the organic cation transporter. The interaction has been described and mechanism verified in vitro (McKinney and Speeg 1982). In a study in 36 hospitalized male patients (aged 65–90 years) who were receiving oral sustained release procainamide every 6 h, the addition of cimetidine 300 mg every 6 h for 3 days resulted in a significant ($P < 0.005$) increase in the steady-state serum concentrations of procainamide (by a mean of 55%), and *n*-acetylprocainamide (mean 36%) (Bauer et al. 1990). In ten of the patients from whom urine was also collected, the ratio of procainamide renal clearance to creatinine renal clearance was reduced by cimetidine by an average of 33% from 2.1 ± 0.6 to 1.4 ± 0.4 ($P < 0.02$). The *n*-acetylprocainamide to creatinine clearance ratio also significantly ($P < 0.05$) declined from 1.4 ± 0.4 to 1.1 ± 0.4. Twelve of these patients experienced mild to severe symptoms relating to procainamide toxicity. Nine of the patients had symptoms of nausea, weakness, malaise and PR interval increases of <20%. Three patients had severe procainamide toxicity from prolongation of the PR interval of >25%, unifocal premature ventricular contractions of <30/h and ventricular rates >150/min for greater than 30s. These latter patients had procainamide serum concentrations of greater than 14mg/l and *n*-acetylprocainamide concentrations of greater than 16mg/l. All patients with adverse effects had cimetidine discontinued, and procainamide was also discontinued in those patients with severe toxicity. The procainamide adverse effects disappeared within 24h after cessation of cimetidine therapy. In a case report (HIGBEE et al. 1984), a 71-year-old male patient was being treated with procainamide (937.5mg every 6h) for frequent multiform premature ventricular depolarizations and had achieved steady-state therapeutic serum procainamide and *n*-acetylprocainamide concentrations of 9 and 7mg/l, respectively. Following the addition of cimetidine 300mg every 6h for a benign gastric ulcer, the patient was re-admitted with increased fatigue, weakness, nausea, anorexia and urinary retention. Examination revealed congestive heart failure and the ECG demonstrated sinus rhythm with first-degree AV block accompanied by a decreased heart rate. The procainamide and *n*-acetylprocainamide serum concentrations were both 15mg/l, in the range associated with toxicity. The dose of procainamide was subsequently decreased to 750mg every 6h and procainamide and *n*-acetylprocainamide serum concentrations declined to 10 and 12mg/l, respectively. Other dosage adjustments were made but the authors concluded by stating that their case report should alert clinicians to the potentially dangerous adverse effects of this combination (cimetidine and procainamide) as well as the need to evaluate carefully the source of gastrointestinal symptoms which may be due either to ulcer disease or procainamide toxicity. They also commented that the changes encountered suggested that older people may be at a greater risk of procainamide toxicity when prescribed cimetidine.

b) Ranitidine

In a study in six male subjects, the administration of cimetidine 400mg 12 hourly resulted in a significant ($P < 0.05$) increase in the area under the plasma ranitidine concentration-time curve from a single 150-mg oral dose of ranitidine. This was associated with a significant reduction in the renal clearance of ranitidine from 326 ± 67 to 244 ± 57 ml/min (VAN CRUGTEN et al. 1986).

c) Triamterene

In a similar design to the above study (MUIRHEAD et al. 1986), cimetidine significantly ($P < 0.03$) increased the area underneath the plasma triamterene concentration-time curve during a 24-h dosing interval of 100mg daily triamterene. This was associated with a reduction in the renal clearance of triamterene from 71 ± 46 to 21 ± 14 ml/min. However, the decrease in renal triamterene clearance occurred only during the first 6h of the triamterene dosing interval, at the time of maximum plasma cimetidine concentrations. When the unbound fraction (0.4) of triamterene in plasma is taken into consideration, the renal clearance due to tubular secretion (minus that due to reabsorption) was reduced by cimetidine from a mean of 22ml/min to minus

28 ml/min. The latter figure indicates that cimetidine completely blocked the renal tubular secretion of triamterene and uncovered significant triamterene passive reabsorption, a mechanism previously reported in vitro in the rat (KAU 1978). Cimetidine had no effect on the renal clearance of the major active metabolite of triamterene, the sulphate conjugate of *p*-hydroxytriamterene (MUIRHEAD et al. 1986).

d) Metformin

Cimetidine significantly ($P < 0.01$) increased the area under the plasma metformin concentration-time curve by an average of 50% in seven young healthy subjects given a 0.25-g daily dose of metformin and a 400-mg twice daily dose of cimetidine. The increase in steady-state plasma metformin concentrations was associated with a significant ($P < 0.01$) reduction (mean 27%) in the renal clearance of metformin from 527 ± 165 ml/min to 378 ± 165 ml/min (SOMOGYI et al. 1987). The reduction in metformin renal clearance only occurred over the first 6 h of cimetidine dosing, a phenomenon consistent with competitive inhibition of metformin renal tubular secretion.

e) Pindolol

In a study to examine the potential stereoselectivity of the inhibition of renal tubular secretion of organic bases by cimetidine, eight healthy young subjects received a single 15-mg oral dose of racemic pindolol with 400 mg cimetidine taken twice daily for 2 days (SOMOGYI et al. 1992). The area under the plasma concentration-time curve for R-(+) pindolol was significantly ($P < 0.01$) increased by cimetidine from 230 ± 90 to 344 ± 78 ng/ml × h and for S-(−) pindolol it was increased from 209 ± 73 to 288 ± 69 ng/ml × h ($P < 0.01$). Cimetidine significantly reduced the renal clearance of R-(+) pindolol from 170 ± 55 ml/min to 104 ± 88 ml/min ($P < 0.01$) and of S-(−) pindolol from 222 ± 66 ml/min to 155 ± 38 ml/min ($P < 0.01$). The enantiomer with the higher renal clearance [S-(−) pindolol] had a smaller (mean 26%) cimetidine-induced reduction in renal clearance compared with R-(+) pindolol (mean 34%) (Fig. 5). Cimetidine has a stereoselective inhibitory effect on the active transport of organic cations in the proximal tubule.

f) Disopyramide

In a study in seven healthy male subjects, co-administration of cimetidine 400 mg twice daily did not alter the disposition or the renal clearance of the enantiomers of disopyramide when given as an intravenous bolus of 150 mg racemic disopyramide (BONDE et al. 1991). The reason for this lack of effect of cimetidine on the renal clearance of a drug (an organic cation) which undergoes extensive tubular secretion (renal clearance of unbound enantiomers of disopyramide is 200–400 ml/min) is not known. However, since the drug exhibits significant concentration-dependent plasma binding, the unbound concentrations need to be measured at all blood sampling times and not just at one

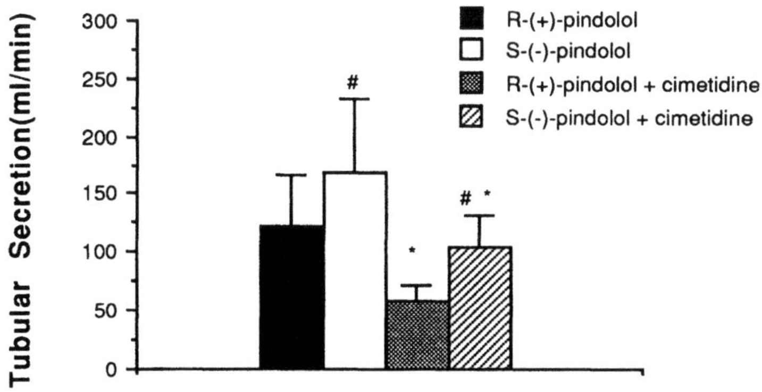

Fig. 5. Net renal clearance via tubular secretion for R-(+) pindolol and S-(−) pindolol when racemic pindolol was administered alone (15 mg) and when co-administered with cimetidine in eight subjects. Values represent mean ± SD. *$P < 0.001$, enantiomer alone vs. enantiomer + cimetidine; #$P < 0.001$, R-(+) pindolol vs. S-(−) pindolol. (From Somogyi et al. 1992)

particular time (or concentration) point. As was the case with triamterene and metformin, the effect of cimetidine may only occur in the first few hours after dosing, particularly with the low dose of cimetidine used. Hence to observe an effect on the renal clearance of disopyramide may have required frequent urine collections and not reliance solely placed on urine collected over a long time interval.

g) Amiloride

In eight healthy subjects given a single dose of 5 mg amiloride, cimetidine 400 mg twice daily reduced the renal clearance of amiloride by an average of 17% from 358 ± 134 to 299 ± 118 ml/min ($P < 0.05$), but there was no effect on the area under the plasma amiloride concentration-time curve (Somogyi et al. 1989). The effect on renal amiloride clearance was small in magnitude and occurred only in the 2- to 4-h time period after cimetidine dosing.

h) Nicotine

In six healthy subjects, cimetidine 600 mg 12 hourly for 2 days significantly ($P < 0.05$) reduced the renal clearance of nicotine (as an intravenous infusion of 1 μg/kg per minute for 30 min) from 1.49 ± 0.30 to 0.79 ± 0.30 ml/min per kilogram, whereas there was no effect ($P > 0.05$) on the renal clearance of the nicotine metabolite cotinine (0.14 ± 0.04 ml/min per kilogram without cimetidine vs. 0.15 ± 0.10 ml/min per kilogram with cimetidine) (Bendayan et al. 1990). Since the renal clearance of nicotine when co-administered with cimetidine was below the renal clearance due solely to filtration (fu × GFR), these results indicate the existence of passive reabsorption of nicotine in addition to active tubular secretion.

i) Quinidine

In six healthy subjects, HARDY et al. (1983) showed that cimetidine given as 1.2 g/day reduced the oral clearance of quinidine (as 400 mg quinidine sulphate) by 37% ($P < 0.05$) with no change in the urinary excretion of unchanged quinidine. As a consequence, calculation of the renal clearance allows one to conclude that cimetidine reduced the renal clearance of quinidine from 5.6 to 3.6 l/h.

j) Cephalothin/Cephalexin

In order to investigate the specificity of cimetidine-induced reductions in the renal tubular secretion of organic cations, VAN CRUGTEN et al. (1986) showed, in six normal healthy subjects, that cimetidine (400 mg twice daily) had no effect ($P > 0.05$) on the renal clearance of the organic anion cephalothin (305 ± 127 ml/min alone vs. 284 ± 141 ml/min with cimetidine), whose renal clearance is inhibited by probenecid (Table 1). However, for the zwitterion cephalexin, cimetidine reduced ($P < 0.01$) the renal clearance from 263 ± 39 to 208 ± 21 ml/min. The renal clearance of this zwitterion, which has both acidic (pK_a 2.6) and basic (pK_a 7.0) functional groups, is also reduced by probenecid (Table 1). These investigators concluded that cimetidine's effect was specific for organic cations, and that, for those zwitterions that undergo tubular secretion, both organic anion and cation transport systems might be involved.

2. Ranitidine

In the light of the inhibitory effect of cimetidine on the renal tubular secretion of other organic cations, studies in humans have determined whether other H_2 antagonists (which are organic cations) behave in a similar manner to cimetidine.

a) Procainamide/n-Acetylprocainamide

In six healthy subjects, the administration of 150 mg twice daily ranitidine 13 h prior to a 1-g oral procainamide dose increased the area under the plasma procainamide concentration-time curve by an average of 14% with a significant ($P < 0.05$) but small (mean 18%) reduction in the renal clearance of procainamide (SOMOGYI and BOCHNER 1984). The area under the plasma *n*-acetylprocainamide concentration-time curve was also reduced by an average of 11% ($P < 0.05$). In the same study, when the dose of ranitidine was increased to 750 mg in three subjects, the area under the plasma concentration-time curve for procainamide was increased by an average of 20% and renal clearance was reduced by an average of 35%, and for *n*-acetylprocainamide, renal clearance was reduced by an average of 38%. This study demonstrated that in the clinical doses used, ranitidine will have little inhibitory effect; however, if larger doses are used or if the patient's renal function declines such that in both cases higher plasma ranitidine concentrations are achieved, then inhibition of renal clearance of other organic cations can be observed. In another study in six healthy male subjects who received 500 mg sustained release procainamide every 6 h, the addition of ranitidine 150 mg twice a day had no effect on the plasma disposition or renal clearance of procainamide or *n*-acetylprocainamide (RODVOLD et al. 1987). These results would be anticipated in the light of the first study (SOMOGYI and BOCHNER 1984) and indicate that any possible effect of ranitidine on the tubular secretion of procainamide or *n*-acetylprocainamide may only occur at plasma ranitidine concentrations in excess of those normally found following standard doses in patients with normal renal function.

b) Triamterene

In eight healthy male subjects, ranitidine 150 mg twice daily reduced the renal clearance of triamterene (100 mg daily for 4 days) by a mean of 51% ($P < 0.05$) from 94 ± 64 to 46 ± 42 ml/min (MUIRHEAD et al. 1988). This was interpreted as competition between triamterene and ranitidine for the renal organic cation transport system. In the same study, the renal clearance of the sulphate conjugate of *p*-hydroxytriamterene was significantly ($P < 0.01$) reduced by ranitidine from 105 ± 35 to 56 ± 20 ml/min. There was no effect of ranitidine on the degree of binding of the sulphate conjugate to plasma components (fu = 0.09). Although the results for triamterene are consistent with inhibition of tubular secretion of triamterene by another organic cation (ranitidine), the inhibition of the renal clearance of the sulphate conjugate raised the possibility of multitransport systems, as cimetidine which reduced triamterene renal clearance did not alter the renal clearance of the sulphate conjugate (see Sect. C.II.1.c).

c) Nicotine

In six healthy subjects, ranitidine 300 mg 12 hourly for 2 days had no effect on the renal clearance of nicotine (intravenous infusion of 1 μg/kg per minute for 30 min) (1.49 ± 0.30 alone and 0.88 ± 0.40 ml/min per kilogram with ranitidine) or cotinine (BENDAYAN et al. 1990).

3. Famotidine

*a) Procainamide/*n*-Acetylprocainamide*

Eight healthy subjects were given a single 5-mg/kg intravenous infusion of procainamide alone and after 5 days of receiving 40 mg famotidine once daily (KLOTZ et al. 1985). Famotidine had no significant effect ($P > 0.05$) on the total body clearance or renal clearance of procainamide and *n*-acetylprocainamide. Although famotidine has a high renal clearance of more than 300 ml/min, its half-life is less than 3 h; hence plasma concentrations from a daily 40-mg dose may have been insufficient to elicit inhibitory effects on the tubular secretion of procainamide and *n*-acetylprocainamide. In addition, it has been observed that the organic anion probenecid inhibits the tubular secretion of famotidine (see Sect. C.I.1.1).

4. Trimethoprim

*a) Procainamide/*n*-Acetylprocainamide*

In eight healthy subjects, procainamide 500 mg orally every 6 h for 3 days was co-administered with and without trimethoprim 200 mg daily for 4 days (KOSOGLOU et al. 1988). Trimethoprim significantly ($P < 0.05$) increased the steady-state plasma concentrations of procainamide by an average of 62% and for *n*-acetylprocainamide by 52%. These were due to significant ($P < 0.05$) reductions in the renal clearances of procainamide (mean 47%) and *n*-acetylprocainamide (mean 13%). These investigators detected small but statistically significant ($P < 0.0001$) increases in the QTc interval when trimethoprim was co-administered, presumably as a result of the elevated plasma concentrations of procainamide and *n*-acetylprocainamide. The same investigators also conducted a study in ten healthy young male subjects who received a single 1-g oral procainamide dose alone and with trimethoprim 100 mg twice daily for 2 days prior to and 200 mg with the procainamide dose (VLASSES et al. 1989). The area under the plasma procainamide concentration-time curve was significantly ($P < 0.05$) increased by trimethoprim by an average of 39% and the renal clearance was reduced from 487 ± 129 to 267 ± 123 ml/min ($P < 0.05$). For *n*-acetylprocainamide, the area under the plasma concentration-time curve was increased by an average of 27% by trimethoprim ($P < 0.05$) and renal clearance was reduced from 275 ± 78 to 192 ± 82 ml/min ($P < 0.05$). Since trimethoprim is actively taken up by the renal cortex via the organic cation transport system, and has a high renal clearance (>220 ml/min for the unbound drug), this interaction is consistent with trimethoprim having a higher affinity than procainamide and *n*-acetylprocainamide for the organic cation transporter and hence inhibiting their tubular secretion.

b) Zidovudine

In nine HIV-infected patients with normal renal function, the effect of trimethoprim (150 mg orally) on the disposition of zidovudine (3 mg/kg intravenous infusion) was evaluated (CHATTON et al. 1992). Trimethoprim significantly ($P < 0.05$) reduced the renal clearance of zidovudine from 0.34 ± 0.15 to 0.17 ± 0.09 l/h per kilogram, an average decrease of 58%. In addition, the apparent renal clearance of the glucuronide conjugate of zidovudine was also significantly ($P < 0.05$) reduced by trimethoprim from 0.70 ± 0.21 to 0.56 ± 0.12 l/h per kilogram. In the same study but on another occasion, the patients received a single dose of trimethoprim 150 mg plus sulphamethoxazole 800 mg. The decrease in zidovudine renal clearance averaged 48% and for the glucuronide conjugate 20%, similar to when trimethoprim was administered alone, indicating that the sulphonamide has no effect on zidovudine disposition. Hence the renal tubular secretion of zidovudine is via the organic cation transport system, whereas that of its glucuronide conjugate is by both organic cation and anion systems as probenecid also reduced its renal clearance (see Sect. C.I.1.g).

5. Amiodarone

*a) Procainamide/*n*-Acetylprocainamide*

The disposition of procainamide was investigated in eight patients who received an intravenous dose of between 6 and 15 mg/kg before and after 1–2 weeks treatment with amiodarone (1.6 g/day) (WINDLE et al. 1987). Amiodarone significantly ($P < 0.01$) reduced the total body clearance of procainamide from 0.43 ± 0.12 to 0.33 ± 0.12 l/kg per hour. In two of the patients, urine was also collected and amiodarone reduced the renal clearance of procainamide from 0.25 and 0.24 l/kg per hour to 0.14 and 0.15 l/kg per hour, respectively. Hence the reduction in total body clearance appears to be predominantly due to renal clearance, most likely a consequence of competition for active tubular secretion. Since the half-life of amiodarone is almost 1 month, the degree of inhibition of procainamide renal tubular secretion might be even higher when steady-state plasma concentrations of amiodarone are reached after about 4 months of chronic dosing.

6. Quinine/Quinidine

a) Amantadine

In a study designed to evaluate the effect of age and gender on inhibition of amantadine renal clearance by quinine and quinidine (GAUDRY et al. 1993), a comparison was made between nine young 27- to 39-year-old subjects (five male and four female) and nine 60- to 72-year-old adults (four male and five female). Amantadine (an organic base undergoing tubular secretion) was administered as a 3-mg/kg oral syrup on the evening before the study and quinine (as quinine sulphate 200 mg) and quinidine (as quinidine sulphate 200 mg) were administered orally approximately 12 h later. Blood and urine samples were collected at 2, 4, 6 and 8 h after quinine/quinidine dosing and all urine was collected in 2 hourly aliquots to 8 h. Quinine and quinidine had no significant ($P > 0.05$) effect on amantadine renal clearance in either young or old subjects. However, when the data were separated according to gender, quinine and quinidine significantly ($P < 0.05$) reduced the amantadine renal clearance in males but not in females. When both male and female data were combined, only quinine reduced the renal clearance of amantadine. The investigators cited animal studies which suggest that testosterone may enhance the renal organic anion and cation transport of xenobiotics. However, it is unclear as to how this would impact on inhibition of transport as there was no gender difference in the renal clearance of amantadine. Clearly, further studies are warranted to elucidate the mechanisms of this finding.

III. Organic Neutral Drugs

1. Digoxin

Digoxin is the most commonly used cardiac glycoside drug for the treatment of congestive heart failure and atrial fibrillation. When administered by the intravenous route, approximately 60% of the dose is recovered unchanged in urine and its renal clearance is in excess of the clearance via glomerular filtration (taking into account the degree of plasma binding), indicating that this chemically neutral compound undergoes tubular secretion (STEINESS 1974). A large number of interactions have been reported for digoxin and several of these involve renal excretory mechanisms.

a) Amiodarone

Six patients with either ventricular or supraventricular tachyarrhythmias were given a 1-mg intravenous dose of digoxin alone and after 2 weeks of amiodarone 1600 mg/day (NADEMANEE et al. 1984). The renal clearance of digoxin was reduced by an average of 22% (from 0.85 ± 0.33 to 0.66 ± 0.39 ml/min per kilogram) but this was not significant ($P > 0.05$). However, in ten healthy subjects who were administered 400 mg amiodarone daily for 3 weeks, there was a mean 22% decrease (from 105 ± 39 to 84 ± 15 ml/min; $P < 0.05$) in the renal clearance of digoxin when administered as a 1-mg intravenous dose (FENSTER et al. 1985). Amiodarone also reduced the non-renal clearance of digoxin in both studies. The reduction in renal clearance was interpreted as a reduction in the tubular secretion of digoxin as there was no effect on the renal clearance of creatinine by amiodarone (NADEMANEE et al. 1984).

b) Spironolactone and Potassium-Sparing Diuretics

It was originally reported by STEINESS (1974) that spironolactone (100 mg daily for 10 days) decreased the renal clearance of digoxin by approximately 30%, in nine patients stabilized on digoxin, to values almost identical to those for inulin clearance. As there was no effect on the glomerular filtration rate, the mechanism remained unexplained. The same group (WALDORFF et al. 1978) also found that spironolactone (100 mg twice daily for 5 days) given to four hospitalized patients with arteriosclerotic heart disease and four healthy subjects reduced the renal clearance of digoxin from 1.52 ± 0.86 to 1.13 ± 0.58 ml/min per kilogram ($P < 0.02$). In a subsequent study in six healthy subjects, spironolactone (100 mg/day for 8 days) was found to reduce the renal clearance of digoxin from 168 ± 22 to 145 ± 17 ml/min ($P < 0.05$), an average decrease of 13% (HEDMAN et al. 1992). The mechanism for the interaction is still unclear as it is not known by which active transport system in the kidney digoxin is secreted (KOREN 1987). It has been reported that amiloride at a dose of 5 mg twice daily for 8 days in six healthy subjects increased ($P < 0.001$) digoxin renal clearance from 1.3 ± 0.1 to 2.4 ± 0.5 ml/min per kilogram following a 15-μg/kg intravenous bolus dose of digoxin without altering inulin clearance (WALDORFF et al. 1981). The mechanism by which amiloride increases the renal clearance of digoxin was attributed to increased potassium in distal tubular cells and therefore increasing digoxin secretion, although this latter mechanism was not proven.

c) Quinidine

Several mechanisms, including a reduction in digoxin renal clearance, are involved in the interaction of quinidine on digoxin disposition. The reduction in the renal clearance of digoxin by quinidine has been reported as: (a) a mean of 35% reduction in three patients on long-term oral digoxin and administered 250 mg twice daily quinidine, with no effect on creatinine clearance (HOOYMANS and MERKUS 1978); (b) a mean of 33% reduction (from 1.64 ± 0.60 to 1.09 ± 0.24 ml/min per kilogram; $P < 0.05$) in six patients given a 1-mg intravenous bolus dose of digoxin with and without 200 mg quinidine 6 hourly for 7 days (HAGER et al. 1979); (c) a mean of 34% reduction (from 53 ± 21 to 35 ± 13 ml/min per 1.73 m^2; $P < 0.001$) in 15 patients prescribed between 0.125 and 0.25 mg/day digoxin and 200–300 mg/day quinidine, with no alteration in creatinine clearance (MUNGALL et al. 1980); (d) a mean of 52% decrease (from 130 ± 32 to 62 ± 10 ml/min; $P < 0.01$) in digoxin renal clearance in five patients given an intravenous dose of tritiated digoxin with and without quinidine 0.6 g twice daily (SCHENCK-GUSTAFSSON and DAHLQVIST 1981); (e) a decrease of 32% in mean renal clearance of digoxin (from 1.93 ± 0.61 to 1.32 ± 0.40 ml/min per kilogram) in seven healthy subjects receiving quinidine 800 mg/day and increasing to a mean 54% reduction in digoxin renal clearance (from 1.47 ± 0.30 to 0.68 ± 0.22; $P < 0.001$) when the dose of quinidine was doubled to 1600 mg/day for 4 days (LEAHEY et al. 1981). Digoxin was administered as a 1-mg intravenous infusion over 60 min on both occasions in both groups. A relationship ($r = 0.60$) was found between the serum quinidine concentration and reduction in digoxin renal clearance. Thus, the reduction in renal clearance of digoxin depends on the concentration of quinidine and is consistent with competition for active tubular secretion; (f) In a comprehensive study in eight healthy young male subjects, digoxin 0.5–0.75 mg/day was given alone and with 400 mg twice daily quinidine (in four subjects). The renal clearance of digoxin was significantly reduced ($P < 0.05$) by an average of 29% from 155 ± 40 to 110 ± 21 ml/min. In the other four subjects, the effect of quinine (the diastereoisomer of quinidine) given as 750 mg/day was investigated. The renal clearance of digoxin was not altered by quinine (177 ± 40 ml/min digoxin alone and 185 ± 53 ml/min with quinine) (HEDMAN et al. 1990).

Since quinidine has no effect on glomerular filtration rate, the results of these studies have been interpreted as a reduction in the tubular secretion of digoxin by quinidine. It is possible that both digoxin and quinidine compete for the same carrier-mediated transport system in the kidney. Quinidine competes for the organic cation transporter; however, it has been shown in animals that digoxin is not eliminated by this transport system (KOREN 1987). Clearly further studies are warranted to delineate the mechanism of the renal tubular secretion of digoxin.

d) Verapamil

Verapamil has been shown to increase plasma digoxin concentrations and produce toxicity and in some cases fatality (as reviewed by STOCKLEY 1994). In eight healthy subjects who received a 1-mg intravenous infusion of digoxin over 15 min, the addition of verapamil 800 mg orally three times daily for 10 days resulted in a reduction in the renal clearance of digoxin from 2.2 ± 0.4 to 1.7 ± 0.5 ml/min per kilogram ($P < 0.025$) (PEDERSEN et al. 1981). Similarly, in seven patients receiving verapamil (80 mg three times daily) for long-term therapy, the renal clearance of digoxin from a dose of 125 μg/day was reduced after 1 week of verapamil therapy by an average of 36% (from 198 ± 46 to 128 ± 27 ml/min; $P < 0.001$), however, after 6 weeks of verapamil there was no reduction in the renal clearance of digoxin (203 ± 71 ml/min) (PEDERSEN et al. 1982). In a study by BELZ et al. (1983) in healthy subjects who were administered digoxin 0.375 mg/day to steady state, the addition of verapamil (80 mg three times/day for 2 weeks) reduced ($P < 0.05$) the digoxin renal clearance from 218 ± 90 to 148 ± 42 ml/min per 1.73 m^2. In a more recent study in six patients with chronic atrial fibrillation, verapamil (240 mg/day) had no effect on the renal clearance of digoxin (dose varied between 0.25 and 0.5 mg/day) (HEDMAN et al. 1991). The renal clearance of digoxin was 153 ± 51 ml/min when given alone and 173 ± 51 ml/min when given with verapamil ($P > 0.05$). This study found a significant association between the renal clearance of digoxin and urine flow rate ($r = 0.75$; $P < 0.001$). At a urine flow rate of 1 ml/min the renal clearance of digoxin was about 100 ml/min and at 4 ml/min the renal clearance had doubled to almost 200 ml/min. There was no difference in urine flow rate between the two study occasions. It would appear that the effect of verapamil on digoxin renal clearance may be transient and may be confounded by alterations in urine flow rate.

e) Other Calcium Antagonists

There have been several studies investigating the effect of other calcium antagonists (diltiazem, nifedipine) on the disposition of digoxin. Most of the studies have resulted in contradictory findings and the magnitude of any change in the disposition of digoxin has usually been small. Effects on the renal clearance of digoxin have usually not been investigated, but based on effects on the total clearance of digoxin (when they occur) it could be concluded that these other calcium antagonists do not alter digoxin renal clearance (RODIN and JOHNSON 1988).

The mechanism for the renal tubular secretion of digoxin remains unknown. Studies in animals have indicated that digoxin is not secreted by the organic cation or anion transport system. There was also no evidence to indicate that binding to the antiluminal Na^+/K^+-ATPase alters the renal clearance of digoxin (KOREN 1987).

D. Tubular Reabsorption

I. Proximal Tubule Site

1. Lithium

Lithium is used for the management of manic depressive psychoses, and has a relatively narrow therapeutic index in which toxicity is manifested by impaired consciousness, disorientation, ataxia, tremor, muscle twitches and vomiting. These adverse effects usually occur at plasma concentrations greater than 2 mmol/l. The kidney is the major clearance organ for the elimination of lithium. After filtration of lithium and sodium at the glomerulus, approximately 70% of the filtered load of sodium and lithium is reabsorbed at the proximal tubule, which does not distinguish between these two inorganic ions. Thus, the renal clearance of lithium is similar and parallels that of sodium and the ratio of the renal clearance of lithium to that of creatinine is approximately 0.2. A number of drugs alter the clearance of lithium through altering its renal clearance. These include non-steroidal anti-inflammatory drugs and thiazide diuretics, which reduce the renal clearance of lithium, and sodium salts and theophylline, which increase the renal clearance of lithium.

a) Theophylline

In ten healthy subjects with normal renal function administered lithium carbonate 900 mg daily for 7 days, the addition of theophylline to achieve steady-state plasma concentrations of between 5 and 12 mg/l significantly ($P < 0.05$) increased the renal clearance of lithium by an average of 42% from 18.3 ± 7.0 to 26.1 ± 11.5 ml/min and shortened the half-life from 28.1 ± 7.5 to 21.8 ± 6.6 h (PERRY et al. 1984). There was also a significant decrease in serum lithium concentrations by an average of 30%. This interaction was dependent on the plasma theophylline concentration, as there was a statistically significant ($P < 0.05$) linear relationship between the plasma theophylline concentration and the percentage change in lithium clearance. The incidence of side effects (restlessness, tremor and anorexia) has been reported to be significantly greater when theophylline was co-administered to patients (COOK et al. 1985). In a single case report of a patient with mania treated with lithium, the addition of theophylline resulted in a rapid onset of relapse of mania. When the dosage of lithium was increased, serum lithium concentrations increased and the patient's mania was controlled (SIERLES and OSSOWSKI 1982). The mechanism for the increase in the renal clearance of lithium by theophylline is unknown. Theophylline does not alter renal blood flow or glomerular filtration rate and therefore appears to reduce the tubular reabsorption of lithium.

b) *Sodium Salts*

THOMSEN and SCHOU (1968) originally reported that administration of sodium bicarbonate with lithium resulted in a 30% increase in the ratio of lithium renal clearance to creatinine clearance, and DEMERS and HENINGER (1971) reported in six patients with mania, receiving lithium 1200–1800 mg/day, that when a high-sodium diet (120–250 mmol/day) was compared to a low-sodium (69–91 mmol/day) diet, there was a decline in the serum lithium concentration from a mean of 0.99 to 0.83 mmol/l, and an increase in manic effects and behaviour ratings. In a case report, a male with depressive illness was prescribed 250 mg lithium carbonate four times daily to achieve a target lithium serum concentration of 0.5 mmol/l. When the frequency of dosage was increased to six times a day his serum lithium concentration was still below 0.6 mmol/l as the patient was also taking sodium bicarbonate. When this was ceased, serum lithium concentrations reached 0.8 mmol/l on the initial dosage (ARTHUR 1975). In a subsequent case study, McSWIGGAN (1978) reported on several patients who failed to achieve and maintain therapeutic serum lithium concentrations. It was subsequently revealed that a nurse had been administering a proprietary saline drink for upset stomachs, which contained about 50% sodium bicarbonate. When this was ceased, two patients developed lithium toxicity. Other studies have shown similar effects in that intake of sodium salts reduces serum lithium concentrations, resulting in a reduced pharmacodynamic effect of lithium (DEMERS and HENINGER 1971).

The mechanism for this interaction is consistent with the known mechanism for the renal elimination of lithium. When plasma sodium concentrations are elevated through the addition of sodium-containing preparations, extracellular fluid is expanded and the kidney reduces its reabsorption of sodium and, as a consequence, lithium, resulting in increased renal clearance and reduced plasma concentrations. This interaction is clinically important and has implications for the management of mania in patients.

c) *Diuretics*

α) Thiazide Diuretics

There has been no systematic pharmacokinetic study investigating the interaction between lithium and thiazide diuretics. Nevertheless, there have been a number of clinically important case reports, which have highlighted a potentially serious interaction between these drugs in which thiazide diuretics increase serum lithium concentrations. For example, in 22 patients receiving hydroflumethiazide or bendrofluazide, the renal clearance of lithium was significantly ($P < 0.05$) reduced from a mean of 20 ml/min to 15 ml/min (PETERSEN et al. 1974) and in a patient treated with lithium carbonate, addition of 500 mg chlorothiazide daily raised the serum lithium concentration from 1.3 to 2.0 mmol/l (LEVY et al. 1973). HIMMELHOCH et al. (1977) showed an association between the percentage reduction in calculated lithium renal clearance and the dose of chlorothiazide (500 mg/day, a 40% decrease, 1000 mg/day, a 68% decrease). KERRY et al. (1980) reported on a patient stabilized on lithium (serum concentration 0.9–1.2 mmol/l) in which the addition of bendrofluazide increased the serum lithium concentration to 2.4 mmol/l and in four normal healthy subjects administered 50 mg/day hydrochlorothiazide there was a significant ($P < 0.01$) increase in serum lithium concentrations from 0.4 to 0.53 mmol/l (JEFFERSON and KALIN 1979).

The mechanism for this interaction is still not completely understood. However, the interaction would appear to be of a long-term nature rather than short term, as in a study in healthy subjects a single dose of bendrofluazide had no effect on the renal excretion of lithium (THOMSEN and SCHOU 1968). It has been proposed (STOCKLEY 1994) that thiazide diuretics, although initially eliciting an increase in sodium excretion through their major effect in the distal tubule, upon chronic administration the proximal tubule compensates for the loss of sodium by retaining sodium. As both sodium and lithium are co-reabsorbed, lithium would also be retained within the body, a result of the reduction in renal clearance and therefore increased plasma concentrations.

β) Loop Diuretics

There have been no comprehensive studies investigating potential interactions between loop diuretics and lithium, although there have been a number of case reports and some studies in normal healthy subjects. In five normal healthy subjects administered 300 mg lithium three times daily, resulting in steady-state serum concentrations of 0.4 mmol/l, addition of 40 mg frusemide daily for 14 days produced no lithium-induced adverse effects or increase in serum lithium concentrations in five of six subjects. However, one subject withdrew from the study because of toxicity which was associated with a 61% increase in serum lithium concentration (from 0.44 to 0.71 mmol/l; JEFFERSON and KALIN 1979). In a case report bumetanide has also been associated with the development of lithium toxicity in a patient (KERRY et al. 1980), in which the serum lithium concentration increased from between 0.8 and 1.1 mmol/l to 1.6 mmol/l.

The mechanism for this interaction is likely to be similar to that described for the thiazide diuretics.

d) Non-steroidal Anti-inflammatory Drugs

Several clinical studies and case reports have demonstrated increases in serum lithium concentrations with the use of various but not all non-steroidal anti-inflammatory drugs.

α) Indomethacin

Although published in abstract from (LEFTWICH et al. 1978), it was reported that in five healthy subjects taking 300–600 mg lithium carbonate daily, addition of indomethacin 50 mg three times a day significantly reduced the renal clearance of lithium by an average of 31% with a significant ($P < 0.05$) 43% increase in serum lithium concentrations. The urinary excretion of the metabolite of prostaglandin (PGE_2) synthesis was reduced by indomethacin. In ten normal healthy female subjects on a balanced sodium diet, addition of 150 mg indomethacin daily to lithium sulphate (330 mg twice daily) significantly increased the serum lithium concentration from 0.68 ± 0.07 mmol/l to 0.84 ± 0.13 mmol/l (REIMANN et al. 1983). This was accompanied by a significant decrease in the renal clearance of lithium (from 26 ± 7 ml/min to 19 ± 5 ml/min; $P < 0.001$) and urinary PGE_2 excretion was reduced from 298 ± 46 to 168 ± 89 ng/24 h ($P < 0.005$). Reports of lithium toxicity associated with increased serum lithium concentrations in patients also taking indomethacin have been described (RAGHEB et al. 1980).

β) Ibuprofen

In a study in 11 healthy young (23–38 years) subjects, ibuprofen 1600 mg daily increased the serum lithium concentration (resulting from a dose of 450 mg lithium sulphate twice daily) by an average of 15%, from 0.67 ± 0.12 to 0.77 ± 0.13 mmol/l ($P < 0.001$) (KRISTOFF et al. 1986), with a significant decrease in renal clearance (24 ± 5 to 19 ± 4 ml/min, $n = 5$). In patients maintained on lithium, the addition of 1800 mg ibuprofen increased serum lithium concentrations by between 12% and 66% (RAGHEB 1987a). There have been reports of lithium toxicity in patients receiving ibuprofen (KHAN 1991).

γ) Diclofenac

Five young healthy women, on a standardized sodium diet of 150 mmol/day, were administered lithium as a sustained release tablet of 330 mg twice daily. Diclofenac 50 mg three times daily was also administered (on another occasion) and it significantly ($P = 0.002$) decreased the renal lithium clearance by an average of 23% and significantly ($P < 0.001$) increased serum lithium concentrations by a mean of 26%. Renal prostaglandin synthesis was measured by the urinary excretion of PGE_2 and was decreased by 53% (mean) compared with the control period (213 ± 81 to 103 ± 36 ng/24 h) (REIMANN and FRÖLICH 1981).

δ) Naproxen

In seven patients stabilized on lithium, the addition of 750 mg naproxen resulted in a 16% (range 0%–42%) increase in serum lithium concentrations (from 0.81 ± 0.11 to 0.94 ± 0.08 mmol/l). The renal clearance of lithium decreased from 17 ± 5 to 13 ± 4 ml/min ($P < 0.05$) (RAGHEB and POWELL 1986). One patient developed signs of toxicity and had an increase in the serum lithium concentration from 0.95 to 1.13 mmol/l.

ε) Piroxicam

A case report documented an increase in serum lithium concentrations in a patient also taking piroxicam which resulted in adverse effects (KERRY et al. 1983).

ζ) Phenylbutazone

Although no systematic pharmacokinetic study has been conducted, there has been a case report in which, in six patients with a bipolar affective disorder, treatment with 300 mg phenylbutazone caused a small increase (range 0%–15%) in serum lithium concentrations, which was associated in some patients with side effects (RAGHEB 1990).

η) Sulindac

In six patients requiring lithium therapy at doses ranging from 600 to 900 mg/day, the addition of sulindac 150 mg twice a day did not significantly ($P > 0.05$) alter serum lithium concentrations (0.84 ± 0.07 lithium alone vs. 0.83 ± 0.10 mmol/l with sulindac) (RAGHEB and POWELL 1986).

θ) Aspirin and Salicylate

Six healthy female subjects aged 22–46 years were placed on a strict sodium diet of 150 mmol/day (REIMANN et al. 1985). Lithium was taken as sustained release tablets of 330 mg twice a day to achieve steady-state plasma concentrations of between 0.6 and 0.8 mmol/l. The subjects received aspirin as an intravenous loading dose of 0.5 g followed by 8.8 mg/min for 170 min. Urinary prostaglandin excretion as measured by PGE_2 was significantly reduced by aspirin from 155 ± 95 to 61 ± 28 ng per 3 h ($P < 0.05$). However, there was no significant effect of aspirin on renal blood flow, glomerular filtration rate, plasma lithium concentration or renal clearance of lithium (38 ± 11 to 35 ± 12 ml/min; $P > 0.05$). In the same study, sodium salicylate (1.8 g over 3 h) had no effect on urinary prostaglandin excretion or serum lithium concentration, but significantly reduced the renal clearance of lithium to 29 ± 5 ml/min ($P < 0.05$). In seven patients stabilized on lithium, the addition of 3.9 g aspirin daily had no effect on serum lithium concentrations or renal lithium clearance (RAGHEB 1987b), and in ten normal female subjects, aspirin (1 g four times daily) had no effect on serum lithium concentrations or renal lithium clearance but did reduce renal prostaglandin excretion by between 65% and 70% (REIMANN et al. 1983).

The mechanism for the interaction between non-steroidal anti-inflammatory drugs and lithium has been attributed to inhibition of the synthesis of renal vasodilatory prostaglandins (which has been documented), resulting in a reduction in renal blood flow and glomerular filtration rate and as a consequence, reduction in the renal clearance of lithium. However, it is unclear as to why aspirin has no effect on renal lithium clearance but inhibits renal prostaglandin synthesis. The renal sparing effect of sulindac would be consistent with its lack of interaction with lithium. This interaction is clinically important as a number of reports of lithium toxicity due to non-steroidal anti-inflammatory drugs excluding sulindac and aspirin/salicylate have been documented.

II. Distal Tubule/Collecting Duct Site

1. Urine pH and Flow Rate

Many studies have been conducted to examine the influence of urine pH or flow rate on the renal clearance of drugs via modifying passive reabsorption (see Sect. B.IV). Two examples will be presented, one for organic anions (chlorpropamide) and one for organic cations (pseudoephedrine).

a) Chlorpropamide

Six healthy subjects received a 250-mg oral dose of chlorpropamide on several occasions, once when urine pH was made acidic (pH 4.7–5.5) with ammonium chloride and once when alkaline (pH 7.1–8.2) with sodium bicarbonate (NEUVONEN and KÄRKKÄINEN 1983). The area under the plasma chlorpropamide concentration-time curve was significantly ($P < 0.001$) larger when urine was acidic (3635 ± 1212 mg/l × h) than when alkaline (700 ± 98 mg/l × h), and there were clear differences in half-life from a mean of 13 h under alkaline urine conditions to 69 h when urine was acidic. These substantial differences were attributed to a marked dependence of renal clearance on urine pH. When all the data were combined, providing 144 sets of urine pH and renal clearance data, it was noted that, at a urine pH of 5, renal clearance was 1 ml/h but, when urine pH was 7, renal clearance was 100 ml/h, a two orders of magnitude difference. This is probably the best example of how urine pH can influence the disposition of a drug. For chlorpropamide, an acidic, lipophilic drug with a pK_a of 4.8, by raising urine pH, the degree of ionization is increased thus preventing passive reabsorption and increasing apparent renal clearance; in contrast, when urine pH is acidic, the degree of unionization is increased, resulting in increased passive reabsorption and negligible renal clearance (0.017 ml/min).

b) Pseudoephedrine

In eight subjects, who received a 5-mg/kg oral dose of pseudoephedrine, BRATER et al. (1980) found a strong relationship between the half-life of pseudoephedrine (a surrogate marker of renal clearance) and urine pH, such that at pH 5.8 the half-life was about 300 min and at pH 7.2 it was 600 min. These investigators also found that for each subject, when urine was alkaline (pH >7), there was a significant correlation ($r = 0.71–0.99$) between renal clearance and urine flow. Thus for this lipophilic organic cation with a pK_a of 9.4, its renal clearance is highly dependent on both urine pH and flow rate. The observations found were in agreement with those predicted, thus confirming the mechanism of passive reabsorption and its modulation by the physiological variables of urine pH and flow rate.

Acknowledgement. This review has been supported by the National Health and Medical Research Council of Australia.

References

Ahern M, Booth J, Loxton A, McCarthy P, Meffin P, Kevat S (1988) Methotrexate kinetics in rheumatoid arthritis: is there an interaction with nonsteroidal antiinflammatory drugs? J Rheumatol 15:1356–1360

Aherne GW, Piall E, Marks V, Mould G, White WF (1978) Prolongation and enhancement of serum methotrexate concentrations by probenecid. Br Med J 1:1097–1099

Andreasen F, Sigurd B, Steiness E (1980) Effect of probenecid on excretion and natriuretic action of furosemide. Eur J Clin Pharmacol 18:489–495

Arthur RK (1975) Lithium levels and "Soda BIC". Med J Aust 2:918

Baber N, Halliday L, Sibeon R, Littler T, Orme M L'E (1978) The interaction between indomethacin and probenecid. Clin Pharmacol Ther 24:298–307

Bauer LA, Black D, Gensler A (1990) Procainamide-cimetidine drug interaction in elderly male patients. J Am Geriatr Soc 38:467–469

Belz GG, Doering W, Munkes R, Matthews J (1983) Interaction between digoxin and calcium antagonists and antiarrhythmic drugs. Clin Pharmacol Ther 33:410–417

Bendayan R, Sullivan JT, Shaw C, Frecker RC, Sellers EM (1990) Effect of cimetidine and ranitidine on the hepatic and renal elimination of nicotine in humans. Eur J Clin Pharmacol 38:165–169

Beyer KJ, Russo HF, Tillson EK, Miller AK, Verwey WF, Gass SR (1951) "Benemid", p-(di-n-propylsulfamyl)-benzoic acid: its renal affinity and its elimination. Am J Physiol 166:625–640

Bloom EJ, Ignoffo RJ, Reis CA, Cadman E (1986) Delayed clearance (CL) of methotrexate (MTX) associated with antibiotics and antiinflammatory agents. Clin Res 34:560A

Bonde J, Pedersen LE, Nygaard E, Ramsing T, Angelo HR, Kampmann JP (1991) Stereoselective pharmacokinetics of disopyramide and interaction with cimetidine. Br J Clin Pharmacol 31:708–710

Borgå O, Larsson R, Lunell E (1986) Effects of probenecid on enprofylline kinetics in man. Eur J Clin Pharmacol 30:221–223

Bourke RS, Cheda G, Bremer A, Watanabe O, Tower DB (1975) Inhibition of renal tubular transport of methotrexate by probenecid. Cancer Res 35:110–116

Brater DC, Kaojarern S, Benet LZ, Lin ET, Lockwood T, Morris RC, McSherry EJ, Melmon KL (1980) Renal excretion of pseudoephedrine. Clin Pharmacol Ther 28:690–694

Brown GR (1993) Cephalosporin-probenecid drug interactions. Clin Pharmacokinet 24:289–300

Burnell JM, Kirby WMM (1951) Effectiveness of a new compound, benemid, in elevating serum penicillin concentrations. J Clin Invest 30:697–700

Chatton JY, Munafo A, Chave JP, Steinhäuslin F, Roch-Ramel F, Glauser MP, Biollaz J (1992) Trimethoprim, alone or in combination with sulphamethoxazole, decreases the renal tubular secretion of zidovudine and its glucuronide. Br J Clin Pharmacol 34:551–554

Christian CD, Meredith CG, Speeg KV (1984) Cimetidine inhibits renal procainamide clearance. Clin Pharmacol Ther 36:221–227

Cook BL, Smith RE, Perry PJ, Calloway RA (1985) Theophylline-lithium interaction. J Clin Psychiatry 46:278–279

Corvaia L, Li SC, Ioannides-Demos LL, Bowes G, Spicer WJ, Spelman DW, Tong N, McLean AJ (1992) A prospective study of the effects of oral probenecid on the pharmacokinetics of intravenous ticarcillin in patients with cystic fibrosis. J Antimicrob Chemother 30:875–878
De Miranda P, Good SS, Yarchoan R, Thomas RV, Blum MR, Myers CE, Broder S (1989) Alteration of zidovudine pharmacokinetics by probenecid in patients with AIDS or AIDS-related complex. Clin Pharmacol Ther 46:494–500
Demers RG, Heninger GR (1971) Sodium intake and lithium treatment in mania. Am J Psychiatry 128:132–136
Elion GB, Yü T-F, Gutman AB, Hitchings GH (1968) Renal clearance of oxipurinol, the chief metabolite of allopurinol. Am J Med 45:69–77
Ellison NM, Servi RJ (1985) Acute renal failure and death following sequential intermediate-dose methotrexate and 5-FU: a possible adverse effect due to concomitant indomethacin administration. Cancer Treat Rep 69:342–343
Fenster PE, White NW, Hanson CD (1985) Pharmacokinetic evaluation of the digoxin-amiodarone interaction. J Am Coll Cardiol 5:108–112
Furst DE, Herman RA, Koehnke R, Ericksen N, Hash L, Riggs CE, Porras A, Veng-Pedersen P (1990) Effect of aspirin and sulindac on methotrexate clearance. J Pharm Sci 79:782–786
Gaudry SE, Sitar DS, Smyth DD, McKenzie JK, Aoki FY (1993) Gender and age as factors in the inhibition of renal clearance of amantadine by quinine and quinidine. Clin Pharmacol Ther 54:23–27
Gilman AG, Rall TW, Nies AS, Taylor P (eds) (1990) The pharmacological basis of therapeutics, 8th edn. Pergamon, New York
Hager WD, Fenster P, Mayersohn M, Perrier D, Graves P, Marcus FI, Goldman S (1979) Digoxin-quinidine interaction. N Engl J Med 300:1238–1241
Hardy BG, Zador IT, Golden L, Lalka D, Schentag JJ (1983) Effect of cimetidine on the pharmacokinetics and pharmacodynamics of quinidine. Am J Cardiol 52:172–175
Hedaya MS, Elmquist WF, Sawchuk RJ (1990) Probenecid inhibits the metabolic and renal clearances of zidovudine (AZT) in human volunteers. Pharm Res 7:411–417
Hedman A, Angelin B, Arvidsson A, Dahlqvist R, Nilsson B (1990) Interactions in the renal and biliary elimination of digoxin: stereoselective difference between quinine and quinidine. Clin Pharmacol Ther 47:20–26
Hedman A, Angelin B, Arvidsson A, Beck O, Dahlqvist R, Nilsson B, Olsson M, Schenck-Gustafsson K (1991) Digoxin-verapamil interaction: reduction of biliary but not renal digoxin clearance in humans. Clin Pharmacol Ther 49:256–262
Hedman A, Angelin B, Arvidsson A, Dahlqvist R (1992) Digoxin-interactions in man: spironolactone reduces renal but not biliary digoxin clearance. Eur J Clin Pharmacol 42:481–485
Hendel J, Nyfors A (1984) Nonlinear renal elimination kinetics of methotrexate due to saturation of renal tubular reabsorption. Eur J Clin Pharmacol 26:121–124
Higbee MD, Wood JS, Mead RA (1984) Procainamide-cimetidine interaction: a potential toxic interaction in the elderly. J Am Geriatr Soc 32:162–164
Himmelhoch JM, Poust RI, Mallinger AG, Hanin I, Neil JF (1977) Adjustment of lithium dose during lithium-chlorothiazide therapy. Clin Pharmacol Ther 22:225–227
Homeida M, Roberts C, Branch RA (1977) Influence of probenecid and spironolactone on furosemide kinetics and dynamics in man. Clin Pharmacol Ther 22:402–409
Honari J, Blair AD, Cutler RE (1977) Effects of probenecid on furosemide kinetics and natriuresis in man. Clin Pharmacol Ther 22:395–401
Hooymans PM, Merkus FWMH (1978) Effect of quinidine on plasma concentration of digoxin. Br Med J 2:1022
Howell SB, Olshen RA, Rice JA (1979) Effect of probenecid on cerebrospinal fluid methotrexate kinetics. Clin Pharmacol Ther 26:641–646

Inotsume N, Nishimura M, Nakano M, Fujiyama S, Sato T (1990) The inhibitory effect of probenecid on renal excretion of famotidine in young, healthy volunteers. J Clin Pharmacol 30:50–56
Israel KS, Black HR, Nelson RL, Brunson MK, Nash JF, Brier GL, Wolney JD (1978) Cinoxacin: pharmacokinetics and the effect of probenecid. J Clin Pharmacol 18:491–499
Jefferson JW, Kalin NH (1979) Serum lithium levels and long-term diuretic use. JAMA 241:1134–1136
Kau ST (1978) Handling of triamterene by the isolated perfused rat kidney. J Pharmacol Exp Ther 206:701–709
Kerry RJ, Ludlow JM, Owen G (1980) Diuretics are dangerous with lithium. Br Med J 2:371
Kerry RJ, Owen G, Michaelson S (1983) Possible toxic interaction between lithium and piroxicam. Lancet I:418–419
Khan IH (1991) Lithium and non-steroidal anti-inflammatory drugs. Br Med J 302:1537–1538
Klotz U, Arvela P, Rosenkranz B (1985) Famotidine, a new H_2-receptor antagonist, does not affect hepatic elimination of diazepam or tubular secretion of procainamide. Eur J Clin Pharmacol 28:671–675
Koren G (1987) Clinical pharmacokinetic significance of the renal tubular secretion of digoxin. Clin Pharmacokinet 13:334–343
Kosoglou T, Rocci ML, Vlasses PH (1988) Trimethoprim alters the disposition of procainamide and N-acetylprocainamide. Clin Pharmacol Ther 44:467–477
Kristoff CA, Hayes PE, Barr WH, Small RE, Townsend RJ, Ettigi PG (1986) Effect of ibuprofen on lithium plasma and red blood cell concentrations. Clin Pharm 5:51–55
Lai MY, Jiang FM, Chung CH, Chen HC, Chao PDL (1988) Dose dependent effect of cimetidine on procainamide disposition in man. Int J Clin Pharmacol Ther 26:118–121
Laskin OL, de Miranda P, King DH, Page DA, Longstreth JA, Rocco L, Lietman PS (1982) Effects of probenecid on the pharmacokinetics and elimination of acyclovir in humans. Antimicrob Agents Chemother 21:804–807
Leahey EB, Bigger JT, Butler VP, Reiffel JA, O'Connell GC, Scaffidi LE, Rottman JN (1981) Quinidine-digoxin interaction: time course and pharmacokinetics. Am J Cardiol 48:1141–1146
Leftwich AB, Walker LA, Ragheb M, Oates JA, Frölich JC (1978) Inhibition of prostaglandin synthesis increases plasma lithium levels. Clin Res 34:291A
Levy ST, Forrest JN, Heninger GB (1973) Lithium-induced diabetes insipidus: manic symptoms, brain and electrolyte correlates, and chlorothiazide treatment. Am J Psychiatry 130:1014–1018
Liegler DG, Henderson ES, Hahn MA, Oliverio VT (1969) The effect of organic acids on renal clearance of methotrexate in man. Clin Pharmacol Ther 10:849–857
Maier WP, Leon-Perez R, Miller SB (1986) Pneumonia during low-dose methotrexate therapy. Arch Intern Med 146:602–603
May DC, Jarboe CH (1981) Inhibition of clearance of dyphylline by probenecid. N Engl J Med 304:791
Mayall B, Poggi G, Parkin JD (1991) Neutropenia due to low-dose methotrexate therapy for psoriasis and rheumatoid arthritis may be fatal. Med J Aust 155:480–484
McKinney TD, Speeg KV Jr (1982) Cimetidine and procainamide secretion by proximal tubules in vitro. Am J Physiol 242:F672–F680
McSwiggan C (1978) Interaction of lithium and bicarbonate. Med J Aust 1:38
Møller JV, Sheikh MI (1982) Renal organic anion transport system: pharmacological, physiological, and biochemical aspects. Pharmacol Rev 34:315–358
Muirhead MR, Somogyi AA, Rolan PE, Bochner F (1986) Effect of cimetidine on renal and hepatic drug elimination: studies with triamterene. Clin Pharmacol Ther 40:400–407

Muirhead M, Bochner F, Somogyi A (1988) Pharmacokinetic drug interactions between triamterene and ranitidine in humans: alterations in renal and hepatic clearances and gastrointestinal absorption. J Pharmacol Exp Ther 244:734–739
Mungall DR, Robichaux RP, Perry W, Scott JW, Robinson A, Burelle T, Hurst D (1980) Effects of quinidine on serum digoxin concentration. Ann Intern Med 93:689–693
Nademanee K, Kannan R, Hendrickson J, Ookhtens M, Kay I, Singh BN (1984) Amiodarone-digoxin interaction: clinical significance, time course of development, potential pharmacokinetic mechanisms and therapeutic implications. J Am Coll Cardiol 4:111–116
Neuvonen PJ, Kärkkäinen S (1983) Effects of charcoal, sodium bicarbonate and ammonium chloride on chlorpropamide kinetics. Clin Pharmacol Ther 33:386–393
Nirenberg A, Mosende C, Mehta BM, Gisolfi AL, Rosen G (1977) High-dose methotrexate with citrovorum factor rescue: predictive value of serum methotrexate concentrations and corrective measures to avert toxicity. Cancer Treat Rep 61:779–783
Overbosch D, Van Gulpen C, Hermans J, Mattie H (1988) The effect of probenecid on the renal tubular excretion of benzylpenicillin. Br J Clin Pharmacol 25:51–58
Pedersen KE, Dorph-Pedersen A, Hvidt S, Klitgaard NA, Nielsen-Kudsk F (1981) Digoxin-verapamil interaction. Clin Pharmacol Ther 30:311–316
Pedersen KE, Dorph-Pedersen A, Hvidt S, Klitgaard NA, Pedersen KK (1982) The long-term effect of verapamil on plasma digoxin concentration and renal digoxin clearance in healthy subjects. Eur J Clin Pharmacol 22:123–127
Perel JM, Dayton PG, Snell MM, Yü TF, Gutman AB (1969) Studies of interactions among drugs in man at the renal level: probenecid and sulfinpyrazone. Clin Pharmacol Ther 10:834–840
Perry PJ, Calloway RA, Cook BL, Smith RE (1984) Theophylline precipitated alterations of lithium clearance. Acta Psychiatr Scand 69:528–537
Peters L (1960) Renal tubular excretion of organic bases. Pharmacol Rev 12:1–35
Petersen V, Hvidt S, Thomsen K, Schou M (1974) Effect of prolonged thiazide treatment on renal lithium clearance. Br Med J 2:143–145
Pitkin D, Dubb J, Actor P, Alexander F, Ehrlich S, Familiar R, Stote R (1981) Kinetics and renal handling of cefonicid. Clin Pharmacol Ther 30:587–593
Ragheb MA (1987a) Ibuprofen can increase serum lithium level in lithium-treated patients. J Clin Psychiatry 48:161–163
Ragheb MA (1987b) Aspirin does not significantly affect patients' serum lithium levels. J Clin Psychiatry 48:425
Ragheb M (1990) The interaction of lithium with phenylbutazone in bipolar affective patients. J Clin Psychopharmacol 10:149–150
Ragheb M, Powell AL (1986) Lithium interaction with sulindac and naproxen. J Clin Psychopharmacol 6:150–154
Ragheb M, Ban TA, Buchanan D (1980) Interaction of indomethacin and ibuprofen with lithium in manic patients under a steady-state lithium level. J Clin Psychiatry 41:397–398
Reimann IW, Frölich JC (1981) Effects of diclofenac on lithium kinetics. Clin Pharmacol Ther 30:348–352
Reimann IW, Diener U, Frölich JC (1983) Indomethacin but not aspirin increases plasma lithium ion levels. Arch Gen Psychiatry 40:283–286
Reimann IW, Golbs E, Fischer C, Frölich JC (1985) Influence of intravenous acetylsalicylic acid and sodium salicylate on human renal function and lithium clearance. Eur J Clin Pharmacol 29:435–441
Rennick BR (1972) Renal excretion of drugs: tubular transport and metabolism. Annu Rev Pharmacol Toxicol 12:141–156
Rennick BR (1981) Renal tubule transport of organic cations. Am J Physiol 240:F83–F89
Roberts DH, Kendall MJ, Jack DB, Welling PG (1981) Pharmacokinetics of cephradine given intravenously with and without probenecid. Br J Clin Pharmacol 11:561–564

Rodin SM, Johnson BF (1988) Pharmacokinetic interactions with digoxin. Clin Pharmacokinet 15:227–244
Rodriguez N, Madsen PO, Welling PG (1979) Influence of probenecid on serum levels and urinary excretion of cinoxacin. Antimicrob Agents Chemother 15:465–469
Rodvold KA, Paloucek FP, Jung D, Gallastegui J (1987) Interaction of steady-state procainamide with H_2-receptor antagonists cimetidine and ranitidine. Ther Drug Monit 9:378–383
Sand TE, Jacobsen S (1981) Effect of urine pH and flow on renal clearance of methotrexate. Eur J Clin Pharmacol 19:453–456
Schenck-Gustafsson K, Dahlqvist R (1981) Pharmacokinetics of digoxin in patients subjected to the quinidine-digoxin interaction. Br J Clin Pharmacol 11:181–186
Sierles FS, Ossowski MG (1982) Concurrent use of theophylline and lithium in a patient with chronic obstructive lung disease and bipolar disorder. Am J Psychiatry 139:117–118
Singh RR, Malaviya AN, Pandey JN, Guleria JS (1986) Fatal interaction between methotrexate and naproxen. Lancet I:1390
Somogyi A, Bochner F (1984) Dose and concentration dependent effect of ranitidine on procainamide disposition and renal clearance in man. Br J Clin Pharmacol 18:175–181
Somogyi A, McLean A, Heinzow B (1983) Cimetidine-procainamide pharmacokinetic interaction in man: evidence of competition for tubular secretion of basic drugs. Eur J Clin Pharmacol 25:339–345
Somogyi A, Stockley C, Keal J, Rolan P, Bochner F (1987) Reduction of metformin renal tubular secretion by cimetidine in man. Br J Clin Pharmacol 23:545–551
Somogyi AA, Hovens CM, Muirhead MR, Bochner F (1989) Renal tubular secretion of amiloride and its inhibition by cimetidine in humans and in an animal model. Drug Metab Dispos 17:190–196
Somogyi AA, Bochner F, Sallustio BC (1992) Stereoselective inhibition of pindolol renal clearance by cimetidine in humans. Clin Pharmacol Ther 51:379–387
St Clair EW, Rice JR, Snyderman R (1985) Pneumonitis complicating low-dose methotrexate therapy in rheumatoid arthritis. Arch Intern Med 145:2035–2038
Steiness E (1974) Renal tubular secretion of digoxin. Circulation 50:103–107
Stewart CF, Fleming RA, Magneson P, Germain B, Evans WE (1990a) Effect of aspirin (ASA) on the disposition of methotrexate (MTX) in patients with rheumatoid arthritis (RA). Clin Pharmacol Ther 47:139
Stewart CF, Fleming RA, Arkin CR, Evans WE (1990b) Coadministration of naproxen and low-dose methotrexate in patients with rheumatoid arthritis. Clin Pharmacol Ther 47:540–546
Stockley IH (1994) Drug interactions. A source book of adverse interactions, their mechanisms, clinical importance and management, 3rd edn. Blackwell, Oxford
Thomsen K, Schou M (1968) Renal lithium excretion in man. Am J Physiol 215:823–827
Thyss A, Milano G, Kubar J, Namer M, Schneider M (1986) Clinical and pharmacokinetic evidence of a life-threatening interaction between methotrexate and ketoprofen. Lancet I:256–258
Tracy TS, Krohn K, Jones DR, Bradley JD, Hall SD, Brater DC (1992) The effects of a salicylate, ibuprofen, and naproxen on the disposition of methotrexate in patients with rheumatoid arthritis. Eur J Clin Pharmacol 42:121–125
Ullrich KJ (1994) Specificity of transporters for "organic anions" and "organic cations" in the kidney. Biochim Biophys Acta 1197:45–62
Ullrich KJ, Fritzsch G, Rumrich G, David C (1994) Polysubstrates: substances that interact with renal contraluminal PAH, sulfate, and NMeN transport: sulfamoyl-sulfonylurea-, thiazide- and benzeneamino-carboxylate (nicotinate) compounds. J Pharmacol Exp Ther 269:684–692
Van Crugten J, Bochner F, Keal J, Somogyi A (1986) Selectivity of the cimetidine-induced alterations in the renal handling of organic substrates in humans. Studies with anionic, cationic and zwitterionic drugs. J Pharmacol Exp Ther 236:481–487

Veenendaal RJ, Brooks PM, Meffin PJ (1981) Probenecid-clofibrate interaction. Clin Pharmacol Ther 29:351–358
Vlasses PH, Kosoglou T, Chase SL, Greenspon AJ, Lottes S, Andress E, Ferguson RK, Rocci ML (1989) Trimethoprim inhibition of the renal clearance of procainamide and N-acetylprocainamide. Arch Intern Med 149:1350–1353
Waldorff S, Andersen JD, Heebøll-Nielsen N, Nielsen OG, Moltke E, Sørensen U, Steiness E (1978) Spironolactone-induced changes in digoxin kinetics. Clin Pharmacol Ther 24:162–167
Waldorff S, Hansen PB, Kjaergård H, Buch J, Egeblad H, Steiness E (1981) Amiloride-induced changes in digoxin dynamics and kinetics: abolition of digoxin-induced inotropism with amiloride. Clin Pharmacol Ther 30:172–176
Waldorff S, Hansen PB, Egeblad H, Berning J, Buch J, Kjaergård H, Steiness E (1983) Interactions between digoxin and potassium-sparing diuretics. Clin Pharmacol Ther 33:418–423
Waller ES, Sharanevych MA, Yakatan GJ (1982) The effect of probenecid on nafcillin disposition. J Clin Pharmacol 22:482–489
Weidekamm E, Portmann R, Suter K, Partos C, Dell D, Lücker PW (1987) Single- and multiple-dose pharmacokinetics of fleroxacin, a trifluorinated quinolone, in humans. Antimicrob Agents Chemother 31:1909–1914
Weiner IM, Mudge GH (1964) Renal tubular mechanisms for excretion of organic acids and bases. Am J Med 36:743–762
Windle J, Prystowsky EN, Miles WM, Heger JJ (1987) Pharmacokinetic and electrophysiologic interactions of amiodarone and procainamide. Clin Pharmacol Ther 41:603–610
Wingender W, Beermann D, Foster D (1985) Mechanism of renal excretion of ciprofloxacin (Bay 9867), a new quinolone carboxylic derivative, in humans. Chemotherapy 4 [Suppl 2]:403–404

Section II
Pharmacodynamic Drug Interactions

CHAPTER 7

Drug-Drug Interactions at Receptors and Other Active Sites

M. SCHORDERET and J.D. FERRERO

A. Introduction

Pharmacodynamic interactions occur at receptor sites or result more broadly from the effect of the drugs on physiological processes. Both antagonism and synergism may be produced. The synergistic interactions, as far as they have the potential to be beneficial and clinically useful, are reviewed separately in this handbook (see Chap. 8).

The present chapter will concentrate on adverse drug interactions resulting in either a diminished therapeutic effect or an increased toxicity. Its aim is to describe the mechanisms by which pharmacodynamic interactions may occur and to illustrate them with clinically relevant examples. It is not a compilation of all possible and known interactions, duplicating standard publications (SCHOU 1982; MCINNES and BRODIE 1988; STOCKLEY 1991; BUGNON et al. 1993; HANSTEN et al. 1993). Accordingly, the classification adopted here is based on sites of action and not on drug classes. Different steps in the same process, e.g., synaptic transmission, or different mechanisms taking part in the regulation of one system or function, e.g., blood pressure, may be affected.

Since they are based on mechanisms of action supposedly known to the prescriptor, pharmacodynamic interactions should be more easily prevented and should have less clinical impact than pharmacokinetic interactions. However, the relevance of such a statement may be questioned in view of the recent advances in the characterization and detection of kinetic interactions.

B. Mechanisms of Pharmacodynamic Interactions

I. Transmitter Systems

1. Noradrenergic Synapse

The various drugs interfering with practically every step of the noradrenergic transmission represent a well-established source of pharmacodynamic interactions (Fig. 1). In the CNS, where adrenaline is a neurotransmitter on its own

(MEFFORD 1988; COOPER et al. 1991), similar interactions can occur. The specific plasma membrane noradrenaline transporter has been cloned and characterized (PACHOLCZYK et al. 1991; USDIN et al. 1991). It plays a role in the mechanism of action and potential interactions of amphetamine and tyramine, which stimulate the outward transport of noradrenaline, as well as of cocaine and imipramine, which block the inward transport (reuptake) of noradrenaline (Fig. 1).

Possible interactions at postsynaptic receptors are not shown in Fig. 1, since they can easily be anticipated. Patients on β-blockers are particularly at risk (Table 1). For example, the blockade of β-receptors makes a patient resistant to adrenaline in the emergency treatment of anaphylaxis. In this

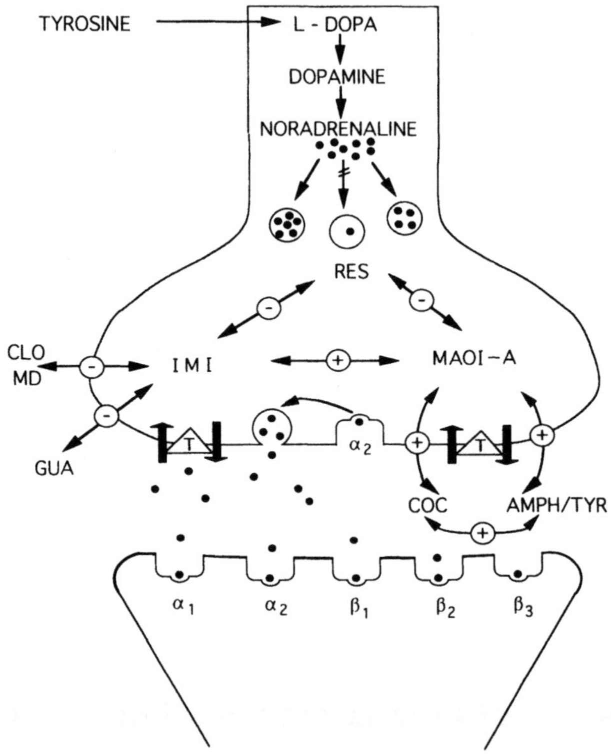

Fig. 1. Noradrenergic synapse. Pharmacodynamic interactions at presynaptic sites and transporters (*T*). *AMPH*, amphetamine; *COC*, cocaine; *CLO*, clonidine; *GUA*, guanethidine; *IMI*, imipramine and imipramine-like antidepressants; *MAOI-A*, monoamine oxidase inhibitor (type A); *MD*, α-methyldopa; *RES*, reserpine; *TYR*, tyramine. The localization of drugs (inside or outside the presynaptic terminal) is purely arbitrary. For the sake of clarity, all subtypes of receptors (α and β) are gathered together on the same postsynaptic membrane. *Symbols* ⊕ *and* ⊖ *on the bidirectional arrows* represent mutual potentiation and inhibition, respectively, of the effects of the compounds *at both ends of the arrows* (see text for additional explanations)

situation, glucagon would still be effective as an inotropic agent (BOCHNER and LICHTENSTEIN 1991). Similarly, the early symptoms of hypoglycaemia may be masked in a diabetic patient treated with insulin or oral hypoglycaemic agents (Table 1; see also Sect. B.III.2). Unopposed α-adrenergic stimulation is the probable mechanism of the increased cardiovascular toxicity (hypertension, coronary constriction, cf. Table 1) of cocaine (LANGE et al. 1990).

As expected, drugs facilitating the noradrenergic transmission [amphetamine, cocaine, monoamine oxidase inhibitors (MAOIs) and imipramine-like antidepressants] will potentiate the effects of α_1- or $\beta_{1/2}$-adrenergic receptor agonists. Indeed, the dosage of adrenaline in the treatment of anaphylaxis

Table 1. Pharmacodynamic interactions with β-blockers

Associated drugs	Mechanisms	Clinical consequences
α_1-Agonists and indirect sympathomimetics (adrenaline, phenylephrine, etilefrine, dopamine, cocaine)	α_1-Receptor-induced vasoconstriction unopposed by β_2-receptor-induced vasodilation	Hypertensive crises, coronary vasoconstriction
Clonidine withdrawal	Increased sympathetic activity: α_1-receptor-induced vasoconstriction unopposed by β_2-receptor-induced vasodilation	Increased rebound hypertension
Ergotamine	α_1- and 5-HT_2-receptor-induced vasoconstriction, unopposed by β_2-receptor-induced vasodilation	Peripheral vasospasm
α_1-Antagonists (prazosin)	Potentiation of the "first-dose effect" of prazosin	Orthostatic hypotension
Digoxin	Additive slowing of conduction	Bradycardia
Disopyramide	Additive cardiodepressant activity	Bradycardia, heart block, hypotension
Verapamil, diltiazem	Additive cardiodepressant activity	Bradycardia, heart block, hypotension, left ventricular failure
NSAID	Inhibition of cyclooxygenases	Decreased antihypertensive effect of β-blockers
Insulin[a]	Hypoglycaemia; masking of hypoglycaemia symptoms	Glycaemia control in diabetic patients made more difficult

NSAID, non-steroidal anti-inflammatory drugs; see Sect. B.IV.2.
[a] See Sect. B.III.2.

should be strongly reduced in a patient receiving tricyclic antidepressants (STOCKLEY 1991). Similarly, sympathomimetic OTC preparations containing, for example, phenylpropanolamine or imidazolines as systemic or topical nasal decongestants can cause potentially serious reactions (CETARUK and AARON 1994). In contrast, the indirect sympathomimetic agents mentioned above will attenuate the effects of sympatholytic drugs such as clonidine and α-methlydopa (agonists at central α_2-adrenergic receptors that inhibit the sympathetic tone), reserpine (a blocker of the vesicular catecholamine transporter) and guanethidine. Since the latter agent must be taken up in the nerve terminal through the membrane noradrenaline transporter in order to be active, it will lose its antihypertensive effect in the presence of imipramine (Fig. 1) or other drugs that block the reuptake of noradrenaline (MICHELL et al. 1970). The inhibition by imipramine of the antihypertensive effect of clonidine or α-methyldopa is based on a different mechanism, possibly a direct interaction at central α_2-receptors.

It has been known for several decades that the combination of MAOIs with foods and beverages containing tyramine may induce the "cheese reaction", characterized by hypertension, headache, sweating, chest pain and even cerebral haemorrhage and death (BLACKWELL and MARLEY 1966). This reaction results from tyramine getting free access, in the absence of MAO activity, to the adrenergic nerve terminals and varicosities, where it stimulates noradrenaline release. It appears, however, that the risk of serious reactions with dietary tyramine is much reduced with new selective and reversible MAOIs-A such as moclobemide (BERLIN et al. 1989; FITTON et al. 1992). Since tyramine is a substrate for MAO-A, the reversibility of MAO inhibition may be a more determinant property than the selectivity (GIESCHKE et al. 1988). Moreover, the simultaneous administration of older irreversible MAOIs with tricyclic antidepressants has been shown to produce a potentially lethal central and cardiovascular syndrome, which is likely to be due to increased amine concentrations at postsynaptic receptors (SPINA and PERUCCA 1994). The association of moclobemide with tricyclic antidepressants, if the treatment is initiated with slowly increasing doses of both drugs, or if imipramine is started before moclobemide, is probably safe, but remains to be fully characterized (FITTON et al. 1992; SPINA and PERUCCA 1994). Finally, possible pharmacodynamic interactions between MAOIs and pethidine (meperidine) or dextromethorphane are described in Sect. B.I.3.

In considering the numerous possible interactions of tricyclic antidepressants, it should be remembered that besides blocking amine membrane transporters, they possess additional (and to some extent paradoxical) antagonistic properties at α_1-adrenergic, H_1-histamine and muscarinic receptors (SUGRUE 1981; REMICK 1988), and that they exert quinidine-like effects on the membrane of excitable cells (GLASSMAN and BIGGER 1981). In this respect, the "second-generation" antidepressants, particularly the serotonin-selective reuptake inhibitors (SSRIs, see Sect. B.I.3), appear to be much safer.

2. Dopaminergic Synapse

Dopamine is a neurotransmitter in the CNS, and as a circulating catecholamine released from the adrenals it also plays multiple roles in the periphery (HORN and MURPHY 1991). In addition, brain dopamine receptors are the primary target in the treatment of schizophrenia, Parkinson's disease and Huntington's chorea (SEEMAN and VAN TOL 1994). A better knowledge of the processes involved in the dopaminergic transmission, as well as of the pharmacological characteristics of the subtypes of receptors (D_1 to D_5), may help to understand and possibly to avoid drug-drug interactions (Fig. 2).

The interaction between L-dopa and pharmacological doses of pyridoxine (vitamin B_6) is due to an increased peripheral decarboxylation of L-dopa by

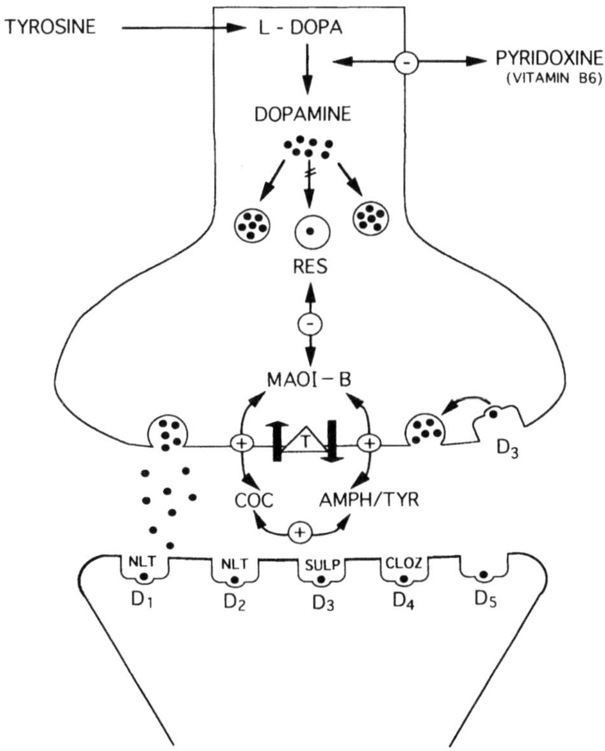

Fig. 2. Dopaminergic synapse. Pharmacodynamic interactions at presynaptic sites and transporter (*T*). *AMPH*, amphetamine; *COC*, cocaine; *MAOI-B*, monoamine oxidase inhibitor (type B); *R*, reserpine; *TYR*, tyramine. The localization of drugs (inside or outside the presynaptic terminal) is purely arbitrary. For the sake of clarity, all subtypes of receptors (D_1–D_5) are gathered together on the same postsynaptic membrane. *CLOZ*, clozapine; *NLT*, typical neuroleptics; *SULP*, sulpiride. *Symbols* ⊕ *and* ⊖ *on the bidirectional arrows* represent mutual potentiation and inhibition, respectively, of the effects of the compounds *at both ends of the arrows* (see text for additional explanations)

the L-amino acid decarboxylase cofactor, which results in a decreased transfer of L-dopa to the CNS and a concomitant reduction in therapeutic efficacy. The interaction does not occur when L-dopa is co-administered with a peripheral inhibitor of the L-amino acid decarboxylase (MARS 1974). The specific plasma membrane transporter of dopamine has been cloned and characterized (GIROS et al. 1991; KILTY et al. 1991; SHIMADA et al. 1991). It is concerned with the mechanism of action and potential interactions of amphetamine and tyramine, which stimulate the outward transport of dopamine, as well as of cocaine, which blocks the inward transport (reuptake) of dopamine (Fig. 2). Possible interactions at postsynaptic receptors are not shown in Fig. 2, since they can easily be anticipated. As expected, the typical neuroleptics represented by chlorpromazine, flupenthixol or haloperidol, which are potent antagonists at D_1 and D_2 receptor subtypes, will inhibit the therapeutic action of the dopamine precursor (L-dopa) and of amantadine, as well as of preferential D_2 agonists (e.g., bromocriptine and analogs) in the treatment of Parkinson's disease. Further interactions with L-dopa can also be expected from the mixed D_2/D_3 antagonist sulpiride. However, D_2 antagonists that exhibit a low incidence of extrapyramidal effects, such as domperidone or metoclopramide, may be of benefit in the treatment of L-dopa- or D_2-agonist-induced emesis (MITCHELSON 1992). Though theoretically founded, interactions implying the D_3- (e.g., selective agonists or antagonists being able to decrease or increase the dopaminergic transmission, respectively) or D_4-receptor subtypes (e.g., the relatively selective D_4 antagonist clozapine) do not yet seem to be clinically relevant.

It should be noted that most neuroleptics, particularly those of the phenothiazine class, may cause additional interactions due, for example, to an antagonism at α-adrenergic and muscarinic receptors (DILSAVER 1993).

3. Serotonergic Synapse

The last decade has been marked by a rapid progress in serotonin (5-hydroxytryptamine, 5-HT) research with the cloning, expression and characterization of numerous subtypes of 5-HT receptors (BOESS and MARTIN 1994), as well as of the specific plasma membrane transporter of serotonin (BLAKELY et al. 1991; HOFFMAN et al. 1991).

The serotonin transporter is concerned with the mechanism of action and possible interactions of amphetamine and 3,4-methylenedioxymethamphetamine (MDMA, "ecstasy"), which stimulate the outward transport of serotonin, as well as of cocaine and the selective serotonin reuptake inhibitors (SSRIs), which block the inward transport of serotonin (Fig. 3). The combined treatment with irreversible MAOIs and SSRIs has been shown to cause a serotonin syndrome (FEIGHNER et al. 1990; GRAM 1994; SPINA and PERUCCA 1994). Moreover, a lethal serotonin syndrome has been reported after the concomitant administration of high-dose moclobemide with citalopram or clomipramine (NEUVONEN et al. 1993; MITCHELL 1994). Thus, the

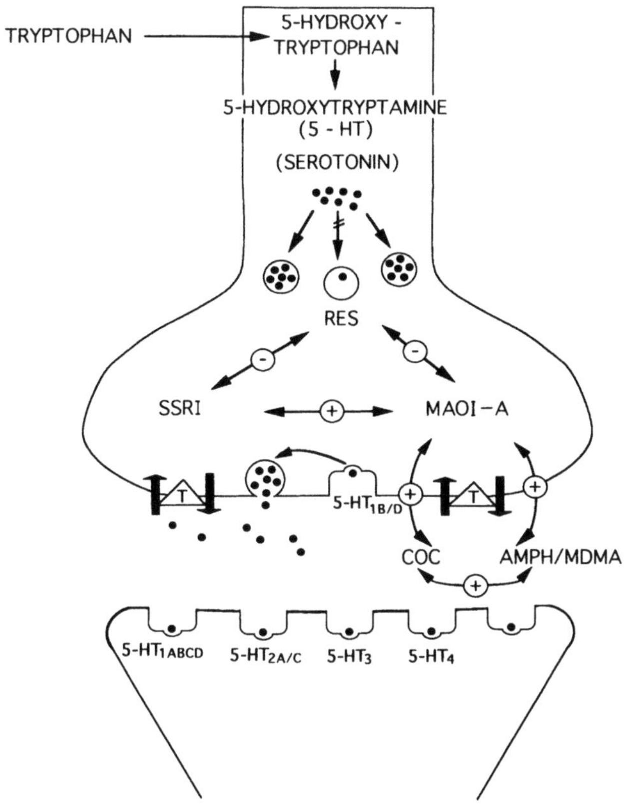

Fig. 3. Serotonergic synapse. Pharmacodynamic interactions at presynaptic sites and transporters (*T*). *AMPH*, amphetamine; *COC*, cocaine; *MDMA*, 3,4-methylenedioxymethamphetamine ("ecstasy"); *MAOI-A*, monomaine oxidase inhibitor (type A); *RES*, reserpine; *SSRI*, selective serotonin reuptake inhibitors, such as citalopram, fluoxetine, fluvoxamine, paroxetine, sertraline and zimeldine. The localization of drugs (inside or outside the presynaptic terminal) is purely arbitrary. For the sake of clarity, different subtypes of receptors (5-HT_1 to 5-HT_4) are gathered together on the same postsynaptic membrane. The unlabelled postsynaptic receptor signals that additional subtypes (5-HT_5 to 5-HT_7), for which ligands remain to be found, have been characterized. *Symbols ⊕ and ⊖ on the bidirectional arrows* represent mutual potentiation and inhibition, respectively, of the effects of the compounds *at both ends of the arrows* (see text for additional explanations)

prospective benefits of combining SSRIs with MAOIs – even with the reversible MAOI-A moclobemide – in the treatment of depression should be seriously weighed against the associated risks (Spina and Perucca 1994).

Like cocaine, the OTC antitussive dextromethorphan, as well as pethidine (meperidine), can also block the neuronal uptake of serotonin (Browne and Linter 1987; Stack et al. 1988; Hill et al. 1992; Cetaruk and Aaron 1994). The association of dextromethorphan or pethidine with MAOIs (including moclobemide) has been shown to produce clinical symptoms very similar to those of the serotonin syndrome (LEJOYEUX et al. 1994), with hypertensive crises, hyperpyrexia, hyperexcitation, muscle rigidity, coma and death (AMREIN et al. 1992; CETARUK and AARON 1994; SPINA and PERUCCA 1994). A serotonin syndrome has also been reported when the 5-HT precursor tryptophan was associated with MAOIs, SSRIs or imipramine-like drugs (LEJOYEUX et al. 1994), as well as during lithium therapy associated with fluvoxamine or fluoxetine (GUY 1992; SPINA and PERUCCA 1994).

Possible pharmacodynamic interactions at the postsynaptic level would result from the erroneous simultaneous administration of agonists and antagonists acting on the same 5-HT receptor subtype (Fig. 3) or from the association of poorly selective or non-selective agonists and antagonists. In view of the increased risk of peripheral vasospasm, for example, ergotamine and sumatriptan should not be administered concomitantly in the therapy of migraine, nor should they be prescribed together with MAOIs or SSRIs (DRUG and THERAPEUTICS BULLETIN 1992).

4. Cholinergic Synapse

In the cholinergic transmission, unlike the monoaminergic transmissions, drugs affecting the synthesis (e.g., choline, lecithins), storage and uptake of the

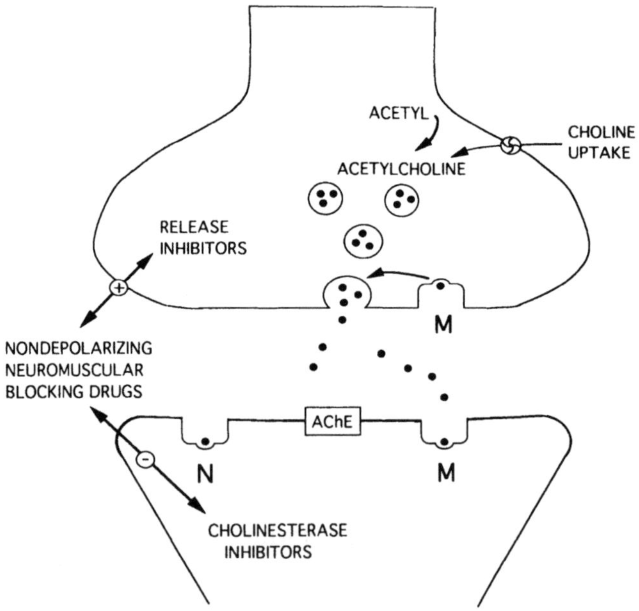

Fig. 4. Cholinergic synapse. Pharmacodynamic interactions of non-depolarizing neuromuscular-blocking drugs with cholinesterase inhibitors and inhibitors of acetylcholine release (see text for details). *Symbols* ⊕ *and* ⊖ *on the bidirectional arrows* represent mutual potentiation and inhibition, respectively, of the effects of the compounds *at both ends of the arrows*. *M*, muscarinic receptors; *N*, nicotinic receptors; *AChE*, acetylcholinesterase

Table 2. Drugs with anticholinergic (antimuscarinic) activity

Therapeutic classes	Examples
Antiarrhythmic drugs	Quinidine
Antiasthmatics	Ipratropium
Antidepressants	Imipramine
Antiemetics (motion sickness)	Scopolamine
Antiepileptics	Carbamazepine
Antiparkinsonians	Trihexyphenidyl
Antiulcer drugs	Pirenzepine
H_1-antihistaminics	Promethazine
Mydriatics	Homatropine
Neuroleptics	Chlorpromazine
Parasympatholytics	Atropine

transmitter have not been found therapeutically useful. A few drugs, however, have been shown to interfere presynaptically with acetylcholine release at the neuromuscular junction (Fig. 4). Apart from the botulinum toxins, some antibiotics (aminoglycosides, polymyxins, lincosamides) (SINGH et al. 1982), quinidine, quinine and procainamide, as well as magnesium salts, play a role in this effect (CLARKE adn MIRAKHUR 1994). The aminoglycosides are not only able to produce weakness on their own or aggravate the muscular symptoms in myasthenia gravis (DRACHMAN 1994), they may also enhance the effect of the neuromuscular blocking agents (CLARKE and MIRAKHUR 1994).

Most interactions occurring at cholinergic receptors involve additive effects between muscarinic receptor blocking drugs, examples of which can be found in many therapeutic classes (Table 2). In the nicotinic domain, additive effects with dangerous cardiovascular consequences have also been reported in patients treated with transdermal nicotine delivery systems who do not stop smoking or who start smoking again (FIORE et al. 1992).

Cholinesterase inhibitors are routinely used in anaesthesiology to reverse curare-induced neuromuscular block (Fig. 4). As the use of cholinesterase inhibitors, notably tacrine, has been proposed in the treatment of Alzheimer's disease, it should be remembered that any agent with antimuscarinic properties which crosses the blood-brain barrier will interact negatively. A more insidious interaction of anticholinergic drugs is the inhibition of the gastrokinetic effect of the 5-HT_4 agonists metoclopramide and cisapride that act by stimulating acetylcholine release in the enteric nervous system (SCHOURKES et al. 1988).

5. GABAergic Synapse

GABA is the main inhibitory transmitter in the mammalian brain. In postsynaptic neurons, it produces either a fast response, through the ligand-gated ion-channel $GABA_A$ receptor, or a slow modulatory response, through the G-protein-coupled $GABA_B$ receptor (MODY et al. 1994). As shown in Table 3, various substances can either facilitate or inhibit the $GABA_A$-mediated increase in Cl^- conductance (SIEGHART 1992).

Table 3. Modulation of GABAergic neurotransmission coupled to $GABA_A$ receptors

Drugs which reinforce GABA-mediated processes	Drugs which inhibit GABA-mediated processes
Benzodiazepines	Flumazenil[a]
Barbiturates	
Gabapentin	Penicillins
Progabide	Quinolones
Vigabatrin	
Ethanol	Bicuculline
General anaesthetics	Picrotoxin
Avermectin insecticides	Cyclodiene insecticides

[a] Benzodiazepine receptor antagonist.

Benzodiazepines, barbiturates and probably ethanol are allosteric modulators that reinforce GABAergic transmission. Potentiation of CNS-depressant effects, with an increased risk of respiratory depression, is a well-known example of pharmacodynamic interaction brought about by the combination of any of these hypnosedative agents. Inversely, the administration of flumazenil, and antagonist at the benzodiazepine site of the $GABA_A$ receptor, will precipitate an acute withdrawal syndrome in a patient physically dependent on benzodiazepines. Convulsions may be triggered, particularly in the epileptic patient treated with benzodiazepines (MARCHANT et al. 1989).

Among other substances of experimental or toxicological interest (Table 3), penicillins in high doses and quinolone antibiotics have been reported to cause CNS stimulation or convulsions, which may be due to the inhibition of GABAergic processes (CURTIS et al. 1972; HALLIWELL et al. 1993). The non-steroidal anti-inflammatory drugs, fenbufen in particular, may increase the incidence of quinolone-induced convulsions. The mechanism involved in this interaction remains unknown (CHRIST et al. 1988).

As a $GABA_B$ receptor agonist, baclofen may increase or decrease the level of synaptic inhibition in the CNS, depending on whether the receptors are located pre- or postsynaptically (MOTT and LEWIS 1994). No clinically important pharmacodynamic interaction has yet been reported with baclofen.

II. Ion Channels

1. Cardiac Ion Channels and Antiarrhythmic Drugs

It is well established that class I antiarrhythmic drugs have sodium channel blocking properties. The affinity for or access to the channel receptor site may depend on specific states of the channel. Most drugs have low affinity for the sodium channel in the resting state and the affinity increases when the channel is in the activated (quinidine) or inactivated (lidocaine) states. Class I agents are further characterized by the association/dissociation kinetics to and from the channel receptor site (WOOSLEY 1991).

The concomitant administration of antiarrhythmic drugs with distinct properties (e.g., class Ia/Ib) may result in either adverse or beneficial interactions (LANGIERI et al. 1995). As an example of the latter, mexiletine (Ib) and quinidine (Ia) in combination have been found to be more effective at a lower dose and to produce fewer adverse effects than each drug given alone in higher dosage (BIGGER 1984; GIARDIA and WESCHSLER 1990). This may be explained by the fact that quinidine-induced prolongation of repolarization prolongs the inactivated state of the sodium channel. As a result, the binding of mexiletine, which possesses higher affinity for the channel in the inactivated state, is increased. Moreover, the competition between "fast" and "slow" sodium channel blockers (including non-antiarrhythmic drugs) may be effective in reversing toxic electrophysiologic effects. Lidocaine has been shown to reverse the QRS prolongation due to cocaine (BAUMAN et al. 1994) or propoxyphene (WHITCOMB et al. 1989) overdoses. Finally, the drugs that cause repolarization abnormalities (see Sect. B.II.2) tend to have greater toxicity at slower heart rate. Increasing heart rate by temporary pacing or isoprenaline infusion is the recommended treatment in this case (BEN-DAVID and ZIPES 1993).

Cardiac calcium channels, particularly in the atrioventricular node, are concerned with a well-known additive interaction between class II (β-blockers) and class IV (verapamil, diltiazem) antiarrhythmic drugs (see also Table 1). The resulting inhibition of atrioventricular conduction is due to the blockade of nodal calcium channels induced both directly by the calcium channel blocker and indirectly, through a reduction in sympathetic drive by the β-blocker (WINNIFORD et al. 1982).

2. Potassium Channels and Drug-Induced Torsades de Pointes

Torsades de pointes is a potentially lethal ventricular arrhythmia, which can be produced by class Ia and class III antiarrhythmic agents, as well as by practically any drug that prolongs repolarization (BEN-DAVID and ZIPES 1993). A delay in the opening of the repolarizing outward-rectifying potassium channels increases the duration of the action potential and of the QT interval on the electrocardiogram. Numerous drugs have been associated with torsades de pointes and the long QT syndrome (Table 4). Synergism is likely to occur when QT-modifying drugs are administered concomitantly. This is, for example, the reason why mefloquine and halofantrine should not be used together (WHITE 1994). Electrolyte disorders, such as the hypokalaemia/hypomagnesaemia caused by the administration of diuretics, can precipitate torsades de pointes and have a synergistic effect with class Ia antiarrhythmic drugs (SISCOVICK et al. 1994). It should be noted that apparently "safe" drugs,

Table 4. Some drugs associated with torsades de pointes

Therapeutic classes	Examples
Antiarrhythmic drugs	
– Class Ia	Quinidine
– Class III	Sotalol
Antihypertensive agents	Ketanserine
Antimalarial drugs	Chloroquine, mefloquine, halofantrine
CNS stimulants	3,4-Methylenedioxy-methamphetamine ("ecstasy")
Diuretics	Hydrochlorothiazide
H_1-antihistaminics	Terfenadine, astemizole
Macrolide antibiotics	Erythromycin
Neuroleptics	Chlorpromazine, haloperidol
Tricyclic antidepressants	Imipramine

like terfenadine and astemizole, can create a life-threatening risk when administered in overdose or in association with macrolide antibiotics, particularly erythromycin. In the latter case, the interaction is largely of a pharmacokinetic nature (see Chap. 6, this volume).

3. ATP-Sensitive Potassium Channels

ATP-sensitive K^+ channels (K_{ATP} channels) of the pancreatic B cell are the target of drugs producing opposite effects on insulin secretion. K_{ATP} channel blockers, such as the sulfonylureas tolbutamide and glibenclamide, reduce membrane polarization. This will increase the opening probability of voltage-dependent Ca^{2+} channels resulting in insulin release (SCHMID-ANTOMARCHI et al. 1987). On the contrary, diazoxide and somatostatin are K_{ATP} channel openers, which hyperpolarize the cell and inhibit insulin release (EDWARDS and WESTON 1993). The other K_{ATP} channel openers, such as minoxidil and pinacidil, have a much lower affinity for the B cell than for smooth muscle channels and are thus much less likely to interact with antidiabetic sulfonylureas.

III. Hormonal Systems

Natural, semisynthetic and synthetic hormones are used as replacement therapy for hormone deficiency states or as drugs to modulate the hypothalamic-hypophyseal regulations of the gonads and other endocrine glands. Most clinically important interactions involving hormonal treatments are of a pharmacokinetic nature. However, a few pharmacodynamic interactions concern the corticosteroids and the antidiabetic agents.

1. Adrenal Corticosteroids

Interactions implying the corticosteroids can be deduced from their broad spectrum of activity. Their metabolic effects, which tend to produce a hyperglycaemic state, will reduce the efficacy of insulin and other antidiabetic agents (see below). Those glucocorticoids possessing mineralocorticoid activity will aggravate the hypokalaemia induced by thiazide or loop diuretics. Finally, the corticosteroids may interfere with the development of the protective immune response to vaccines (BUTLER and ROSSEN 1977; ZIELENSKI et al. 1986). Moreover, in view of altered defense mechanisms, live vaccines should not be administered to patients treated with systemic corticosteroids (GROSS et al. 1985).

2. Glycaemic Regulation

The association of insulin and oral hypoglycaemic agents has been advocated in the treatment of type II diabetes (GERICH 1989; LEWITT et al. 1989). Clearly enough, the additive effects of such a combination would increase the risk of severe hypoglycaemia. Similarly, the α-glucosidase inhibitors (e.g., acarbose) may aggravate insulin- and sulphonylurea-induced hypoglycaemia (BALFOUR and MCTAVISH 1993). Inversely, the therapeutic efficacy of oral antidiabetic agents would be reduced by the simultaneous administration of thiazide diuretics or of diazoxide (see Sect. B.II.3).

Many drugs affect glycaemic regulation and can thus interfere with antidiabetic agents. It has been mentioned previously (see Sect. B.I.1) that β-blockers may mask the early signs of hypoglycaemia and should be avoided in diabetic patients. In addition, non-selective β-blockers have been reported to cause severe hypoglycaemia in both diabetic and non-diabetic patients (KOTLER et al. 1966; ANGELO-NIELSON 1980). Unexpectedly, the salicylates in

Table 5. Drugs affecting glycaemic regulation[a]. Possible interactions with insulin and oral hypoglycaemic agents

Hyperglycaemia	Hypoglycaemia
β-Blockers	β-Blockers
	β_2-Adrenergic agonists
Cyclosporine	Disopyramide
Corticosteroids	Angiotensin-converting enzyme inhibitors (ACEIs)
Diazoxide	Ethanol
Nicotinic acid	Fibric acid derivatives
	Octreotide
Pentamidine	Pentamidine
Phenothiazines	Quinine
Phenytoin	Salicylates
Thiazide diuretics	Streptozotocin
Thyroid hormones	

[a] See the review by PANDIT et al. (1993) for an exhaustive list and evaluation of medications influencing plasma glucose levels.

high doses (4–6 g/day) have been shown to decrease blood glucose concentrations (SELTZER 1989). Salicylates increase insulin secretion in non-insulin-dependent diabetes, increase insulin sensitivity and inhibit lipolysis (MICOSSI et al. 1978; PANDIT et al. 1993). They can thus enhance the hypoglycaemic effects of insulin and oral antidiabetic agents (RICHARDSON et al. 1986). Additional pharmacokinetic interactions with the sulphonylureas (e.g., displacement from protein-binding sites) may also play a role (see Chap. 4, this volume). If anti-inflammatory doses of aspirin are to be prescribed to a diabetic patient, they should be started low and increased progressively with frequent blood glucose monitoring.

As mentioned in Sect. B.III.1, the glucocorticoids represent another cause of interaction with antidiabetic agents. By stimulating hepatic gluconeogenesis and decreasing peripheral utilization of glucose (JACKSON and BRESSLER 1981; GERICH 1989), they tend to produce a diabetes-like state (PANDIT et al. 1993).

This survey is limited to a few classical examples of pharmacodynamic interactions that implicate insulin and other antidiabetic agents. Many additional drugs (including ethanol) can interfere with glycaemic regulation (Table 5).

IV. Homeostatic Regulations

1. Renal Haemodynamics and Drug-Induced Acute Renal Failure

Decreased renal perfusion is a major factor that predisposes to nephrotoxicity. Besides various causes unrelated to drugs, renal perfusion can be decreased by drug-induced sodium and volume depletion (diuretics) or systemic hypotension (antihypertensive agents). In the face of reduced renal perfusion, the glomerular hydrostatic pressure and filtration rate are maintained through a balance between angiotensin II-mediated efferent arteriole constriction and prostaglandin-effected afferent arteriole dilation (Fig. 5). In such a condition, drugs interfering with either prostaglandin synthesis (e.g., the non-steroidal anti-inflammatory drugs, NSAIDs) or the renin-angiotensin system (e.g., angiotensin-converting enzyme inhibitors, ACEIs) will precipitate acute renal failure (HRICIK and DUNN 1990; WHELTON et al. 1990).

The role of prostaglandins in preserving glomerular filtration under conditions of elevated circulating angiotensin II has recently been demonstrated in man (MOTWANI and STRUTHERS 1994). In the clinical context, it should be remembered that prolonged diuretic treatment stimulates the renin-angiotensin system, which in turn necessitates renal prostaglandins for preserved glomerular filtration. In this situation, even the trivial administration of an NSAID for minor musculoskeletal disorders may have serious consequences. It is still not clear whether all NSAIDs are equi-effective in causing this kind of nephrotoxicity or whether some of them, notably sulindac, may be considered as safer drugs (WHELTON et al. 1990; JOHNSON et al. 1994).

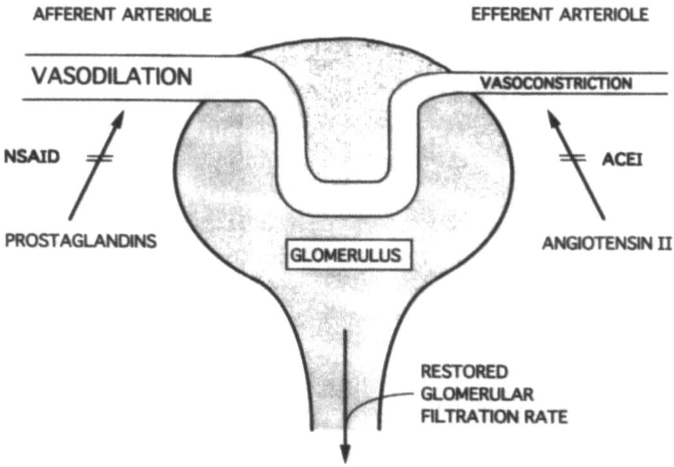

Fig. 5. Compensatory mechanisms to decreased renal perfusion and drug interactions. Decreased renal perfusion due, for example, to intravascular volume depletion (diuretics) or systemic hypotension (antihypertensive agents) results in decreased intraglomerular hydrostatic pressure and filtration rate. To compensate, both the renin-angiotensin system and the production of prostaglandins are stimulated, mainly in response to sympathetic system activation. Afferent arteriole vasodilation by prostaglandins and efferent arteriole vasoconstriction by angiotensin II restore glomerular hydrostatic pressure and filtration rate. In this condition, drugs that interfere with the compensatory mechanisms, such as the non-steroidal anti-inflammatory drugs (*NSAIDs*) and angiotensin-converting enzyme inhibitors (*ACEIs*), may, either or both, precipitate acute renal failure (see text for additional explanations)

2. Prostaglandins, Natriuresis and Antihypertensive Therapy

By interfering with the production of renal prostaglandins, the NSAIDs induce hydrosaline retention (see Sect. B.IV.1) and may reduce the natriuretic effect of concurrently administered diuretics, particularly furosemide (Favre et al. 1983). Moreover, the NSAIDs can decrease the antihypertensive effect of most antihypertensive drugs (Zahler et al. 1990). This has been confirmed in a recent meta-analysis of randomized trials studying the effect of NSAIDs on blood pressure: among the NSAIDs, piroxicam had the greatest effect, whereas sulindac and acetylsalicylic acid were the least effective (Johnson et al. 1994). In view of the fact that some NSAID preparations have now become available over the counter, the possibility of a drug interaction should be considered when unexpected blood pressure elevation is detected in a previously stabilized hypertensive patient.

3. Potassium Homeostasis

All thiazide and loop diuretics increase the renal excretion of potassium and reduce kalaemia. Hypokalaemia augments the risk of arrhythmias, particularly in the course of therapy with antiarrhythmic drugs (see Sect. B.II.2). It is also well known to increase the toxicity of cardiac glycosides (STEINESS and OLESEN 1976). In order to treat hypokalaemia, certain patients will require the prescription of potassium supplements or of a potassium-sparing diuretic. In these patients, the co-prescription of ACE inhibitors creates a serious risk of hyperkalaemia (BURNAKIS and MIODUCH 1984). By interfering with renin release, the NSAIDs tend to produce hyperkalaemia (TAN et al. 1979), which would also be aggravated by the simultaneous prescription of potassium-sparing diuretics or of potassium salts (ZIMRAN et al. 1985).

References

Amrein R, Güntert TW, Dingemanse J, Lorscheid T, Stabl M, Schmid-Burgk W (1992) Interactions of moclobemide with concomitantly administered medication: evidence from pharmacological and clinical studies. Psychopharmacology (Berl) 106:S24–S31

Angelo-Nielson K (1980) Timolol topically and diabetes mellitus. JAMA 244:263

Balfour JA, McTavish D (1993) Acarbose. An update of its pharmacology and therapeutic use in diabetes mellitus. Drugs 46:1025–1054

Bauman JL, Grawe JJ, Winecoff AP, Hariman RJ (1994) Cocaine-related sudden cardiac death: a hypothesis correlating basic science and clinical observations. J Clin Pharmacol 34:902–911

Ben-David J, Zipes DP (1993) Torsades de pointes and proarrhythmia. Lancet 341:1578–1582

Berlin I, Zimmer R, Cournot A, Payan C, Peclariosse AM, Puech AJ (1989) Determination and comparison of the pressor effect of tyramine during long-term moclobemide and tranylcypromine treatment in healthy volunteers. Clin Pharmacol Ther 46:344–351

Bigger JT (1984) The interaction of mexiletine with other cardiovascular drugs. Am Heart J 107:1079–1085

Blackwell B, Marley E (1966) Interactions of cheese and its constituents with monoamine oxidase inhibitors. Br J Pharmacol Chemother 26:120–141

Blakely RD, Berson HE, Fremeau RT, Caron MG, Peek MM et al. (1991) Cloning and expression of a functional serotonin transporter from rat brain. Nature 354:66–70

Bochner BS, Lichstenstein LM (1991) Anaphylaxis. N Engl J Med 324:1785–1790

Boess FG, Martin IL (1994) Molecular biology of 5-HT receptors. Neuropharmacology 33:275–317

Browne B, Linter S (1987) Monoamine oxidase inhibitors and narcotic analgesics. Br J Psychiatry 151:210–212

Bugnon O, Dommer J, Mesnil M, Schorderet M (1993) Detection and management of drug interactions (in French). Swiss Society of Pharmacy, Bern (Basic formation, vol 1)

Burnakis TG, Mioduch HJ (1984) Combined therapy with captopril and potassium supplementation. A potential for hyperkalemia. Arch Intern Med 144:2371–2372

Butler WT, Rossen RD (1977) Effects of corticosteroids on immunity in man. I. Decreased serum IgG concentration caused by 3 or 5 days of high doses of methylprednisolone. J Clin Invest 30:1451–1455

Cetaruk EW, Aaron CK (1994) Hazards of nonprescription medications. Emerg Med Clin North Am 12:483–510

Christ W, Lehnert T, Ulbrich B (1988) Specific toxicologic aspects of the quinolones. Rev Infect Dis 10 [Suppl 1]:S141–S146

Clarke RSJ, Mirakhur RK (1994) Adverse effects of muscle relaxants. Adv Drug React Toxicol Rev 13:23–41

Cooper JR, Bloom FE, Roth RH (1991) The biochemical basis of neuropharmacology, 6th edn. Oxford University Press, New York
Curtis DR, Game CJA, Johnston GAR, McCulloch RM, MacLachlan RM (1972) Convulsive action of penicillin. Brain Res 43:242–245
Dilsaver SC (1993) Antipsychotic agents: a review. Am Fam Physician 47:199–204
Drachman DB (1994) Myasthenia gravis. N Engl J Med 330:1797–1809
Drug and Therapeutics Bulletin (1992) Sumatriptan: a new approach to migraine. Drug Ther Bull 30:85–87
Edwards G, Weston AH (1993) The pharmacology of ATP-sensitive potassium channels. Annu Rev Pharmacol Toxicol 33:597–637
Favre L, Glasson P, Riondel A, Vallotton MB (1983) Interaction of diuretics and non-steroidal anti-inflammatory drugs in man. Clin Sci 64:407–415
Feighner JP, Boyer WF, Tyler DL, Neborsky RJ (1990) Adverse consequences of fluoxetine-MAOI combination therapy. J Clin Psychiatry 51:222–225
Fiore MC, Jorenby DE, Baker TB, Kenford SL (1992) Tobacco dependence and the nicotine patch; clinical guidelines for effective use. JAMA 268:2687–2694
Fitton A, Faulds D, Goa KL (1992) Moclobemide. A review of its pharmacological properties and therapeutic use in depressive illness. Drugs 43:561–596
Gerich JE (1989) Oral hypoglycemic agents. N Engl J Med 321:1231–1245
Giardia E, Weschsler M (1990) Low dose quinidine-mexiletine combination therapy versus quinidine monotherapy for treatment of ventricular arrhythmias. J Am Coll Cardiol 15:1138–1145
Gieschke R, Schmid-Burgk W, Amrein R (1988) Interaction of moclobemide, a new reversible monoamine oxidase inhibitor with oral tyramine. J Neural Transm [Suppl 26]:97–104
Giros B, Mestikawy L, Bertrand L, Caron MG (1991) Cloning and functional characterization of a cocaine-sensitive dopamine transporter. FEBS Lett 295:149–154
Glassman AH, Bigger JT Jr (1981) Cardiovascular effects of therapeutic doses of tricyclic antidepressants. Arch Gen Psychiatry 38:815–820
Gram LF (1994) Fluoxetine. N Engl J Med 331:1354–1361
Gross PA, Gould A, Brown AE (1985) Effect of cancer chemotherapy on the immune response to influenza virus vaccine: review of published studies. Rev Infect Dis 7:613–618
Guy EJ (1992) Selective serotonin reuptake inhibitors. Br Med J 304:1655–1658
Halliwell RF, Davey PG, Lambert JJ (1993) Antagonism of $GABA_A$ receptors by 4-quinolones. J Antimicrob Chemother 31:457–462
Hansten PD, Horn JR, Koda-Kimble MA, Young LY (eds) (1993) Drug interactions and updates quarterly. Applied Therapeutics, Vancouver
Hill S, Yau K, Whitwam J (1992) MAOIs to RIMAs in anaesthesia – a literature review. Psychopharmacology (Berl) 106 [Suppl]:S43–S45
Hoffman BJ, Mezby E, Brownstein MJ (1991) Cloning of a serotonin transporter affected by antidepressants. Science 254:79–80
Horn PT, Murphy MB (1991) Dopamine receptor agonists in cardiovascular medicine. Trends Cardiovasc Med 1:103–107
Hricik DE, Dunn MJ (1990) Angiotensin-converting enzyme inhibitor-induced renal failure: causes, consequences, and diagnostic uses. J Am Soc Nephrol 1:845–858
Jackson EJ, Bressler R (1981) Clinical pharmacology of sulphonylurea hypoglycemic agents. Drugs 22:211–245, 295–320
Johnson AG, Nguyen TV, Day RO (1994) Do nonsteroidal anti-inflammatory drugs affect blood pressure? A meta-analysis. Ann Intern Med 121:289–300
Kilty JE, Lorang D, Amara SG (1991) Cloning and expression of a cocaine-sensitive rat dopamine transporter. Science 254:578–579
Kotler MN, Berman L, Rubenstein AH (1966) Hypoglycaemia precipitated by propranolol. Lancet II:1389–1390

Lange RA, Cigarro RG, Flores ED, McBride W, Kim AS, Bedotto JB, Danziger RS, Hillis LD (1990) Potentiation of cocaine-induced coronary vasoconstriction by beta-adrenergic blockade. Ann Intern Med 324:1485–1490
Langieri G, Marinchak RA, Rials SJ, Kowey PR (1995) Antiarrhythmic drug interactions. Curr Op in Cardiol 10:26–28
Lejoyeux M, Adès J, Rouillon F (1994) Serotonin syndrome. Incidence, symptoms and treatment. CNS Drugs 2:132–143
Lewitt MS, Yu VK, Rennie GC, Carter JN, Marel GM, Yue DK, Hooper MJ (1989) Effects of combined insulin-sulfonylurea therapy in type II patients. Diabetes Care 12:379–383
Marchant B, Wray R, Leach A, Nama M (1989) Flumazenil causing convulsions and ventricular tachycardia. Br Med J 299:860
Mars H (1974) Levodopa, carbidopa and pyridoxine in Parkinson's disease. Arch Neurol 30:444–447
McInnes GT, Brodie MJ (1988) Drug interactions that matter. A critical reappraisal. Drugs 36:83–110
Mefford IN (1988) Epinephrine in mammalian brain. Prog Neuropsychopharmacol Biol Psychiatry 12:365–388
Michell JR, Cavanaugh JH, Arias L, Oates JA (1970) Guanethidine and related agents. III. Antagonism by drugs which inhibit the norepinephrine pump in man. J Clin Invest 49:1596–1604
Micossi P, Pontiroli AE, Baron SH, Tamayo RC, Lengel F, Bevilacqua M, Raggi U, Norbiato G, Foà PP (1978) Aspirin stimulates insulin and glucagon secretion and increases glucose tolerance in normal and diabetic subjects. Diabetes 27:1196–1204
Mitchell PB (1994) Selective serotonin reuptake inhibitors: adverse effects, toxicity and interactions. Adv Drug React Toxicol Rev 13:121–144
Mitchelson F (1992) Pharmacological agents affecting emesis; a review (part I). Drugs 43:295–315
Mody I, De Koninck Y, Otis TS, Soltesz I (1994) Bridging the cleft at GABA synapses in the brain. Trends Neurosci 17:517–525
Mott DD, Lewis DV (1994) The pharmacology and function of central $GABA_B$ receptors. Int Rev Neurobiol 36:97–223
Motwani JG, Struthers AD (1994) Interactive effects of indomethacin, angiotensin II and frusemide on renal haemodynamics and natriuresis in man. Br J Clin Pharmacol 37:355–361
Neuvonen PJ, Pohjola-Sintonen S, Tacke U (1993) Five fatal cases of serotonin syndrome after moclobemide-citalopram or moclobemide-clomipramine overdoses. Lancet 342:1419
Pacholczyk T, Blakely RD, Amara SG (1991) Expression cloning of a cocaine- and antidepressant-sensitive human noradrenaline transporter. Nature 350:350–353
Pandit MK, Burke J, Gustafson AB, Minocha A, Peiris AN (1993) Drug-induced disorders of glucose tolerance. Ann Intern Med 118:529–539
Remick RA (1988) Anticholinergic side effects of tricyclic antidepressants and their management. Prog Neuropsychopharmacol Biol Psychiatry 12:225–231
Richardson T, Foster J, Mawer GE (1986) Enhancement by sodium salicylate of the blood glucose lowering effect of chlorpropamide-drug interaction or summation of similar effects? Br J Clin Pharmacol 22:43–48
Schmid-Antomarchi H, De Weille J, Fosset M, Lazdunski M (1987) The receptor for antidiabetic sulfonylureas controls the activity of the ATP-modulated K^+ channel in insulin-secreting cells. J Biol Chem 262:15840–15844
Schou JS (1982) Drug interactions at (pharmacodynamically active) receptor sites. Pharmacol Ther 17:199–210
Schourkes JA, Van Bergen PJ, Van Nueten JM (1988) Prejunctional muscarinic (M_1)-

receptor interactions on guinea-pig ileum: lack of effect of cisapride. Br J Pharmacol 94:228–234
Seeman P, Van Tol HM (1994) Dopamine receptor pharmacology. Trends Pharmacol Sci 15:264–270
Seltzer HS (1989) Drug-induced hypoglycemia. A review of 1418 cases. Endocrinol Metab Clin North Am 18:163–183
Shimada S, Kitamaya SK, Lin CL, Patel A, Nanthakumar P, Gregor P, Kuhar M, Uhl G (1991) Cloning and expression of a cocaine-sensitive dopamine transporter complementary DNA. Science 254:576–578
Sieghart W (1992) $GABA_A$ receptors: ligand-gated Cl^- ion channels modulated by multiple drug-binding sites. Trends Pharmacol Sci 13:446–450
Singh YN, Marshall IG, Harvey AL (1982) Pre- and postjunctional blocking effects of aminoglycosides, polymyxin, tetracycline and lincosamide antibiotics. Br J Anaesth 54:1295–1305
Siscovick DS, Raghunathan TE, Psaty BM, Koepsell TD, Wicklund KG, Lin X, Cobb L, Rautaharju PM, Copass MK, Wagner EH (1994) Diuretic therapy for hypertension and the risk of primary cardiac arrest. N Engl J Med 330:1852–1857
Spina E, Perucca E (1994) Newer and older antidepressants. A comparative review of drug interactions. CNS Drugs 2:479–497
Stack CG, Rogers P, Linter SPK (1988) Monoamine oxidase inhibitors and anaesthesia. A review. Br J Anaesth 60:222–227
Steiness E, Olesen KH (1976) Cardiac arrhythmias induced by hypokalaemia and potassium loss during maintenance digoxin therapy. Br Heart J 38:167–172
Stockley IH (1991) Drug interactions. Blackwell, Oxford
Sugrue MF (1981) Current concepts on the mechanisms of action of antidepressant drugs. Pharmacol Ther 13:219–247
Tan SY, Shapiro R, Franco R, Stockard H, Mulrow PJ (1979) Indomethacin-induced prostaglandin inhibition with hyperkalemia. A reversible cause of hyporeninemic hypoaldosteronism. Ann Intern Med 90:783–785
Usdin TB, Mezey E, Chen C, Brownstein MJ, Hoffman BJ (1991) Cloning of the cocaine-sensitive bovine dopamine transporter. Proc Natl Acad Sci USA 88:11168–11171
Whelton A, Stout RL, Spilman PS, Klassen DK (1990) Renal effects of ibuprofen, piroxicam, and sulindac in patients with asymptomatic renal failure. Ann Intern Med 112:568–576
Whitcomb DC, Gilliam FR III, Starmer CF, Grant AO (1989) Marked QRS complex abnormalities and sodium channel blockade by propoxyphene reversed with lidocaine. J Clin Invest 84:1629–1636
White NJ (1994) Mefloquine in the prophylaxis and treatment of falciparum malaria. Br Med J 308:286–287
Winniford MD, Markham RV Jr, Firth BG, Nicod P, Hillis LD (1982) Hemodynamic and electrophysiologic effects of verapamil and nifedipine in patients on propranolol. Am J Cardiol 50:704–710
Woosley RL (1991) Antiarrhythmic drugs. Annu Rev Pharmacol Toxicol 31:427–455
Zahler SM, al-Kiek R, Ivanovich P, Mujais SK (1990) Nonsteroidal anti-inflammatory drugs and antihypertensives: cooperative malfeasance. Nephron 56:345–352
Zielenski CC, Stuller I, Dorner F, Pötzi P, Müller C, Eibl MM (1986) Impaired primary, but not secondary, immune response in breast cancer patients under adjuvant chemotherapy. Cancer 58:1648–1652
Zimran A, Kramer M, Plaskin M, Herschko C (1985) Incidence of hyperkalaemia induced by indomethacin in a hospital population. Br Med J 291:107–108

CHAPTER 8

Synergistic Drug Interactions

J.P. GRIFFIN and P.F. D'ARCY

A. Introduction: "Mithridatium"

Synergy is derived from the Greek word συνεργόζ, meaning working together. In pharmacological terms it has a precise meaning, which is distinct from interaction between two or more active substances where the interaction may be on the absorption, distribution, metabolism or excretion of one or more of the substances. For a synergistic therapeutic effect to occur there need not necessarily be an interaction of this type but the effect should be to the patient's therapeutic advantage.

Historically, the seeking for the universal panacea or cure for all ills has to some extent hinged on the philosophy of empirical polypharmacy. This can be traced back to philosophies such as that evolved by Mithridates VI, King of Pontus 133 B.C.–63 B.C. (see GRIFFIN 1994). Pontus abounded in medicinal plants and Mithridates acquired considerable knowledge of them. Like every despot of that period he lived in fear of being assassinated by poisoning, in consequence of which he sought the universal antidote to all poisons. Mithridates proceeded along a simple line of reasoning. Having investigated the powers of a number of single ingredients, which he found to be the antidote to various venoms and poisons individually, he evaluated them experimentally on condemned criminals. He then compounded all the effective substances into one antidote hoping thereby to produce universal protection. A daily dose was taken prophylactically to give the immunity he sought.

Pliny wrote: "by his unaided efforts Mithridates devised the plan of drinking poison daily after first taking remedies in order to achieve immunity by sheer habituation. He was the first to discover the various antidotes, one of which is even known by his name". So effective was Mithridates' formulation that he tried unsuccessfully to commit suicide by poisoning and finally killed himself with a "Celtic sword".

Galen writing in the second century A.D. at a time when he was physician to the Roman Emperor, Marcus Aurelius, refers to "Mithridatium" and a formulation derived from it by one Andromachus, Nero's physician. It is said that Andromachus removed some ingredients from Mithridates' formulation and added others particularly viper's flesh. To this new product he gave the name "Galene", which means "tranquillity". Galene later became known as theriac. Details of various theriacs, including Mithridatium and Galene, were given in Galen's *Antidotes I* and *Antidotes II*.

In Galen's *Antidotes I* he distinguished three kinds of antidote, those that countered poisons, those that countered venoms and those that countered ailments. Some would counter all three and Galen claimed that Mithridatium and Galene belonged to this latter class. According to Galen, Mithridatium contained 41 ingredients and the Galene of Andromachus 55 components.

The use of Mithridatium persisted in Britain until the 1746 London Pharmacopoeia; it was removed from the 1788 edition. The Edinburgh Pharmacopoeia included it in its 1699 edition but dropped it from the 1759 edition. However, Mithridatium persisted in the German Pharmacopoeia of 1872 and the French Pharmacopoeia of 1884.

WILLIAM HEBERDEN in 1745 described Mithridatium and theriacs as: "this medley of discordant samples . . . made up of a dissonant crowd collected from many countries, mighty in appearance, but in reality, an ineffective multitude that only hinder one another".

What was written by WILLIAM HEBERDEN two and a half centuries ago could equally well be applied to a considerable number of the multi-ingredient products that persist in use today, particularly in many of the traditional OTC remedies and in many of the unlicensed Indian and Chinese medicines which, although unlicensed, are freely available throughout Europe through various ethnic outlets.

In current medical practice a number of examples of co-formulations and/or co-prescribing of active ingredients on the basis of a believed synergistic action or interaction can be cited. For many of these the evidence for the combined use has been derived on the basis of theory which may or may not stand up to critical analysis in terms of practical therapeutics.

A brief account of examples of the more important treatments with believed synergistic drug combinations is presented in this chapter. The examples start with the early use of probenecid to slow the renal elimination of penicillin. They follow on with the demonstration of synergistic drug effects in the use of diuretics, with the synergistic design of co-trimoxazole, with synergism in the use of antiepileptics, antitubercular agents, antileprosy drugs, and drugs to treat peptic and duodenal ulcers, and finally with synergism in the control of diabetes and in cancer chemotherapy.

B. Early Use of Probenecid with Penicillin

Many acidic drugs and drug metabolites are actively secreted by the proximal renal tubular active transport mechanism and interactions may arise from competition for this system. Drugs actively secreted include the penicillins and probenecid; thus the plasma half-life of penicillin may be prolonged by probenecid. The newer penicillins have rendered this combination obsolete although the inhibition of the urinary excretion of penicillin and cephalosporins has been used as a device to increase the biliary excretion of these agents, thereby raising the antibiotic concentrations in the biliary tract. This has been used to improve the efficacy of antibiotic treatment of cholecystitis (SALES et al. 1972) (see also, Chap. 6, this volume).

C. Diuretic Combinations

Commonly prescribed fixed combinations that can be cited as examples of believed benefit are combined diuretics. The BRITISH NATIONAL FORMULARY No. 27 (1994) cites over 25 products comprising a loop or thiazide diuretic combined with either potassium or a potassium-sparing diuretic. The MeRec Bulletin (MEDICINES RESOURCE CENTRE 1994) asks the question: Are such combinations necessary? and concludes that for the vast majority of patients there is no need for such combinations, and that the routine approach should be to give a thiazide or loop diuretic alone.

D. Co-trimoxazole

In 1948, HITCHINGS et al. demonstrated that nearly all 2,4-diaminopyrimidines and related compounds inhibited the growth of *Lactobacillus casei* by virtue of their properties as folic acid antagonists. Four years later the precise mode of action was shown to involve inhibition of the enzyme dihydrofolate reductase. This is the enzyme which catalyses the transformation of folic acid into a form that can be utilised in the process leading to thymine formation prior to its incorporation into DNA. It was eventually shown that the degree of inhibition of the enzyme by different antimetabolites varied according to the species from which the enzyme was isolated. This led to the development of species-specific dihydrofolate reductase inhibitors amongst a series of 5-benzyl derivatives of 2,4-diaminopyrimidine. The antimalarial drug pyrimethamine is one such example. By varying the substituents on the benzene and pyrimidine rings, it was found that methoxyl groups enhanced the inhibitory potency against cultures of *Proteus vulgaris*, whilst not markedly inhibiting the enzyme derived from mammalian sources.

This discovery was fully exploited with trimethoprim, the trimethoxy derivative, which is 50000 times more potent as an inhibitor of bacterial dihydrofolate reductase as it is of human dihydrofolate reductase (see SNEADER 1985). Trimethoprim was marketed in fixed combination with sulphamethoxazole as Septrin and Bactrim by Wellcome and Roche, respectively. This combination was based on the fact that sulphonamides prevent the bacterial synthesis of dihydrofolate. It was theoretically conceived that if a bacterium was resistant to the sulphonamide it could be hoped that the dihydrofolate acid reduction process would be inhibited by trimethoprim or vice versa. The choice of sulphamethoxazole as the sulphonamide in the combination was based purely on the fact that it had approximately the same duration of action as trimethoprim.

The eventual choice of the name co-trimoxazole by the British Pharmacopoeia Commission gave a form of sanctity to this combination and it was a sanctity that time has shown to be ill-deserved.

The BRITISH NATIONAL FORMULARY No. 27 states that "trimethoprim alone can be used for the treatment of urinary and respiratory-tract infections and for prostatitis, shigellosis and invasive salmonella infections. Side effects are less than with co-trimoxazole especially in older patients; therefore it should be used in place of co-trimoxazole for most infections (except *Pneumocystis carinii* infections and respiratory infections associated with AIDS)".

The side effects of co-trimoxazole are the same as those of sulphonamides, e.g., rashes, Stevens-Johnson syndrome, toxic epidermal necrolysis, renal failure and blood dyscrasias notably bone-marrow depression and agranulocytosis. In short, the serious side effects of co-trimoxazole are due to the sulphonamide component which, apart from a possible but dubious synergistic role in *Pneumocystis carinii* infections, is not serving any useful purpose.

This example of co-trimoxazole is one where a synergistic role was conceived on good theoretical grounds but was shown after some 12 years to be unnecessary and undesirable and unsafe for many patients, particularly the elderly. The sulphonamide component contributes little to the efficacy of the product but causes most of the combination's adverse reactions.

Despite the fact that the undesirability of this fixed combination has been known for over 15 years and that repeated comments to this effect have been made in the BRITISH NATIONAL FORMULARY in every issue since 1981, the product remains widely used by British general practitioners. Some 18 million prescriptions per year are written in general practice for co-trimoxazole, and approximately the same number for trimethoprim. However, in hospital practice, the use of co-trimoxazole has virtually disappeared.

E. Combination Treatment in Control of Epilepsy

For many decades epileptic patients were treated with combinations of anticonvulsant drugs, one of the more common combinations being phenobarbitone and phenytoin. This combination is still one of the most widely used medications for epileptic patients. GRIFFIN and WYLES (1991) state that "until the 1970s polytherapy (more than one anticonvulsant) was the most widely used and accepted practice. Monotherapy is now generally recognised as being superior. The basis of monotherapy is accurate diagnosis followed by the selection of a single appropriate anticonvulsant".

Table 1 shows the drugs of choice for epilepsy (after CHADWICK 1993).

The BRITISH NATIONAL FORMULARY No. 1 in 1981 stated "therapy with several drugs at once should be avoided. Patients are best controlled with one antiepileptic. Combinations of drugs often used to be given on the grounds

Table 1. Antiepileptic drugs: single therapy by epilepsy type. (After CHADWICK 1993)

Epilepsy type	First choice	Second choice
Idiopathic		
Simple absences	Sodium valproate or ethosuximide	Benzodiazepines or lamotrigine
Juvenile myoclonic	Sodium valproate	Phenobarbitone or lamotrigine
Awakening time/clonic	Sodium valproate	Carbamazepine, phenytoin or lamotrigine
Symptomatic	Sodium valproate or benzodiazipines	Carbamazepine, phenytoin or phenobarbitone
Partial epilepsy	Carbamazepine or sodium valproate	Phenytoin, phenobarbitone or vigabatrin
Unclassified	Sodium valproate	Lamotrigine

that their therapeutic effects were additive while their individual toxicity was reduced, but there is no evidence for this." The wording in the BNF No. 27 (1994) is identical.

The continuing high usage of the phenobarbitone/phenytoin combinations shows how hard it is to break an established but undesirable prescribing practice. The usage of the new anticonvulsants (e.g., GABA derivatives, vigabatrin) remains low. The popularity of the ingrained phenobarbitone/ phenytoin combination is, however, made more attractive by the annual low price for treatment (see Table 2). Downward pressures by Government on prescribing costs are an undoubted factor in perpetuating the use of two agents once thought to have a useful synergistic action. Conversely, the use of new single-therapy regimens is being discouraged by cost.

F. Antitubercular Drugs

The synergistic use of antibiotics and antibacterial agents in the treatment of tuberculosis is well established. The treatment is in two phases, an "initial phase" with *at least* three drugs and a "continuation phase" using two drugs. The following regimen is recommended by the British Thoracic Society:

Initial Phase. The concurrent use of three antitubercular agents is designed to reduce the population of viable bacteria as rapidly as possible. The usual regimen is the daily use of rifampicin, isoniazid and pyrazinamide; ethambutol may be added if resistant organisms are suspected. Streptomycin is rarely used in the United Kingdom, although it is an effective agent for use in the "initial phase" and can be used instead of isoniazid. The "initial phase" of treatment should last at least 2 months.

Continuous Phase. This should last a minimum of 4 months using a combination of isoniazid and rifampicin.

Table 2. Comparison of daily dose and annual cost of anticonvulsants. (From CRIFFIN and WYLES 1991)

Anticonvulsant	Annual cost	Commonest daily dose
Carbamazepine	£65.00	600 mg
Clonazepam	£101.00	6 mg
Ethosuximide	£94.00	1000 mg
Phenytoin	£28.00	300 mg
Phenobarbitone	£13.00	180 mg
Primidone	£27.00	1000 mg
Sodium valproate	£96.00	800 mg
Vigabatrin	£168.00	3000 mg

Treatment regimens for Third World countries may well differ from that recommended in the United Kingdom and will largely depend on the local pattern of resistance and the cost and availability of antitubercular drugs.

Dosage regimens can be found in the BRITISH NATIONAL FORMULARY and the current edition should always be consulted.

G. Anti-Leprosy Treatments

For more than 20 years the use of dapsone alone was commonplace but multidrug regimens have had to be developed due to the emergence of drug-resistant strains of the leprosy bacillus. Thus, the principle for the use of multi-drug treatments in leprosy is the same as that of tuberculosis. For therapeutic purposes, leprosy can be subdivided into, firstly, multibacillary leprosy, for which a three-drug regimen is recommended using dapsone, rifampicin and clofazimine; secondly, paucibacillary leprosy, for which a two-drug regimen of rifampicin and dapsone are recommended. A 2-year period of treatment is recommended in both grades of severity and advice from the Panel of Leprosy Opinion from the Department of Health should be sought before initiating treatment.

H. Peptic Ulcer Therapy

For many years the treatment of gastric and duodenal ulcer was based on the use of antacids such as aluminium hydroxide and magnesium hydroxide. This was supplemented by bed rest and milk alkali drip feeds through a nasogastric tube in conjunction with aspiration of acid. Surgical intervention with partial gastrectomy, Polya gastrectomy, Billroth gastrectomy or vagotomy and pyloroplasty were common until the introduction of the H_2-antagonist cimetidine in the late 1970s, shortly followed by other H_2-antagonists ranitidine, famotidine and nizatidine in the early 1980s. Misoprostol, a synthetic prostaglandin analogue, and proton pump inhibitors such as omeprazole revolutionised the treatment of peptic ulcer and virtually eliminated the need for the heroic gastric surgery conducted in earlier decades.

Treatment of peptic ulcer is due to undergo a further revolution with the discovery of *Helicobacter pylorii* as the causative organism. In a press release following the VIIth Workshop on Gastroduodenal Pathology and *Helicobacter pylorii* held in Texas in September 1994, the detailed results of a trial by Dr. R.P.H. Logan of The Queen's Medical Centre, Nottingham, UK, were presented in detail; they are given below:

Dr. Logan presented 1-year follow-up data on a multicenter study of 154 patients with duodenal ulcers and *H. pylorii* infection. Six months data from the same study were presented in May during Digestive Disease Week in New Orleans.

Participants in the trial received either 500 mg clarithromycin (three times a day) or placebo along with 40 mg omeprazole (once a day) during the first 2 weeks of the study; all study participants received 40 mg omeprazole for an additional 2 weeks.

According to Dr. Logan's 6-months interim results, ulcers healed after 4 weeks in 100% (64/64) of patients receiving clarithromycin and omeprazole and 99% (71/72) of patients receiving omeprazole and placebo. However, a significant difference was observed between the two study arms in the eradication of *H. pylorii* at 4–6 weeks after the conclusion of therapy. Patients who were treated with the clarithromycin/omeprazole combination showed *H. pylorii* rates of 83% – or 50 out of 60 evaluable patients – at 4–6 weeks post-treatment versus 1% (1/74) in the group receiving placebo/omeprazole.

Furthermore, 6 months after treatment only 4% (2/53) of patients who received clarithromycin and omeprazole experienced ulcer recurrence, versus 55% (36/60) of those who received placebo and omeprazole.

One year after treatment, the final results indicate incidence of ulcer recurrence was 6% (3/51) in patients treated with clarithromycin and omeprazole, versus 76% (47/62) of patients who received a placebo and omeprazole. These data suggested that dual therapy with clarithromycin and omeprazole might provide an effective regimen to heal and prevent occurrence of ulcers (LOGAN 1994).

The clarithromycin and omeprazole combination appeared to be well tolerated according to researchers. Dr. Logan reported that only four patients receiving clarithromycin and omeprazole and one patient receiving omeprazole with placebo discontinued therapy because of adverse events. Adverse events that occurred more frequently in the clarithromycin/omeprazole-treated patients were tongue discoloration and taste perversion. In this study, no patient experienced a serious adverse event.

At the Xth World Congress of Gastroenterology held in Los Angeles (2–7 October 1994), a paper by GUSTAVSON et al. was presented. This dealt with the pharmacokinetic interactions between omeprazole and clarithromycin in 20 male healthy volunteers. The results of their study led the authors to draw three salient conclusions, namely:

1. Administration of omeprazole with clarithromycin resulted in higher and more prolonged plasma concentrations of omeprazole than when omeprazole was administered alone, but there was no significant difference in the effect of omeprazole on gastric pH.
2. Administration of clarithromycin with omeprazole resulted in somewhat higher plasma concentrations of clarithromycin and 14(R)-hydroxy-clarithromycin than when clarithromycin was administered with placebo.
3. Good penetration of clarithromycin and 14(R)-hydroxy-clarithromycin into gastric tissues and mucus was observed. Tissue and mucus concentrations were higher when clarithromycin was administered with omeprazole than when clarithromycin was administered with placebo.

Other regimens combining an acid control and antibiotic element are currently under clinical evaluation, for example:

1. Ranitidine bismuth citrate with clarithromycin
2. Omeprazole plus amoxycillin plus metronidazole
3. Omeprazole plus amoxycillin plus clarithromycin
4. Bismuth plus amoxycillin plus metronidazole

In addition experimental work from Japan suggested that degrading the cell wall of *Helicobacter pylorii* with a proteolytic enzyme in combination with the proton pump inhibitor lansoprazole in combination with bismuth, amoxycillin and metronidazole indicated that there might be an improved antibacterial effect with the use of pronase.

The work on *H. pylorii* elimination by the synergistic action of various combinations indicated that the clinical evaluation of synergistic actions should and could be evaluated. The "armchair" combination of treatments based on sound theoretical considerations are hopefully a thing of the past. The first combination approved both in the United Kingdom and Sweden for *H. pylorii* infections is omeprazole and amoxycillin. It would be reasonable to expect that regulatory approval for recommendation to use omeprazole and clarithromycin in combination can be expected shortly.

Most clinical researchers now believe that adding another antibacterial to the currently accepted dual therapy would improve the *H. pylorii* cure rate; metronidazole for 1 week appears to be the frontrunner for this role.

A paper from Hong Kong by HOSKINGS et al. (1994) demonstrated that a 1-week cource of bismuth, tetracycline and metronidazole in patients with duodenal ulcer and *H. pylorii* infection led to ulcer healing in 92% of patients. The addition of omeprazole in a similar-sized group of patients led to ulcer healing in 95%. The authors of this report concluded that the addition of omeprazole greatly reduced ulcer pain but had no effect on ulcer healing.

I. Non-Insulin-Dependent (Maturity Onset) Diabetes

Oral antidiabetic drugs are used in the control of non-insulin-dependent (type 2, obesity or maturity onset) diabetes. They should not be prescribed until patients have been shown not to respond adequately to 3 months restriction of calorie and carbohydrate intake. They should be used to augment attempts to control the diabetes by diet.

There are two classes of oral hypoglycaemics: firstly, the sulponylureas, which act mainly by augmenting insulin secretion and consequently depend on the patient having some remaining pancreatic β-cell function. In long-term therapy there is also thought to be an extrapancreatic action. The second group are the biguanides; metformin is the only available biguanide currently on the British market; phenformin was withdrawn due to its propensity to cause lactic acidosis. Metformin can also induce lactic acidosis but this occurs almost exclusively in patients with renal impairment, in whom it is contraindicated. Metformin exerts its effect by a different mode of action from the sulphonylureas, mainly by decreasing gluconeogenesis and by increasing the peripheral utilisation of glucose. Since it acts only in the presence of insulin it is effective only in those diabetics with some functional pancreatic islet B cells.

In 1970, the University Group Diabetes Program (UGDP) in the United States described the results from the first 8½ years of study in 12 centres in non-insulin-requiring diabetics. These patients were randomly allocated to five treatment groups: placebo; a standard dose of the sulphonylurea tolbutamide; a standard dose of insulin; a variable dose of insulin; or a standard dose of phenformin. The study findings indicated that the combination of diet plus tolbutamide was no more effective than diet alone in prolonging life. Moreover, it was suggested that tolbutamide and diet may be less effective than diet alone or diet with insulin as far as cardiovascular mortality was concerned. A year later a similar conclusion was reached regarding phenformin (Knatterud et al. 1971).

Prout (1975) demonstrated in his analysis that diabetic patients under active therapy with tolbutamide at the time of myocardial infarction were less likely to survive than if no sulphonylurea was being administered.

Shanks (1979) held that these observations suggested that the increased mortality in patients receiving tolbutamide was directly related to an effect of the drug rather than to effects which might result, for example, on unrestricted diet. The mechanism by which tolbutamide might influence the outcome of an acute myocardial infarction was unclear.

A retrospective study was carried out in Birmingham in 1974 of 184 known diabetics who sustained a myocarial infarction (Soler et al. 1974). The mortality rate was higher in patients treated with insulin or oral hypoglycaemic agents than in non-diabetics, but earlier deaths were commoner in the patients on oral therapy and these patients had a higher incidence of primary ventricular fibrillation.

Such effects may account for the increased mortality among patients treated with tolbutamide who develop myocardial infarction. The factors which increase the incidence of primary ventricular fibrillation in patients receiving tolbutamide are not known although in vitro studies have shown that sulphonylureas have an ionotropic effect on heart muscle and increase the automaticity of Purkinje fibres; these factors may increase the extent of myocardial damage and lead to arrhythmias (LASSETER et al. 1972). It is clear that this subject required further investigation.

It should be borne in mind that the UGDP study also showed an increased mortality with the biguanide phenformin. Despite the recommendations of SHANKS (1979), the authors know of no further investigations to shed light on the clinical observations of the UGDP study. In fact the findings of the UGDP study appear to have been forgotten over the last 2 decades and the only real effect on practice has been to try and control maturity onset diabetic patients with diet alone and only after this has failed to resort to the use of a sulphonylurea. If diet plus a sulphonylurea fails to achieve control of the diabetic then metformin may be added to the regimen (ALBERTI and HOCKADAY 1987; BRITISH NATIONAL FORMULARY 1994).

ALBERTI and HOCKADAY (1987) state that "the biguanide's principal role in treatment is as a synergist to a sulphonylurea when this together with dietary advice has failed to produce satisfactory glucose control". These authors then go on to state that "failure is often an indication to transfer to insulin". Perhaps this would, in general, be the preferred course of treatment.

More than 2 decades ago, a new area of research in diabetes involving aldose reductase inhibitors (ARIs) was introduced. They were a new class, of drugs aimed at the consequences of hyperglycaemia (neuropathy, nephropathy, retinopathy and cateracts) rather than the control of hyperglycaemia per se. They were envisaged as synergistic additions to standard diabetic treatments and were a group of structurally different compounds that bound to aldose reductase, thus inhibiting the conversion of glucose to sorbitol. This inhibition circumvents the build-up of sorbitol in the cellular milieu and the resulting decrease in myoinositol uptake. It is this decreased myoinositol uptake in nerve tissue that appears to be responsible for a variety of metabolic and electrophysiologic alterations including alterations in nerve conduction (BOULTON et al. 1990; GREEN and JASPAN 1990; KIRCHAIN and RENDELL 1990; MASSON and BOULTON 1990; ZENON et al. 1990; VAN GERVEN et al. 1992).

Tolrestat is the only one of the original ARIs still undergoing clinical trial. Results have been encouraging, but by no means definitive, for some aspects of diabetic neuropathy. Improvements have occurred in paraesthesia and neuropathy, but unfortunately not in pain symptoms. Ongoing studies continue (see review TSAI and BURNAKIS 1993).

A more recent and possibly synergistic treatment with established oral hypoglycaemic agents is acarbose, an α-glucosidase inhibitor. This is a new type of antidiabetic agent, which because of its similarity in structure to dietary disaccharides inhibits the enzymes responsible for the digestion of carbohydrate. The effect is to delay the digestion of starch and sucrose. In diabetic patients this results in a lowering of postprandial hyperglycaemia and smoothing of daily blood glucose fluctuations. The antidiabetic effect does not involve the stimulation of insulin secretion. Gastrointestinal problems are the most commonly reported adverse effects in clinical trials. These include flatulence, abdominal distension and diarrhoea. They are caused by fermentation of unabsorbed carbohydrate in the bowel (CLISSOLD and EDWARDS 1988; NEW MEDICINES 1994).

Currently, acarbose is licensed in the United Kingdom solely for the treatment of non-insulin-dependent diabetes inadequately controlled by diet alone or diet and oral hypoglycaemic agents. Although it has well-documented antidiabetic properties, its precise place in therapy remains to be established. It is not clear which patients should receive acarbose in place of, or in addition to, established oral hypoglycaemic agents. It has been suggested that acarbose may be useful for patients receiving an oral hypoglycaemic agent at the top of the dosage range as an alternative strategy to commencing insulin. Further clinical experience is required to establish whether it should be used in place of traditional oral agents or whether it will act synergistically in combination with them.

J. Cancer Chemotherapy

It has become standard practice to treat various malignant diseases with cocktails of intensive combination regimens. For example in the treatment of myeloma a combination of melphalan, cyclophosphamide, carmustine, vincristine and prednisolone (M2) has been used. Other regimens combining vincristine, doxorubicin and a corticosteroid, either dexamethasone (VAD) or methylprednisolone (VAMP) have been developed.

In Ewing's sarcoma a protocol combining chemotherapy with cyclophosphamide, vincristine, dactinomycin and doxorubicin (VACA) with local radiotherapy has achieved a 5-year survival rate. Other combination regimens have been used for special tumours, e.g., CHAP 5, cisplatin, cyclophosphamide, altretamine and doxorubicin combined with surgery for ovarian carcinoma. Two of the better known cocktails are the MOPP regimen of mustine, vincristine, procarbazine and prednisolone used in Hodgkin's disease, and the COAP combination of cyclophosphamide, vincristine, ara-AC (fazarabine) and prednisolone used in acute leukaemias.

Cocktails for use in malignant conditions have been developed by specialists in the field, partly on good therapeutic experience and partly empirically. Where there is synergy, the mechanisms are often obscure. In many cases the regimens have been developed from the philosophy of throwing into the treatment everything in the pharmacopoeia that logically might work, and avoiding two agents with the same mechanism of action on the malignancy. From that initial experimental philosophy over the last 20 years there has developed a rational approach to the evolution of the chemotherapeutic cocktails.

K. Beneficial Interactions: A Philosophical Approach

The examples of beneficial drug-drug interactions that have been described in the previous sections of this chapter emphasise that synergistic drug interactions when they occur can be of clinical value in drug therapy. Apart from these major interactions that have already been cited, there are many others that have the potential for clinical advantage. Food-drug interactions can be beneficial in some instances, for example, taking propranolol or metoprolol with food improves their bioavailability (MELANDER et al. 1977) and the absorption of cefuroxime axetil is improved in the presence of food (RIDGWAY et al. 1991; REYNOLDS 1993). Improvements are also seen in the absorption of the antileprosy agent clofazimine (HOLDINESS 1989; REYNOLDS 1993), the urinary antiseptic nitrofurantion (ROSENBERG and BATES 1976), the antifungal itraconazole (VAN PEER et al. 1989) and the antifungal griseofulvin in the presence of fatty foods (CROUNSE 1961) or milk (GINSBURG et al. 1983).

There are other examples where drug-drug interactions can improve the efficacy of therapy, for example, the treatment of erythropoietin resistance with cyclosporin (ALMOND et al. 1994). Cimetidine binds to microsomal P450 enzymes and inhibits the oxidative phase of hepatic drug metabolism, thus potentiating the effect and/or duration of a variety of drugs. These drugs include anticoagulants (e.g., warfarin), theophylline and aminophylline, some benzodiazepines, anti-arrhythmic agents (e.g., lignocaine and quinidine), anticonvulsants (e.g., carbamazepine and phenytoin), some beta-blockers, narcotic analgesics, and tricyclic antidepressants (e.g., nortriptyline). Cimetidine also increases the bioavailability of fluorouracil by over 70% without evidence of increased toxicity (see GRIFFIN et al. 1988).

It is understandable to think that increasing the plasma concentrations of a primary drug, for example by cimetidine, will automatically and consequently increase the probability of increased toxicity of that drug. However, this may not always occur and if it does not then the dose-sparing effect on the primary drug would have both a clinical and cost advantage. There could therefore be much advantage in searching for a chemical substance which would reversibly inhibit the P450 oxidase systems without itself having a distinct pharmacological or toxic action. However, examination of this possibility lies in the future.

Today, in practical terms, the "eldorado" of synergy is very often, but fortunately not always, as elusive as it was at the time of Mithridates. A healthy trend towards scepticism with a touch of optimism probably summarises our current viewpoint.

References

Alberti KGMM, Hockaday TDR (1987) Diabetes mellitus. In: Weatherall DJ, Ledingham JGG, Warrell DA (eds) Oxford textbook of medicine vol 1, 2nd edn. Oxford Medical, Oxford, pp 9.51–9.101

Almond MK, Tailor D, Kelsey SM, Cunningham J (1994) Treatment of erythropoietin resistance with cyclosporin. Lancet 343:916–917
Boulton AJM, Levin S, Comstock J (1990) A multicenter trial of the aldosereductase inhibitor, tolrestat, in patients with symptomatic diabetic neuropathy. Diabetologia 33:431–437
British National Formulary (1981) No 1. British Medical Association and Royal Pharmaceutical Society of Great Britain, London
British National Formulary (1994) No 27. British Medical Association and Royal Pharmaceutical Society of Great Britain, London
Chadwick D (1993) Seizures, epilepsy and other episodic disorder. In: Walton J (ed) Brain's diseases of the nervous system, 10th edn. Oxford Medical, Oxford, p 718
Clissold SP, Edwards C (1988) Acarbose: a preliminary review of its pharmacodynamic and pharmacokinetic properties, and therapeutic potential. Drugs 35:214–243
Crounse RG (1961) Human pharmacology of griseofulvin: the effect of fat intake on gastrointestinal absorption. J Invest Dermatol 37:529–533
Ginsburg CM, McCracken GH Jr, Petruska M, Olsen K (1983) Effect of feeding on bioavailability of griseofulvin in children. J Pediatr 102:309–311
Green A. Jaspan J (1990) Treatment of diabetic neuropathy with inhibitors of the aldose reductase enzyme. J Diabetic Complications 4:138–144
Griffin JP (1994) Mithridates VI of Pontus, the first experimental toxicologist. Adv Drug React Toxicol Rev 14:1–6
Griffin JR, Wyles M (1991) Epilepsy, towards tomorrow. Office of Health Economics, London
Griffin JP, D'Arcy PF, Speirs CJ (1988) A manual of adverse drug interactions, 4th edn. Wright, (Bulterworth) London
Gustavson LE, Kaiser JF, Mukherjee DX, Schneck DW (1994) Evaluation of the pharmacokinetic drug interactions between clarithromycin and omeprazole (Abstr). 10th World Congress of Gastroenterology, Oct 2–7, Los Angeles
Heberden W (1745) Antitherica, essay on Mithridatium and Theriac. London
Hitchings GH, Elion GB, Vanderwerff H, Falco EA (1948) Pyrimidine derivatives as antagonists of P.G.A. J Biol Chem 174:765–766
Holdiness MR (1989) Clinical pharmacokinetics of clofazimine: a review. Clin Pharmacokinet 16:74–85
Hoskings SW, Ling TKW, Chung S, Yung MY, Cheng AFB (1994) Duodenal ulcer healing by eradication of Helicobacter pylori without anti acid treatment. Lancet 343:508–510
Kirchain W, Rendell M (1990) Aldose reductase inhibitors. Pharmacotherapy 10:326–336
Knatterud GL, Meinert CL, Klimt C, Osborne RK, Martin DB (1971) Effects of hypoglycemic agents on vascular complications in patients with adult-onset diabetes. IV. A preliminary report on phenformin results. JAMA 217:777–784
Lasseter KC, Levey GS, Palmer RF, McCarthy JS (1972) The effects of sulfonylurea drugs on rabbit myocardial contractility, canine Purkinje fiber automaticity, and adenylcyclase activity from rabbit and human hearts. J Clin Invest 5:2429–2432
Logan RPH (1994) Clarithromycin may prevent ulcer recurrence. 7th Workshop on Gastroduodenal Pathology and Helicobacter pylori, Sept 30, Houston
Masson EA, Boulton AJM (1990) Aldose reductase inhibitors in the treatment of diabetic neuropathy: a review of the rationale and clinical evidence. Drugs 39:190–202
Medicines Resource Centre (1994) Combination diuretics. MeRec Bull 5(10)
Melander A, Danielson K, Schersten B, Wahlin E (1977) Enhancement of the bioavailability of propranolol and metoprolol by food. Clin Pharmacol Ther 22:108–112
New Medicines (1994) Acarbose – an α-glucosidase inhibitor. Int Pharm J 8:11–12
Prout TE (1975) A progress report on the University Group Diabetes Program. Int J Clin Pharmacol 12:244–251

Reynolds JEF (ed) (1993) Martindale, the extra pharmacopoeia. Pharmaceutical Press, London, pp 137–138, 151–152
Ridgway E, Stewart K, Rai G, Kelsey MC, Bielawska C (1991) The pharmacokinetics of cefuroxim axetil in the sick elderly patient. J Antimicrob Chemother 27:663–668
Rosenberg HA, Bates TR (1976) The influence of food on nitrofurantoin bioavailability. Clin Pharmacol Ther 20:227–232
Sales JEL, Sutcliffe M, O'Grady F (1972) Cephalexin levels in human bile in presence of biliary tract disease. Br Med J 3:441–443
Shanks RG (1979) Cardiac dysfunction. In: D'Arcy PF, Griffin JP (eds) Iatrogenic diseases, 2nd edn, Oxford Medical, Oxford, pp 116–131
Sneader W (1985) Drug discovery: the evolution of modern medicines. Wiley, Chichester
Soler NG, Pentecost BL, Bennett MA, Fitzgerald MG, Lamb P, Malins JM (1974) Coronary care for myocardial infarction in diabetes. Lancet I:457–477
Tsai SC, Burnakis TG (1993) Aldose reductase inhibitors: an update. Ann Pharmacother 27:751–754
University Group Diabetes Program (1970) A study of the effects of hyperglycemic agents on vascular complications in patients with adult onset diabetes. II. Mortality results. Diabetes 19 [Suppl]:789–830
Van Gerven J, Lemkes H, van Dijk J (1992) Long-term effects of tolrestat in symptomatic diabetic sensory polyneuropathy. J Diabetes Complications 6:45–48
Van Peer A, Westenborghs R, Heykants J, Gasparini R, Gauwenbergh G (1989) The effect of food and dose on the oral systemic availability of itraconazole in healthy subjects. Eur J Clin Pharmacol 36:423–426
Zenon G, Abobo C, Carter B, Ball D (1990) Potential use of aldose reductase inhibitors to prevent diabetic complications. Clin Pharm 9:446–457

CHAPTER 9

In Vitro Drug Interactions

J.C. McElnay and C.M. Hughes

A. Introduction

In vitro drug interactions may be defined as those interactions which occur outside the body. Drug interactions which will be discussed in this chapter are those which occur between drugs due to reasons of incompatibility (e.g., drug-drug interactions in an intravenous infusion), due to interaction of a drug with its packaging (e.g., drug binding to an infusion bag), due to loss of drugs during laboratory analyses (e.g., binding to laboratory equipment) or due to changes in the bioavailability of drugs when the formulation is altered (McElnay and D'Arcy 1980).

B. Incompatibility Interactions

This form of interaction is often described as an incompatibility between two agents and may be physical or chemical in nature. The interactions often occur in solution, e.g., following addition of drugs to intravenous fluid containers, after mixing drugs in syringes or in liquid preparations for oral or topical administration. Physical incompatibility may be due to immiscibility or insolubility. For example, addition of high concentrations of electrolytes to mixtures in which the vehicle is a saturated aqueous solution of a volatile oil causes the oil to separate and collect as a surface layer. This is the case with potassium citrate mixture BPC, in which the large quantity of soluble solids salts out the lemon oil; quillaia tincture is therefore added as a suspending and emulsifying agent to prevent this (Carter 1975).

Chemical incompatibility may be caused by pH change, decomposition or complex formation. A prime example of this type of interaction involves the barbiturates, e.g., pentobarbitone and phenobarbitone. Solutions of the salts of these compounds are very alkaline and incompatible with acids, acidic salts and acidic syrups, e.g., lemon syrup, which may precipitate the corresponding insoluble barbituric acid derivative (Carter 1975). The British Pharmaceutical Codex (Lund 1994) documents a number of physical and chemical interactions between drugs and other medicinal products; these are summarised in Table 1.

Many such incompatibilities are no longer viewed as being particularly problematic as far as oral formulations are concerned since easy solutions to

Table 1. Examples of in vitro drug-drug incompatibilities from the British Pharmaceutical Codex (Lund 1994)

Mechanism of interaction	Example
Precipitation of drug on dilution of solutions which contain cosolvents	Cosolvents may be used to increase the solubility of drugs which are poorly aqueous-soluble, e.g., digoxin and diazepam. These cosolvents may, however, reduce the solubility of other constituents such as gums, polymers and sugars. Dilution may result in precipitation of drug unless the extent of dilution is such that the final concentration is below its solubility in water.
Precipitation by salting-out	The solubility of an organic compound or excipient may be reduced by addition of a salt. Sodium chloride and or potassium chloride decrease the solubility of benzoic acid; reconstitution of erythromycin lactobionate with saline rather than water may result in precipitation or gel formation. Such phenomena may be attributed to the effects of the ionic size and valency of salts on the structure of water or to competition between salts and organic solutes for water molecules.
Ionic interactions	Examples of cationic-anionic interactions resulting in possible precipitation from solution are promethazine hydrochloride-thiopentone sodium, kanamycin sulphate-sulphadiazine sodium.
Formation of complexes	Tetracycline derivatives form stable complexes with metal ions, e.g. calcium, magnesium and ferric iron, resulting in reduced absorption of tetracycline.
Adsorption on solid particles	Antidepressants, antihistamines, β-blockers and benzodiazepines may be adsorbed in vitro on clays such as kaolin or suspended antacids. Such interactions may arise due to physical adsorption (drug is bound by weak Van der Waal forces) or chemical adsorption (drug is bound by strong valence forces).

the problems can usually be found. There are often, however, many contemporary reports in the pharmaceutical literature on drug interactions due to incompatibility in intravenous fluids. This can either be due to an interaction between two or more drugs added to intravenous fluids prior to administration to patients, e.g., in the case of cancer chemotherapy or due to a single added drug being incompatible with the intravenous vehicle. Griffin and D'Arcy (1988) have prepared a list of such incompatibilities, for example, ampicillin should not be mixed with amino acids [constituents of total parental nutrition (TPN) regimens] as immunogenic and allergic conjugates could be formed; amphotericin is incompatible with all electrolytes and, thus, must be reconstituted with water and infused with 5% glucose (Griffin and D'Arcy 1988). Allwood (1994) described the problem of drug-drug co-precipitation, which may be particularly problematic in intravenous therapy. Such interactions arise due to the mixing of organic anions and cations, which form ion pairs. Examples of drugs at risk of forming ion pairs with other drugs include gentamicin and other aminoglycosides, with heparin and some cephalosporins.

One of the areas of medicine where simultaneous co-administration of a number of intravenous medicines is a clinical necessity is cancer chemotherapy. FOURNIER et al. (1992) studied the effect of a commercial formulation of 5-fluorouracil on the stability of cisplatin and carboplatin to determine whether the drugs could be mixed in containers or intravenous lines. When cisplatin was incubated with 5-fluorouracil, high-performance liquid chromatography (HPLC) studies demonstrated a rapid disappearance of the parent platinum compound, the extent of degradation being 75% after 3.5 h. It was found that the degradation was in fact caused by an interaction between cisplatin and trometamol, the excipient used in the 5-fluorouracil formulations in some countries to buffer the solution at pH 8.2. This interaction resulted in complete inhibition of the antitumour activity of cisplatin in mice. When cisplatin was incubated at the same pH in trometamol-free sodium hydroxide solutions, the parent compound was transformed into an active compound that was toxic to mice. Degradation of carboplatin-trometamol admixtures was similar to that found for cisplatin, but occurred at a slower rate. As in the case of cisplatin, incubation of carboplatin in a sodium hydroxide solution resulted in toxic effects. It was concluded that both carboplatin and cisplatin were incompatible with 5-fluorouracil formulations containing trometamol.

Due to the possibility of incompatibility with constituents, as a general rule no drugs should be added to TPN fluids. To avoid direct contact between TPN fluids and drugs administered by the intravenous route, COLLINS and LUTZ (1991) have examined the co-administration of phenytoin and TPN mixtures in multilumen catheters. These devices can be used to reduce the need for frequent venipuncture and provide a method of administering incompatible drugs. The study utilised an in vitro model flow system to examine the physicochemical phenomena that occurred when these two incompatible entities (phenytoin and TPN fluid) were simultaneously administered through multilumen catheter systems into a circulating fluid (sodium chloride, sodium bicarbonate and albumin in solution). Double- and triple-lumen catheters were used as shown in Fig. 1. Video recordings were made of the drug flow and assays of phenytoin concentration were performed on samples of the circulating fluid. White clouds of phenytoin precipitation were seen near the tip of the double-lumen catheter, but not the triple-lumen device. Infusion through the former resulted in an average of 6% loss of phenytoin which, on microscopic examination, appeared as spindle-shaped crystals. In some cases, millimetre-size fragments of phenytoin precipitate were seen to dislodge from the tip of the double-lumen catheter. It was suggested that the interaction was due to the close proximity of the two orifices at the end hole of the double-lumen catheter, which permitted mixing of the two effusing streams of the incompatible agents. The staggered orifices of the triple-lumen catheter reduced this interaction greatly; no crystal fragments were observed, although a thin coating of

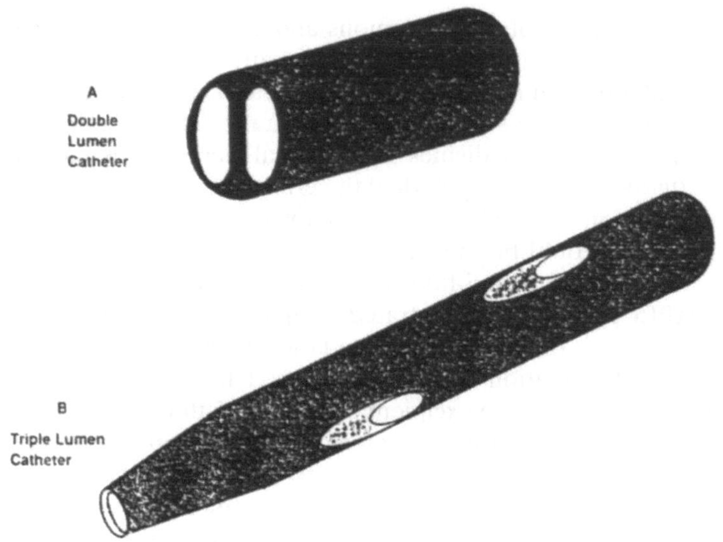

Fig. 1. Drawings of tip of two types of catheters used in study of concomitant administration of phenytoin and TPN fluid. (After Collins and Lutz 1991)

white film did form near the opening of the proximal and middle side holes of the catheter. Although the authors recognised the value of these multilumen devices, they did advise caution in their use, particularly in the clinical setting, with combinations of incompatible drugs. To be forewarned is to be forearmed and indeed the best approach to prevention of drug-drug incompatibility interactions is to use published information on the physical and chemical properties of medicinal agents which are to be used concomitantly and avoid their coming into direct contact.

C. In Vitro Drug Interactions with Pharmaceutical Packaging and Intravenous Administration Equipment

One of the criteria which has been designated as essential in relation to pharmaceutical packaging is that it should not interact with a drug product and, conversely, that the drug product should not interact with the packaging. Careful consideration should therefore be given to the materials used in pharmaceutical packaging. Incompatibilities between drugs and such materials have been well documented in the literature. The two materials which are most widely used in packaging, particularly in intravenous therapy, are plastics and, to a lesser extent, glass.

I. General Properties of Plastics

Plastics are defined as a wide range of solid composite materials which are largely organic, usually based upon synthetic resins or upon modified polymers of natural origin and possessing mechanical strength. The physical, chemical and mechanical properties of a plastic material are determined by its chemical structure, molecular weight range and alignment of the resin, and the type and concentration of additives. Selection of the monomer, for example, ethylene, propylene, glucose, vinyl chloride, determines the chemical structure and the type of substituents in and/or on the polymer chain (AUTAIN 1963; COOPER 1974; BIRELY and SCOTT 1982; BRYDSON 1982). Table 2 lists a number of

Table 2. Polymer compounds which are widely used in plastic pharmaceutical packaging and intravenous administration equipment. (After YAHYA et al. 1986)

Chemical name	Plastic type	Repeating unit of the polymeric structure of the plastic
Cellulose acetate	Amorphous thermoplastic	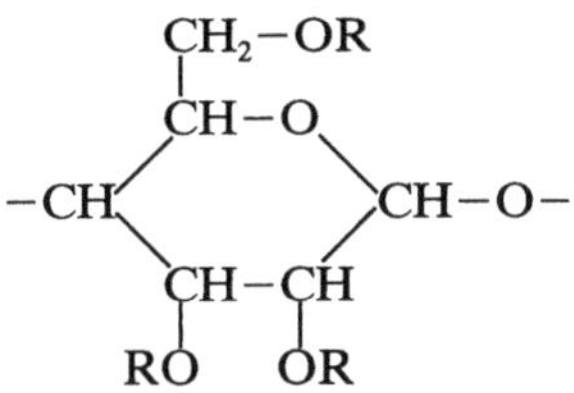 R = H
Cellulose propionate	Amorphous thermoplastic	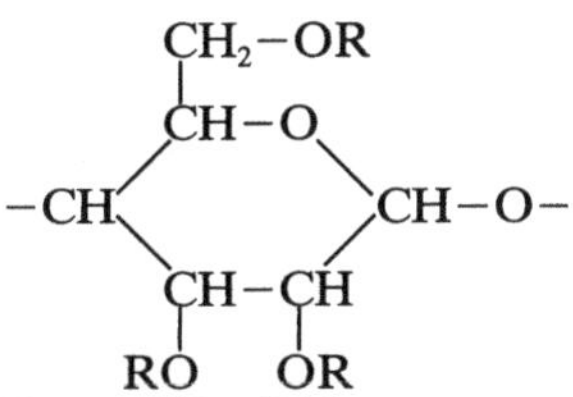 $R = -CO-C_2H_5$
Ethylvinyl acetate	Amorphous thermoplastic	$-CH_2-CH(-O-CO-CH_2-CH_3)-$
Methacrylate butadiene styrene	Amorphous thermoplastic	Co-polymer
Polyethylene	Partially crystalline thermoplastic	$-CH_2-CH_2-$
Polypropylene	Partially crystalline thermoplastic	$-CH_2-CH(CH_3)-$
Polystyrene	Crystalline thermoplastic	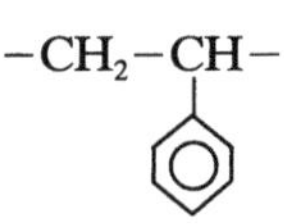
Polyvinylchloride	Amorphous thermoplastic	$-CH_2-CH(Cl)-$

polymer compounds which are widely used in plastic pharmaceutical packaging and intravenous administration equipment.

II. General Properties of Glass

Glass containers are particularly useful for liquid preparations owing to their rigidity, their superior protective qualities and their ability to allow easy inspection of the contents. Glass is impermeable to air and moisture, inert to most medicinal products and can be coloured to protect the contents from light of certain wavelengths (Lund 1994).

For most drug storage purposes, ordinary so-called soda glass or white flint glass containers are used. Amber glass is used for light-sensitive drugs and its composition differs little from white flint glass, except that a small proportion of iron oxide is added under strongly reducing conditions. Although glass is more resistant to chemical attack than other packaging materials, it is not entirely inert and, when it is in prolonged contact with water, alkali tends to be extracted. Alkaline solutions attack glass more rapidly than water and the higher the alkali content of the glass the more rapid is this effect. Thus for preparations which are liable to attack glass, a glass of low alkali content must be used.

Glass completely devoid of alkali is not practical but in borosilicate glass such as Pyrex the soda content is minimal at about 3.5%, the silica content is around 80% and the boric acid content is approximately 13%. Such glass is neutral and has very high chemical and thermal resistance. It is, however, hard and more difficult to process than glass with a higher alkali content. Borosilicate glass is widely used in laboratories where inert glass is required.

III. Mechanisms of Interaction with Pharmaceutical Packaging

There are a number of mechanisms which are considered central to any interpretation of drug-packaging interactions. These are:

1. Sorption – a term which describes both adsorption of a drug to a surface and the absorption of a drug into the matrix of the material under examination.
2. Leaching – this is the process by which components of the packaging material migrate into the medicine.
3. Permeation – this process involves the transfer of drug through the packaging material to the external surface. This can be followed by evaporation of the drug.
4. Polymer modification – in some cases, drugs can interact with the polymer in such a way as to modify the chemical structure of the polymer which may lead to changes in its physical properties.

Each mechanism will be discussed and illustrated with an appropriate example. Table 3 summarises a number of documenter drug-packaging interac-

Table 3. Examples of drug-packaging interactions

Drug	Example of interaction	Management of interaction	Reference
Antineoplastic agents			
Carmustine	Sorbed to PVC	Use glass containers	Benvenuto et al. (1981)
Bleomycin	Sorbed to PVC	Use glass containers	
Mitomycin	Sorbed to PVC and glass		
Doxorubicin	Sorbed to glass		
Fluorouracil	Sorbed to glass		
Methotrexate	Sorbed to glass when prepared in methanol or ethanol	Use plastic containers and protect from light	Chen and Chiou (1982)
Vinblastine	Sorbed to PVC and cellulose propionate	Use alternative plastic materials, e.g., methacrylate butadiene styrene	McElnay et al. (1988)
Immunosuppressant agents			
FK 506	Sorbed to PVC and leaching of plasticiser from PVC	Use polyolefin or glass containers	Taormina et al. (1992)
Hypnotic and psychotropic drugs			
Diazepam	Sorbed to PVC	Use glass or alternative plastic materials, e.g., polyethylene and polypropylene	Parker and McCara (1980)

Table 3. *Continued*

Drug	Example of interaction	Management of interaction	Reference
Chlopromazine	Sorbed to PVC	Use alternative form of plastic for administration, e.g., Stedim infusion bags with polyethylene lining	Airaudo et al. (1993)
Clomipramine	Sorbed to PVC		
Chlorazepate dipotassium salt	Sorbed to PVC		
Vitamins			
Vitamin A	Sorbed to PVC	Use vitamin A palmitate which does not sorb to PVC	Chiou and Moorhatch (1973)
Vitamin E	Sorbed to plastic materials in infusion sets (material not specified)	Use glass	Drott et al. (1991)
Vitamin D (calcitriol)	Sorbed to PVC	Increase amount of surfactant used in formulation as this reduces amount of free drug available for sorption. Avoid PVC	Li et al. (1992)
Miscellaneous agents			
Miconazole	Sorbed to PVC	Use glass; if PVC must be used, drug must be prepared and administered immediately	Holmes and Aldous (1991)
Bupivacaine	Sorbed to glass and plastic after alkalinistation (prolongs release of bupivacaine and thus increases duration of action)	Avoid alkalinisation as sorption effects conteract any clinical benefits in adjusting pH	Bomhomme et al. (1992)
Warfarin	Sorbed to PVC	Use glass	Kowaluk et al. (1981)

tions, the underlying mechanism of action and potential solutions to overcoming these problems.

1. Sorption

Sorption is probably the most important underlying mechanism responsible for a number of drug interactions associated with pharmaceutical packaging. This term describes loss of drug to the interacting material and includes adsorption to the surface plus absorption of the drug into the body of the material. Adsorption is characterised by rapid and substantial initial drug loss associated with binding to the surface, leading to surface saturation. Adsorption can occur to all solid surfaces irrespective of the nature of the material (ALLWOOD 1994). Absorption is defined as the migration of the drug into the matrix of the interacting material (usually plastic) and an equilibrium becomes established between liquid and solid phases. The process develops more slowly than adsorption, but saturation occurs eventually, although it may take many hours (ALLWOOD 1994). The plastic that is most often implicated in such interactions is polyvinyl chloride (PVC), although other materials can also be involved.

When examining sorption interactions, the physicochemical properties of the drug in question must be considered: the extent of ionisation and lipid solubility have been proposed as two determining factors of sorption by plastic infusion bags (KOWALUK et al. 1981). A number of models have been developed to predict solute sorption by PVC. ROBERTS et al. (1991) estimated the time course of the sorption of drugs using a diffusional model in which the plastic was assumed to act as an infinite sink. This model appeared to be suitable for estimation of storage times relevant to clinical usage and enabled the magnitude of the uptake in a specified time to be described by a single parameter known as the sorption number. This sorption number was defined by the plastic-infusion solution partition coefficient, the diffusion coefficient in the plastic, the fraction of drug unionised in solution, the volume of the infusion solution and the exposed surface area of the plastic. This parameter (sorption number) could be used to predict the effects of a number of factors, e.g., time, plastic surface area, solution volume and solution pH on fractional drug loss. JENKE (1993) developed a similar model in which storage time proved to be a critical factor in predicting sorption; at longer storage times, it was found that the infinite sink approach as described by ROBERTS et al. (1991) was inaccurate, while at shorter storage times the sorption profile for a solute can be estimated from its hexane-water and octanol-water partition coefficients via a simple diffusion-based expression. It was noted that differences between observed and predicted behaviour illustrated the problems in a theoretical approach; thus, it is only in the experimental or clinical environment that accurate assessment relating to drug sorption can be made.

Not all sorption interactions will be clinically significant. The schematic representation of an arbitrary time scale (Fig. 2) illustrates that those inter-

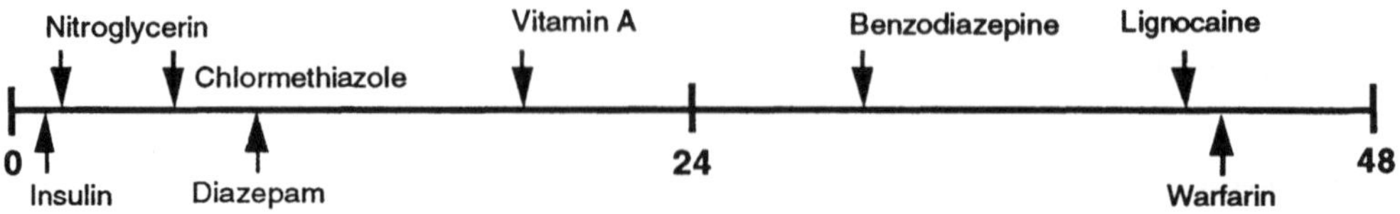

Fig. 2. Time-scale for clinically significant losses of drug due to sorption to plastics materials (Allwood 1995, personal communication)

actions occurring close to time 0 are clinically significant, whereas those occurring at the other end of the scale may be considered insignificant. Insulin and the nitrates are examples of drugs which are sorbed to a clinically significant degree; interactions involving these drugs are discussed below.

a) Insulin

The non-specific surface binding of insulin from dilute solutions was first reported by Ferrebee et al. (1951), who showed that the polypeptide was strongly adsorbed by laboratory glassware. Since then, further work has shown that insulin also binds to siliconised glassware, paper and polypropylene (Newerly and Berson 1957; Hill 1959; Petty and Cunningham 1974). Wideroe et al. (1983) evaluated the degree of insulin adsorption to plastic containers during continuous ambulatory peritoneal dialysis (CAPD). Approximately 65% of insulin which was added to the dialysis fluid was sorbed to the plastic container.

A series of studies by Twardowski et al. (1983a–c) examined this sorption phenomenon in greater depth. Binding of insulin to the plastic of dialysis solution containers was instantaneous and apparently not influenced by time of contact. Further experimental work revealed that insulin binding could be partly reversed by simple washing with water (Twardowski et al. 1983b). A third study demonstrated that insulin sorption was influenced by temperature: more insulin was bound at 37°C than at 24°C (Twardowski et al. 1983c).

McElnay et al. (1987) studied the binding of human insulin to three types of burette intravenous administration sets – a Standard set (cellulose propionate burette and PVC tubing), an Amberset (cellulose propionate burette and PVC tubing) and a Sureset (methacrylate butadiene styrene burette and polybutadiene tubing) over a 6-h period. The latter administration set had been previously shown to resist sorption of a number of drugs (Lee 1986). However, in this case, the Sureset bound insulin more extensively than the Standard set or Amberset as illustrated in Fig. 3. It was suggested that, in the clinical setting, this adsorption would present a particular problem since during intravenous infusion the concentration of drug is relatively low and the surface area available for binding is relatively high. This non-specific adsorption of insulin can best be approached by administering insulin in a small volume via a syringe pump as the surface area is much reduced compared with the total amount of insulin present or, alternatively, more drug than is required is added to the infusion container and blood sugar levels closely moni-

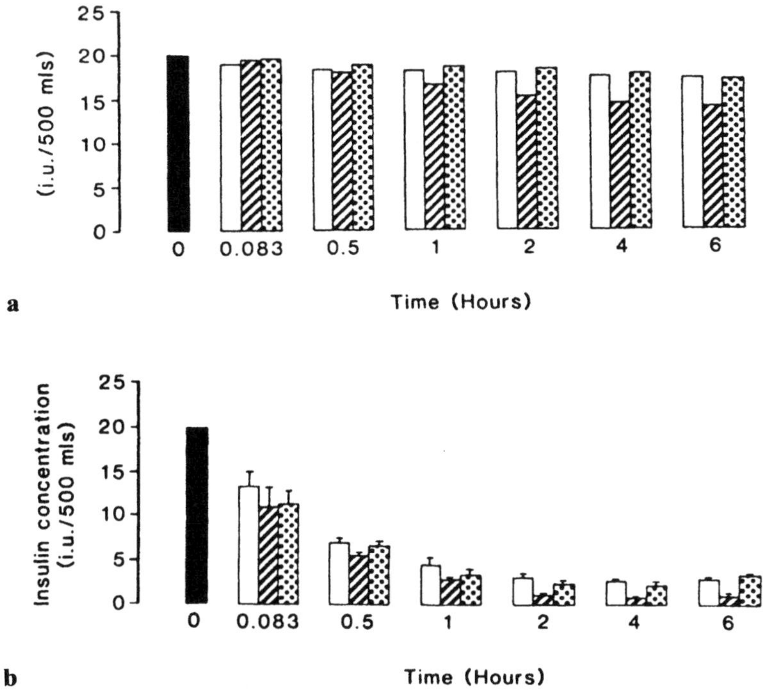

Fig. 3. a Concentration versus time profile for human insulin when stored in burette chambers of a standard set (□), Amberset (▩) and Sureset (▨) burette administration sets. ■ represents mean control data at zero time. (Maximum coefficient of variation within replicate data points was 3.2%.) **b** Concentration (± SD) versus time profile for human insulin when stored in administration tubing from a standard set (□), Amberset (▩) and Sureset (▨) burette administration sets. ■ represents mean control data at zero time. (After McElnay et al. 1987)

tored. It has also been suggested that human serum albumin or hydrolysed gelatin (polygeline) can be used as a carrier for insulin, since the binding to these agents exceeds that of the binding to surfaces (Schildt et al. 1978). This approach is, however, not used routinely.

b) Nitrates

A number of studies have reported interaction of a range of nitrates, e.g., glyceryl trinitrate and isosorbide dinitrate with packaging and intravenous delivery equpiment. A number of mechanisms of interaction appear to take place simultaneously with the nitrates including adsorption, absorption and permeation.

c) Glyceryl Trinitrate (Nitroglycerin)

The physical instability of glyceryl trinitrate can be partly attributed to sorption of drug from solution to plastic containers, Yuen et al. (1979) published the results of a study in which the equilibrium and kinetic techniques were used to examine the loss of drug efficacy. It was concluded that glyceryl trinitrate was removed from aqueous solution by plastic via an absorptive process, suggesting a degree of penetration into the plastic. It is also likely, due to the volatile nature of glyceryl trinitrate, that it penetrates through the plastic and is lost by evaporation from the plastic air interface.

The loss of glyceryl trinitrate was confirmed by Baaske et al. (1980), who studied the effect of intravenous filters, containers and administration sets on glyceryl trinitrate efficacy. Filters decreased glyceryl trinitrate concentrations by 2%–55%, whereas storage in glass bottles over 48 h had no effect. Storage in plastic intravenous infusion bags led to substantial concentration decreases that were related to surface contact area and temperature. Glyceryl trinitrate solution was also infused through an administration set, resulting in an immediate and substantial reduction in drug concentration. Using a theoretical approach, Malick et al. (1981) described a model of glyceryl trinitrate loss in which the drug is adsorbed onto the material surface and then partitioned into the plastic.

Work has continued on the glyceryl trinitrate sorptive phenomenon. Loucas et al. (1990), for example, examined the sorptive properties of glyceryl trinitrate in relation to PVC, with particular reference to the role of the admixture vehicle. Initially (first 10 min of sampling) the loss of glyceryl trinitrate to the PVC administration set was greatest from dextrose admixtures, intermediate for water admixtures and least for saline admixtures. Between 15 and 20 min of the experimental period, the greatest loss of glyceryl trinitrate was noted for saline. It was concluded that glyceryl trinitrate availability in admixtures in contact with PVC was dependent on the ionic strength of the vehicle and the time at which measurements were made.

The loss of glyceryl trinitrate during passage through a PVC or a polyethylene infusion set has also been investigated in relation to concentration and flow rate (Hansen and Spillum 1991). The concentration of glyceryl trinitrate had no influence on the loss of drug, although the greatest loss occurred at the slowest rate of infusion. A loss of 70% of glyceryl trinitrate during infusion using PVC apparatus was found, whereas the loss when using the polyethylene set did not exceed 15% over 8 h.

As PVC appears totally unsuitable as a material in which to store glyceryl trinitrate, other forms of plastics have been used. Salomies et al. (1994) measured the sorption of glyceryl trinitrate, diazepam and warfarin sodium (in normal saline) by a new polypropylene-lined infusion bag (Softbag) and compared this to sorption in glass bottles and PVC bags. The containers were stored at room temperature without protection from light for 24 h (diazepam) and 120 h (glyceryl trinitrate and warfarin). The three drugs did not demonstrate any sorption to glass bottles or the Softbag containers although sorption was high to PVC (Fig. 4). The polypropylene-lined Softbag therefore provides a useful approach in the avoidance of sorption of these agents.

Altavela et al. (1994) viewed that loss of glyceryl trinitrate onto administration sets may have little impact on the care of patients as infusions of the

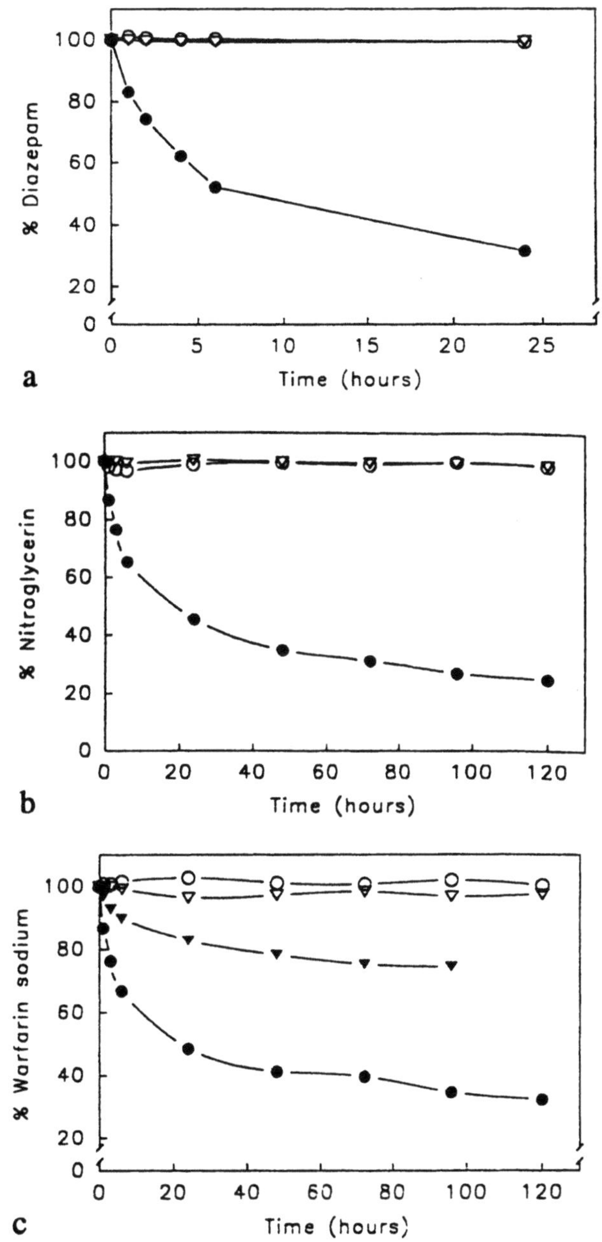

Fig. 4. **a** Concentration of diazepam (measured by UV spectrophotometry) as a function of time in glass bottles (○) and in PVC (●) and Softbag (▽) infusion bags. Values expressed as means of six determinations. Coefficient of variation (CV) values varied between 0.50% and 3.26%. **b** Concentration of nitroglycerin (measured by HPLC) as a function of time. Values expressed as means of six determinations. CV values varied between 0.71% and 2.21%. Symbols as in **a**. **c** Concentrations of warfarin sodium (measured by UV spectrophotometry) as a function of time in glass bottles (○) and in PVC pH 4.9 (●), PVC pH 5.6 (▼) and Softbag (▽) infusion. Values expressed as means of two (pH 5.6) or four (pH 4.9) determinations. CV values of the latter varied between 0.38% and 3.09%. (After SALOMIES et al. 1994)

drug are adjusted for the patient's response and thus drug loss may be of little significance. Another consideration was the expense of polyethylene sets compared to PVC sets; thus this group randomly assigned patients with ischaemic heart disease to receive glyceryl trinitrate solution through either a polyethylene administration set or a PVC set. It was found that patients who received glyceryl trinitrate through the PVC set had the same clinical response as patients who had received the drug through the polyethylene set and there were no differences between the initial dosage adjustment time, the number of dosage adjustments and indications for them, the duration of the infusion, the time of tubing changes and total dose of glyceryl trinitrate.

d) Isosorbide Dinitrate

This long-acting nitrate has been used orally, sublingually and intravenously in the treatment and prevention of angina and in congestive heart failure therapy. As stated in its current United Kingdom data sheet, isosorbide dinitrate is incompatible with PVC containers, but may be used with glass and polyethylene (Association of British Pharmaceutical Industry 1994). Cossum and Roberts (1981) investigated the availability of this drug from intravenous delivery systems. Appreciable loss of the drug was found following storage in PVC infusion bags and/or in burettes (cellulose propionate) of administration sets. This degree of loss was also related to the flow rate of the infusion; lower rates of recovery of the nitrate were found at slower flow rates. No loss of isosorbide dinitrate was found during the storage of the drug in glass for 200h.

Sautou et al. (1994) evaluated the compatibility of isosorbide dinitrate and heparin with polypropylene, polyethylene and PVC under simulated clinical conditions. Results indicated that both drugs were compatible with polypropylene syringes; preparing the syringes 8h in advance did not alter the stability of the drugs, provided that the nitrate was not refrigerated. The concentration of isosorbide dinitrate infused through PVC/polyethylene tubing (PVC on the outside and polyethylene on the inside) was unaffected, irrespective of whether it was administered alone or together with heparin. As was expected, both drugs were sorbed by the PVC. The nitrate was rapidly fixed to the PVC and then released again; this release occurred earlier when heparin was co-administered. It was concluded by the authors that adsorption and release of the drugs appeared to be very random.

2. Leaching

The term leaching is applied to the migration of a substance from packaging material into a medicine. It is influenced in rate and extent by the solvent system used in the preparation, and pH and temperature conditions during processing and storage (Cooper 1974). Examples are the leaching of alkali from soda-glass bottles and the leaching of zinc salts, used as activators, from rubber closures. Barium ions leached from borosilicate glass may react with sulphates of drugs such as kanamycin, atropine or magnesium to form barium sulphate crystals; the presence of sulphurous acid salts as antioxidants in solutions may result in the formation of crystals (LUND 1994).

Leaching of plastics components has received a great deal of attention, especially in relation to parenteral preparations. MOORHATCH and CHIOU (1974) examined the leaching of chemicals from PVC bags containing various solutions. Distilled water, normal saline, 5% dextrose solution, 5% alcohol solution and aqueous buffers produced insignificant leaching from PVC bags. Solutions of the surfactants Tween 20 and Tween 80 produced significant leaching of the plasticiser diethylhexyl phthalate (DEHP) from PVC bags. Leaching was found to increase with the surfactant concentration and storage time. VENKATARAMANAN et al. (1986) studied the leaching of DEHP from flexible PVC containers into intravenous cyclosporin A solutions. As found in the MOORHATCH and CHIOU study (1974), the amount of DEHP leached into solutions was dependent on storage time. By 48h, 33mg DEHP had leached into the solution. The leaching effect was attributed to the presence of a non-ionic surfactant, polyoxyethylated castor oil, the injection vehicle for cyclosporin A. PEARSON and TRISSEL (1993) found that DEHP was leached from PVC containers by polysorbate 80, polyoxyethylated castor oil, cyclosporin, miconazole, teniposide, chlordiazepoxide hydrochloride, etoposide and the vehicles used in paclitaxel and taxotere formulations. The importance of the paclitaxel vehicle as a causative agent of leaching was further confirmed by a later study (TRISSEL et al. 1994). Paclitaxel, used in metastatic ovarian cancer, is formulated in 49% dehydrated alcohol and polyoxyethylated castor oil, the latter being known to leach the plasticiser (DEPH) from PVC infusion containers and intravenous administration sets (PEARSON and TRISSEL 1993). It is recommended that paclitaxel is not to be used with polyvinyl chloride equipment; a short PVC inlet or outlet on a filter may be acceptable (BRITISH NATIONAL FORMULARY 1995). Glass, polypropylene and polyolefin containers are recommended for use with this formulation. This recommendation was largely supported by TRISSEL's study, in which the compatibility of the paclitaxel injection vehicle (drug was not used), corresponding to paclitaxel injection 0.3 and 1.2mg/ml, was evaluated with a variety of intravenous administration and extension sets by measuring the amount of DEHP leached from the sets (TRISSEL et al. 1994). All the extension sets were compatible with the injection vehicle; however, two administration sets were incompatible with the vehicle simulating the low concentration (0.3mgl/ml) injection, and five sets were incompatible with the vehicle simulating the high concentration (1.2mg/ml) injection. In addition, leaching of DEHP occurred, despite some of the incompatible administration sets being labelled as not containing PVC. Careful selection of administration and extension sets was suggested as one way of overcoming potential leaching when using paclitaxel injection.

In earlier work, ALLWOOD (1986) demonstrated that the Intralipid component of a TPN mixture could also extract DEHP from PVC administration sets. In this study, the sets were primed according to the manufacturer's recommendations and contained the fat emulsion alone (10% or 20%), a TPN regimen containing fat emulsion (20%) or a TPN regimen without added Intralipid; storage was at 4°–6°C or at ambient temperature for 24h. Extraction depended on the storage temperature; refrigeration (4°–6°C) reduced the leaching of DEHP. Slightly more DEHP was extracted into the 10% compared to 20% Intralipid, although the difference was not significant. The rate of extraction of the plasticiser into a TPN regimen containing the fat emulsion was substantially lower than into undiluted Intralipid; quantities released into the TPN regimen without Intralipid were negligible. It was suggested that patients receiving Intralipid infused through a standard PVC-containing administration set would receive a small dose of DEHP. Although it was considered that this would not be a problem in most patients, pregnant mothers and neonates would be at greater risk and, thus, it is now generally accepted that non-DEHP-containing sets (made from ethyl vinyl acetate) should be used to administer Intralipid-containing TPN fluids.

3. Permeation

Permeation is another physicochemical mechanism of drug binding associated with some forms of pharmaceutical packaging. It is a combination of absorption followed by migration to the outer surface, from which the drug is released by evaporation. Losses may be substantial and they continue throughout the period of administration since saturation of the plastic is never achieved (ALLWOOD 1994). Permeation may also occur due to passage of gases, vapours and liquids through packaging; water vapour permeates through silicone rubber, and to varying degrees through plastics, particularly PVC and polystyrene (LUND 1994).

In a study conducted by KOWALUK et al. (1981) a number of hypnotic agents were investigated in relation to incompatibility with PVC infusion bags. Chlormethiazole and diazepam were found to be substantially lost after 24h of storage (33% and 20%, respectively). The chlormethiazole was shown to penetrate through the plastic as evidenced by a distinctive odour of the drug in the immediate vicinity of the bag. COSSUM and ROBERTS (1981) also reported a loss in drug concentration when chlormethiazole was stored and infused through PVC bags and infusion sets; the average loss from the infusion fluid approximated to 15%–30%. Loss was negligible when infused from glass syringes through polyethylene tubing, thereby presenting an alternative to PVC materials. As mentioned above the loss of glyceryl trinitrate from intravenous fluid containers involves a degree of permeation through plastic materials followed by loss by evaporation (particularly in the case of PVC). This phenomenon is not limited to intravenous preparations of the drug, i.e., sublingual glyceryl trinitrate should be stored in glass bottles.

4. Polymer Modification

In the work of KOWALUK et al. (1981) described in the previous section, it was also reported that the PVC infusion bags containing chlormethiazole became more pliable and softer, particularly at higher concentrations of the drug. It would seem that chlormethiazole may be responsible for this effect, rather than a surfactant or cosolvent as it is formulated in an aqueous vehicle. It was suggested that at high concentrations chlormethiazole caused greater PVC polymer-chain flexibility through plasticisation. In this way, it enhanced its own sorption and permeation through the plastic. Such effects were partly attributed to its high lipid solubility.

IV. Drug Interactions with Contact Lenses

Although this phenomenon is somewhat different to the interactions already discussed, it is still appropriate to discuss drug interactions with contact lenses in this section. A number of drugs are known to cause problems for those who wear contact lenses. Reports have suggested that some drugs can enter into surface interactions with the lens: rifampicin is associated with a permanent and damaging discoloration of the lens plastic (LYONS 1979). Adenochrome pigmentation associated with the use of adrenaline eye drops has been reported, involving the hydrophilic lenses of three elderly patients; in each case the lens had dense brown coloration (SUGAR 1974). Sulphasalazine used in the treatment of colitis has been implicated in the staining of a soft contact lens (RILEY et al. 1986). The patient wore the soft contact lens in one eye and a gas-permeable lens in the other. During the first 48 h of sulphasalazine treatment, the lenses were not worn; after this time, the lenses were refitted and yellow staining of the soft lens was noted. It is now commonplace for those patients undergoing topical treatment with eye drops and who also wear lenses to be advised to remove their lenses for the duration of treatment, unless eye drops are specifically indicated as safe to use with contact lenses. It is more difficult, however, to predict possible interactions in those patients undergoing systemic therapy, and it is only as a result of isolated reports in the literature that clinicians are in a position to advise patients accordingly.

D. In Vitro Drug Interactions in Therapeutic Drug Monitoring and Drug Analysis

The examples discussed above represent interactions which may occur in the clinical setting. However, there are also examples of in vitro drug interactions occurring in the laboratory setting which may have implications for therapeutic drug monitoring and drug assays, and in turn may have clinical consequences.

I. Cyclosporin

This cyclic peptide is widely used in organ and tissue transplantation and is associated with a number of side effects, most notably nephrotoxicity (CALNE et al. 1978). This has led to the development of a number of therapeutic monitoring techniques, e.g., radioimmunoassay (RIA) and high-performance liquid chromatography (HPLC). However, a number of these techniques gave erroneous cyclosporin levels due to analytical problems (Shaw et al. 1987). The use of an inappropriate pipette for manipulating cyclosporin standards has led to low measured concentrations of the drug. This was due to the pipette having a plastic tip (rather than a glass tip) to which cyclosporin adsorbed. Sorption was attributed to the lipophilic nature of cyclosporin (McMillan 1989) and it is now recommended that serum level monitoring techniques should use glass where possible.

II. Chloroquine

Geary et al. (1983) showed that in the laboratory setting preparations of chloroquine in various solutions showed decreases in concentration of up to 40% when stored in glass containers and that the drug was extensively bound to cellulose acetate filters. Yahya et al. (1985) examined the binding of chloroquine to different grades of laboratory glassware at different concentrations and different pH conditions. Storage in soda glass (test tubes and glass wool) showed a decrease in original drug concentration of up to 60% and 97%, respectively. The highest degree of binding was recorded at physiological pH (7.4) and at low concentrations. Borosilicate glass did not exhibit any chloroquine binding. A second study investigated the binding of this antimalarial agent to a range of plastic materials under conditions of varying pH and concentration (Yahya et al. 1986). Chloroquine was sorbed by cellulose propionate, ethylvinyl acetate, methacrylate butadiene styrene, polyethylene, polypropylene and PVC, but not by high-density polystyrene. Passing chloroquine solutions through cellulose acetate filters once resulted in a high loss of drug. The unionised form of the drug was preferentially sorbed. These studies indicate that care must be taken when quantifying chloroquine in the laboratory since the drug will come into contact with a range of materials. This is particularly the case in kinetic studies and malarial sensitivity testing, when chloroquine may be bound to plastic syringes or pipettes during sampling; binding to membrane filters used to clarify and/or to sterilise chloroquine solutions prior to use or analysis could also give rise to erroneous results (Yahya et al. 1986). Although this problem has been highlighted for chloroquine, it is well known that care must be taken in the manipulation of other drugs in the laboratory setting, e.g., some benzodiazepines are highly bound to glassware and this necessitates the silylation of glassware prior to use.

E. In Vitro Drug Interactions as a Result of Formulation Changes

The final category of in vitro drug interactions which will be discussed in this chapter are those which occur as a result of changes in a product formulation;

this may result in changes in bioavailability of the active drug constituent and there have been a number of reported cases where this has led to adverse clinical consequences for patients (McElnay and D'Arcy 1980). Other reports are from experimental studies which stress the importance of formulation in relation to bioavailability.

The two classical examples of this type of in vitro effect involve the drugs phenytoin and digoxin. Anticonvulsant intoxication resulted from subtle changes to the formulation of phenytoin capsules (Tyrer et al. 1970); lactose replaced calcium sulphate as an excipient, resulting in increased dissolution and higher serum levels of phenytoin as illustrated in Fig. 5. Particle size changes to digoxin also led to marked changes in digoxin levels (Johnson et al. 1973). Tablets manufactured after May 1972 contained a smaller particle size digoxin, which resulted in a higher degree of absorption of the drug and overdigitilisation of patients. This led to much stricter statutory controls in relation to formulation in developed countries. However, a later study by Fletcher and Summers (1980) revealed that these controls did not extend to all countries. Five brands of digoxin, which were available in South Africa at the time, were compared in relation to physical and dissolution characteristics and bioavailability. Only one brand satisfied all requirements, based on official specifications. One brand did not satisfy diameter specifications, was soft and crumbled easily, and showed only 34.82% dissolution within the period examined. A second brand was viewed as a very low grade product which neither dispersed readily nor complied with the specifications for digoxin content. It is this type of report which has given rise to concerns over the quality of some generic products available in developing countries. For most products, this is

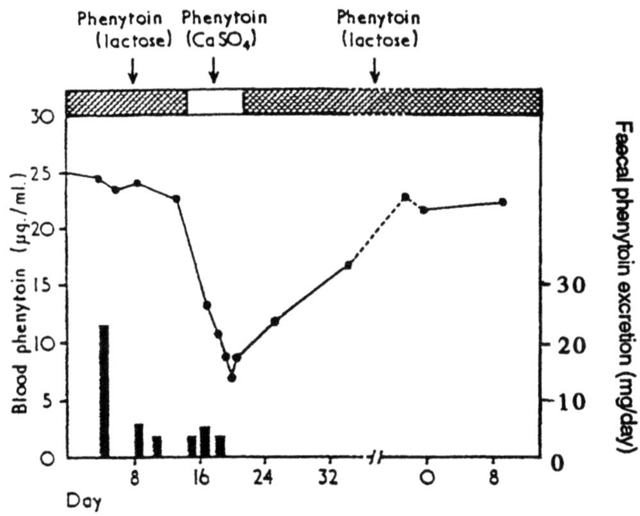

Fig. 5. Blood phenytoin concentrations in a patient taking phenytoin (400 mg/day), with excipients respectively as shown (lactose, calcium sulphate, lactose). *Vertical columns* represent daily faecal excretion of phenytoin when measured. (After Tyrer et al. 1970)

not a problem and slight changes in bioavailability are of no clinical importance; however, many authorities recommend that products such as phenytoin and lithium, which have a narrow therapeutic index, should be prescribed by proprietary (brand) name to ensure that the patient receives the same formulation each time.

I. Problems with Sustained-Release Formulations

One other drug for which prescription by brand name is strongly recommended and whose formulations and kinetics has been widely investigated is theophylline. Weinberger et al. (1978), in a study of the relation of product formulation to absorption, found that the absorption of theophylline from both a solution and uncoated tablets was rapid and complete. Three of six sustained-release formulations were completely and consistently absorbed, although the absorption was slower, as would be expected, than with the solution and uncoated tablets. Absorption of theophylline from the other three sustained-release products appeared to be more erratic and less complete as shown in Fig. 6. Further work by this group examined the absorption characteristics of a once-daily dosage formulation, Theo-24, compared to a theophylline reference solution, in healthy volunteers (Hendeles et al. 1985). Absorption of Theo-24 after an overnight fast was very slow, with only 71% ± 6% (±SE) of the dose ultimately absorbed. In contrast, food caused precipitous "dose-dumping", resulting in peak levels of drug in the serum that averaged 2.3 times higher than after a fasting dose. About half of the dose was absorbed in a 4-h period, generally beginning 6–8 h after postprandial administration, and complete absorption was then attained within 24 h. Theophylline toxicity was evident in four subjects when they took the dose postprandially whereas no toxic effects occurred during the fasting regimen. Although this example is not strictly an in vitro drug interaction as previously defined, it does illustrate the influence of formulation on clinical outcome. The pattern of absorption following breakfast was consistent with a greatly increased dissolution of the formulation (encapsulated coated beads which are designed to dissolve slowly) following exposure to the alkaline bile salts or pancreatic enzymes (or both) that are secreted following post-prandial gastric emptying.

Theophylline is available worldwide as a number of modified-release formulations, e.g., Nuelin, Slo-Phyllin, Theo-Dur and Uniphyllin. Despite a wide availability of branded products, work is still continuing on the development of other modified-release theophylline preparations. Modification of the formulation components gives rise to changes in the dissolution, binding to and release from different formulations. Numerous publications in the pharmaceutical literature explore the influence of formulation on drug release and consequent bioavailability. Illustrative of this work is a recent study by DiLuccio et al. (1994), who examined crystalline polyvinyl alcohol (PVA) polymer and low-crystallinity polyvinyl alcohol-methyl acrylate copolymer (PVA-MA) as sustained-release excipients with theophylline. By blending of different pro-

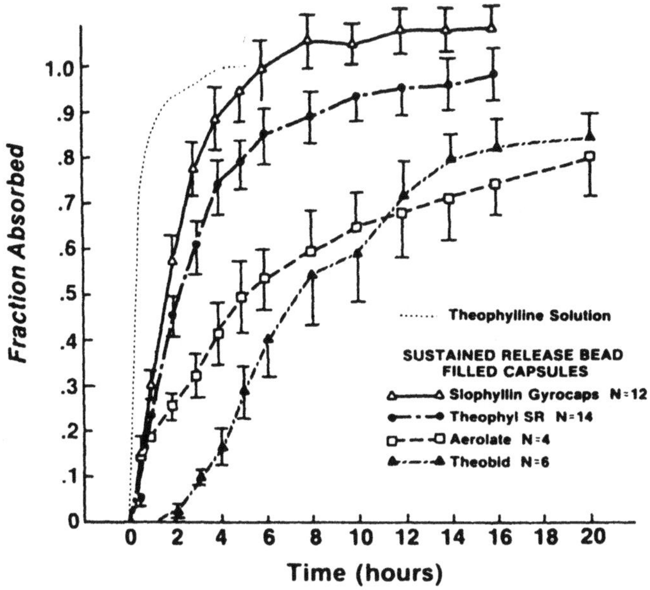

a

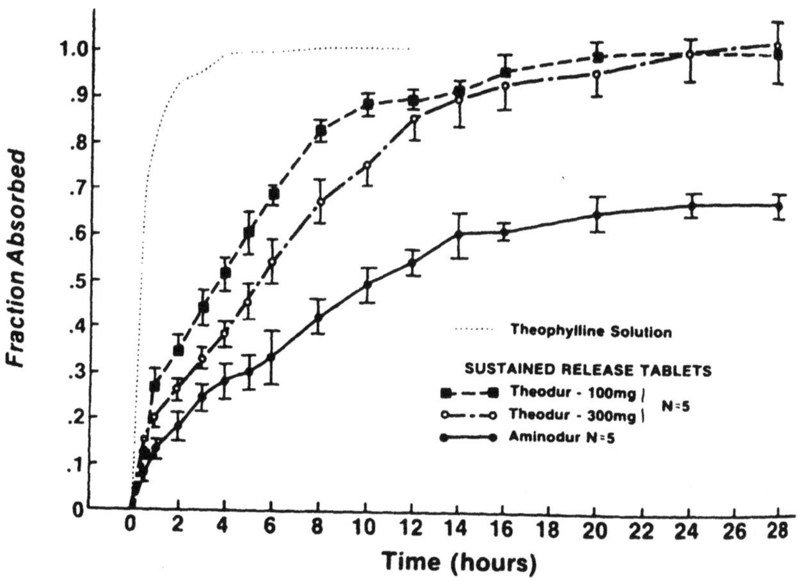

b

Fig. 6. **a** Absorption of sustained-release bead-filled capsules of theophylline. **b** Absorption of sustained-release tablets of theophylline. Aminodur, Aerolate and Theobid are seen to have suboptimal absorption. (After WEINBERGER et al. 1978)

portions of PVA and PVA-MA it was possible to affect the release characteristics of the drug. Tablets made with crystalline PVA provided instant release of theophylline in vitro, whereas those made with a larger proportion of PVA-MA relative to PVA demonstrated a very prolonged release profile. It was concluded that the crystallinity of the polymer played a major role in the release of the drug, as high-crystallinity polymers (PVA) are less permeable to diffusion than low-crystallinity polymers (PVA-MA); the latter are known to dissolve more slowly in water and hence drug release is sustained. The bioavailability (in dogs) of a formulation containing PVA:PVA-MA:theophylline in the ratio of 1:9:10 was equivalent to that obtained after administration of a commercially available product Theo-Dur.

As previously stated, theophylline is a drug which is normally prescribed by brand name to ensure that the patient receives the same release/absorption patterns of the drug each time it is administered. A recent study by HENDELES et al. (1995) evaluated the relative bioavailability and clinical efficacy of a generic slow-release theophylline tablet and Theo-Dur in 14 adults with asthma. The results of the study indicated that the generic slow-release theophylline tablet and Theo-Dur were not bioequivalent. Although they were absorbed to the same extent, as indicated by similar areas under the serum concentration-time curves and peak concentrations, the generic product was absorbed more rapidly. This was demonstrated by the shorter time to peak concentration, larger fluctuations between peak and trough concentrations and slightly lower mean trough concentration. Therapeutic equivalence was also assessed using attenuation of the response to exercise as a surrogate for clinical effect (exercise challenge was 17h after last dose of theophylline). Although neither product effectively attenuated exercise-induced bronchospasm, the authors considered that despite the differences in absorption rates both formulations were therapeutically equivalent, but not bioequivalent.

As a result of stricter statutory controls with regard to changes in product formulation, there are now relatively few clinical reports of drug interactions arising from formulation modification.

II. Drug Excipient Interactions

The BRITISH PHARMACEUTICAL CODEX (LUND 1994) documents a number of drug-excipient interactions. Ionic interactions between cationic and anionic entities may be particularly problematic. Carmellose sodium, an anionic polymer, reacts with large cations such as neomycin and chlorpromazine in solution; formation of a precipitate depends upon the conditions and also the concentrations of each substance. Benzalkonium chloride and other cationic antimicrobial preservatives are inactivated to varying degrees in the presence of carbomer and other anionic polymers. Complex formation between drug and excipient may also be common. Phenolic preservatives may be partly inactivated in complexes formed with macrogol derivatives. The non-ionic excipient povidone may complex with anionic or cationic dyes, and drugs such as chlorpromazine and chloramphenicol. Gels made with povidone may increase in viscosity as a result of complexation with 8-hydroxyquinolone sulphate or thiomersal sodium. Excipients (or another drug) may also affect the stability of a drug, leading to reduced bioavailability and changed therapeutic efficacy as reported by the BRITISH PHARMACEUTICAL CODEX (LUND 1994): sodium metabisulphite rapidly inactivates cisplatin and propylene glycol and macrogols catalyse the degradation of benzoyl peroxide to benzoic acid and carbon dioxide.

An early study by MCGINITY and LACH (1976), using dissolution and dialysis techniques, examined the adsorption of various drugs to montmorillonite clay; this colloidal magnesium aluminium silicate clay has been used as a disintegrant, binder and lubricant. Chlorpheneramine maleate, amphetamine sulphate and propoxyphene hydrochloride were all found to be strongly bound to montmorillonite due to their cationic nature; amphoteric compounds such as theophylline and caffeine were found to bind moderately to the clay, whereas sodium salicylate and sulphanilamide (anionic in nature) were released rapidly from montmorillonite, i.e., binding was minimal.

Although these latter binding interactions are largely regarded as detrimental in that drug absorption will be reduced, the binding can be used to advantage in the formulation of sustained-release products. The interaction of piroxicam, for example, with various synthetic polymers (polyethylene glycol 400, 600, 1000, 4000, 6000 and 20000, polyvinylpyrrolidone 30000, 40000 and 64000, and dextran 40000 and 75000) has been determined by ELSHATTAWY et al. (1994). The binding of piroxicam in different concentrations with the different polymers showed that increasing drug concentration was accompanied by an increase in the amount of bound drug to the polymer, until near saturation. The binding capacity of piroxicam was found to be greatest for the polyvinylpyrrolidones and least for dextran.

CHOWHAN and CHI (1986a) compared the two lubricants magnesium stearate and sodium stearyl fumarate under identical conditions, in order to study their roles in drug-excipient interactions; the drug in question was ketorolac. After prolonged mixing, sodium stearyl fumarate did not affect the drug of excipient (crospovidone) and, as a result, the disintegration time and drug dissolution from hand-filled capsules were not adversely affected. In contrast, magnesium stearate did demonstrate a drug-crospovidone interaction which resulted in an increased disintegration time and dissolution time. This resulted from particle-particle interaction whereby lamination and flaking of magnesium stearate occurred. These flakes adhered to drug particles and caused a significant reduction in the dissolution rate. In a further study, using prednisone, the effect of powder mixing on drug-excipient interactions and their effect on in vitro dissolution from capsules was studied (CHOWHAN and CHI 1986b). Two powder formulations contained dibasic calcium phosphate dihydrate as a filler and potato starch or sodium starch glycolate as a disintegrant were studied. The third powder formulation contained pregelatinised starch as a disintegrant/filler. The lubricant in all formulations was magnesium stearate. When prednisone, calcium filler and disintegrant were mixed thoroughly and hand-filled into capsules, only 70% of the drug dissolved in 30 min. The decrease in the rate of drug dissolution resulted from formation of agglomerates (calcium filler and disintegrant) and the inclusion of drug particles in these agglomerates. When a mixture of prednisone and pregelatinised starch was used, complete dissolution of the drug was achieved after 30 min; formation of agglomerates did not occur with this formulation, and complete dissolution resulted.

The compatibility of pyridoxine hydrochloride (vitamin B_6) with various tableting excipients has been investigated using a variety of techniques, e.g., isothermal stress testing and differential scanning calorimetry (Durig and Fassihi 1993). The experiments revealed chemical incompatibilities between the drug and mannitol, lactose and corn starch. Various cellulose derivatives and colloidal silicone dioxide were identified as strong stabilising factors in formulations. It appeared that these excipients acted as moisture scavengers, thereby decreasing the amount of free water available for interaction with pyridoxal hydrochloride.

In a more recent study, cellulose derivatives were found to have an effect on the stability of diethylpropion both in terms of stability and bioavailability (Gomez et al. 1993). In vitro results showed that sodium carboxymethylcellulose and methylcellulose caused a marked degradation of diethylpropion. Dissolution assays revealed that the availability of the drug was also vastly reduced by these cellulose derivatives.

III. Formulation Effects on Rectal Bioavailability

Bioavailability/formulation interactions are not restricted to those preparations which are administered orally. This was illustrated by McElnay and Nichol (1984), who compared the performance of a novel flow-through bead-bed dissolution apparatus for suppositories with that of the more traditional rotating basket technique. Similar dissolution profiles were obtained for fatty-based benzocaine suppositories using the two methods at 37°C. A second set of experiments measured the dissolution profiles, using these two techniques, of two commercially available indomethacin suppository formulations (Indocid and Imbrilon). Although both products contained the same dose of indomethacin in a polyethylene glycol base, clear differences were noted between the dissolution profiles of the two products; a markedly decreased rate of dissolution was noted for Indocid suppositories and this difference was particularly marked when using the flow-through bead-bed apparatus (Fig. 7). A later study examined the serum indomethacin profiles after rectal administration of these two brands of indomethacin in healthy volunteers (McElnay et al. 1986). Statistical analysis indicated that there were no significant differences between the two formulations in relation to any of the pharmacokinetic parameters which were compared, although the time taken to attain maximal

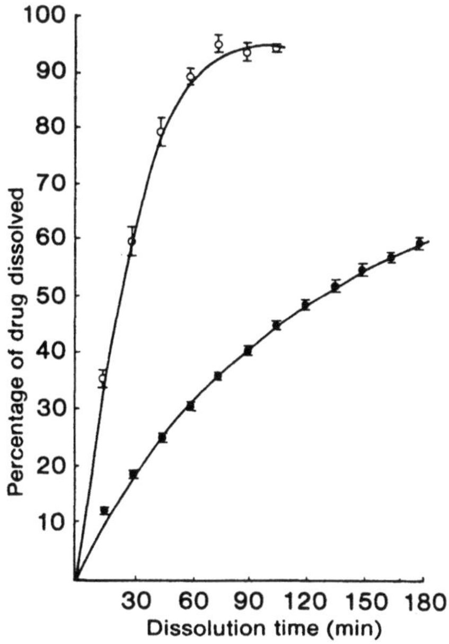

a

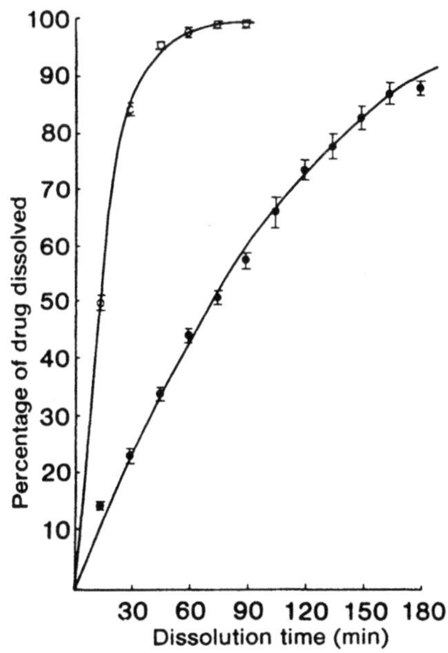

b

Fig. 7. a Dissolution profiles of indomethacin suppositories obtained using the flow-through bead-bed apparatus: ●, Indocid; ○, Imbrilon. Mean ± SE data presented. **b** Dissolution profiles of indomethacin suppositories obtained using the rotating basket apparatus: ●, Indocid; ○, Imbrilon. Mean ± SE data presented. (After McElnay and Nicol 1984)

plasma concentration was somewhat greater with the Indocid suppositiories (1.75 ± 0.16 h) than with the Imbrilon formulation (1.42 ± 0.17 h).

Morgan et al. (1992) investigated the release of morphine (15 mg) from a number of suppository bases using the United States Pharmacopeia rotating basket dissolution apparatus. Morphine hydrochloride in polyethylene glycol (a hydrophilic suppository base), morphine alkaloid in polyethylene glycol, and morphine hydrochloride in Novata BBC (a lipophilic base) completely released the drug within 25 min, whereas morphine alkaloid in Novata BBC released the drug over 10 h. This prolonged release was also seen in vivo when all formulations were used in a clinical trial; morphine alkaloid in Novata BBC was viewed as a useful long-acting preparation of the opioid and warranted further investigation. Clearly, the choice of suppository base was critical in relation to bioavailability of the drug.

F. Conclusions

Interpretation of in vitro drug interactions involves a number of basic principles (Allwood 1994):

1. An understanding of the chemistry of drugs in the salt and/or free form and the effects of pH
2. A knowledge of the basis for physicochemical interactions between solutes and solids
3. Implementing practical solutions to overcome an in vitro drug interaction
4. Evaluation of the clinical effect of the in vitro drug interactions

Many of the interactions which have been discussed in this chapter may be explained in chemical and physical terms. Once the interaction has been elucidated a number of approaches can be used to overcome or avoid unwanted sequelae. Ways of overcoming the interactions include:

1. Avoidance of direct contact of interacting agents, e.g., via use of multiuse cannulae for intravenous administration
2. Adding increased doses of drugs to infusion fluids when the drug is bound to the infusion apparatus
3. Ensuring that oral formulations are adequately tested for bioavailability prior to clinical use
4. Use of glass or suitable plastics which do not interact with drugs in question, e.g. use of polypropylene-lined infusion bags and administration sets for drugs which are highly bound to PVC

References

Airaudo CB, Gaytesorbier A, Bianchi C, Verdier M (1993) Interactions between six psychotherapeutic drugs and containers: influence of plastic material and infusion solutions. Int J Clin Pharmacol Ther 31:261–266

Allwood MC (1986) The release of phthalate ester plasticizer from intravenous administrations sets into fat emulsion. Int J Pharm 29:233–236
Allwood MC (1994) Drug stability and intravenous administration, 2nd edn. Clinical pharmacy practice guide. UKCPA, University of Leeds Media Service, Leeds
Altavela JL, Haas CE, Nowak DR, Powers J, Gacioch GM (1994) Clinical response to intravenous nitroglycerin infused through polyethylene or polyvinyl chloride tubing. Am J Hosp Pharm 51:490–494
Association of British Pharmaceutical Industry (ABPI) (1994) Data sheet compendium 1994/95. Datapharm, London
Autain J (1963) Plastics in pharmaceutical practice and related fields, part I. J Pharm Sci 52:1–23
Baaske DM, Amann AH, Wagenknecht DM, Mooers, M, Carter JE, Hoyt HJ, Stoll RG (1980) Nitroglycerin compatibility with intravenous fluid filters, containers, and administration sets. Am J Hosp Pharm 37:201–205
Benvenuto JA, Anderson RW, Kerkof K, Smith RG, Loo TL (1981) Stability and compatibility of antitumor agents in glass and plastic containers. Am J Hosp Pharm 38:1914–1918
Birely AW, Scott MJ (1982) Plastic materials, properties and applications. Blackie, Glasgow, pp 1–18
Bonhomme L, Benhamou D, Beugre T (1992) Slow-release effect of pH-adjusted bupivacaine: in vitro demonstration. Int J Pharm 84:33–37
British National Formulary (1994) No 28. British Medical Association and Pharmaceutical Press, London
British National Formulary (1995) No 29. British Medical Association and Pharmaceutical Press, London
Brydson JA (1982) Plastics materials. Butterworth, London, pp 18–143
Calne RY, White DRG, Thiru S (1978) Cyclosporin A in patients receiving renal allografts from cadaver donors. Lancet II:515–524
Carter SJ (1975) Cooper and Gunn's dispensing for pharmaceutical students, 12th edn. Pitman, London, pp 253–264
Chen ML, Chiou WL (1982) Adsorption of methotrexate onto glassware and syringes. J Pharm Sci 71:129–131
Chiou WL, Moorhatch P (1973) Interaction between vitamin A and plastic intravenous bags. JAMA 223:328
Chowhan ZT, Chi LH (1986a) Drug-excipient interactions resulting from powder mixing. IV. Role of lubricants and their effect on in vitro dissolution. J Pharm Sci 75:542–545
Chowhan ZT, Chi LH (1986b) Drug-excipient interactions resulting from powder mixing. III. Solid state properties and their effect on drug dissolution. J Pharm Sci 75:534–541
Collins JL, Lutz RJ (1991) In vitro study of simultaneous infusion of incompatible drugs in multilumen catheters. Heart Lung 20:271–277
Cooper J (1974) Plastic containers for pharmaceuticals testing and control. WHO, Geneva, pp 1–13
Cossum PA, Roberts MS (1981) Availability of isosorbide dinitrate, diazepam and chlormethiazole from i.v. delivery systems. Eur J Clin Pharmacol 19:181–185
DiLuccio RC, Hussain MA, Coffin-Beach D, Torosian G, Shefter E, Hurwitz AR (1994) Sustained-release oral delivery of theophylline by use of polyvinyl alcohol and polyvinyl alcohol-methyl acrylate polymers. J Pharm Sci 83:104–106
Drott P, Meurling S, Meurling L (1991) Clinical adsorption and photodegradation of the fat-soluble vitamin A and vitamin E. Clin Nutr 10:348–351
Durig T, Fassihi AR (1993) Identification of stabilizing and destabilizing effects of excipient-drug interactions in solid dosage form design. Int J Pharm 97:161–170
Elshattawy HH, Ahmed EI, Bayomi MA, Abdelfatah IM (1994) Piroxicam synthetic polymers interactions. Pharm Ind 56:396–399
Ferrebee JW, Johnson BB, Mithoefer JC, Gardella JW (1951) Insulin and adrenocorticotropin labelled with radio-iodine. Endocrinology 48:277–283

Fletcher S, Summers RS (1980) Physical and dissolution characteristics and bioavailability of five brands of digoxin tablets. S Afr Med J 57:530–533
Fournier C, Hecquet B, Bastian G, Khayat D (1992) Modification of the physicochemical and pharmacological properties of anticancer platinum compounds by commercial 5-fluorouracil formulations – a comparative study using cisplatin and carboplatin. Cancer Chemother. Pharmacol 29:461–466
Geary TG, Akood MA, Jensen JB (1983) Characteristics of chloroquine binding to glass and plastic. Am J Trop Med Hyg 32:19–23
Griffin JP, D'Arcy PF (1988) Drug interactions in vitro. In: A manual of adverse drug interactions, 4th edit. Wright, Bristol, pp 6–16
Gomez M, Alvarez A, Cid E (1993) Influence of cellulosic derivatives on in vitro stability and availability of diethylpropion in prescription orders. Rev Med Chile 121:1013–1016
Hansen HC, Spillum A (1991) Loss of nitroglycerin during passage through two different infusion sets. Acta Pharm Nord 3:131–136
Hendeles L, Weinberger M, Milavetz G, Hill M, Vaughan L (1985) Food-induced "dose-dumping" from a once-a-day theophylline product as a cause of theophylline toxicity. Chest 87:758–765
Hendeles L, Breton AL, Beaty R, Harman E (1995) Therapeutic equivalence of a generic slow-release theophylline tablet. Pharmacotherapy 15:26–35
Hill JB (1959) Adsorption of insulin to glass. Proc Soc Exp Biol Med 102:75–77
Holmes SE, Aldous S (1991) Stability of miconazole in peritoneal dialysis fluid. Am J Hosp Pharm 48:286–290
Jenke DR (1993) Modelling of solute sorption by polyvinyl chloride plastic infusion bags. J Pharm Sci 82:1134–1139
Johnson BF, Fowle ASE, Lader S, Fox J, Munro-Faure AD (1973) Biological activity of digoxin from Lanoxin produced in the United Kingdom. Br Med J 4:323–326
Kowaluk EA, Roberts MS, Blackburn HD, Polack AE (1981) Interactions between drugs and polyvinyl chloride infusion bags. Am J Hosp Pharm 38:1308–1314
Lee MG (1986) Reduced absorption of drugs onto polybutadiene administration sets. Am J Hosp Pharm 43:1945–1950
Li LC, Zhang H, Pecosky DA (1993) The effect of surfactant on drug-plastic sorption phenomenon in solution. J Pharm Pharmacol 45:748–749
Loucas SP, Maager P, Mehl B, Loucas ER (1990) Effect of vehicle ionic strength on sorption of nitroglycerin to a polyvinyl chloride administration set. Am J Hosp Pharm 47:1559–1562
Lund W (ed) (1994) British pharmaceutical codex. Pharmaceutical Press, London, pp 311–322
Lyons RW (1979) Orange contact lenses from rifampicin. N Engl J Med 300:372–373
Malick AW, Amann AH, Baaske DM, Stoll RG (1981) Loss of nitroglycerin from solutions to intravenous plastic containers: a theoretical treatment. J Pharm Sci 70:798–800
McElnay JC, D'Arcy PF (1980) Sites and mechanisms of drug interactions. I. In vitro, intestinal and metabolic reactions. Int J Pharm 5:167–185
McElnay JC, Nichol AC (1984) The comparison of a novel continuous flow dissolution apparatus for suppositories with the rotating basket technique. Int J Pharm 19:89–96
McElnay JC, Taggart AJ, Kerr B, Passmore P (1986) Comparison of serum indomethacin profiles after rectal administration of two brands of indomethacin suppositories in healthy volunteer subjects. Int J Pharm 33:195–199
McElnay JC, Elliott DS, D'Arcy PF (1987) Binding of human insulin to burette administration sets. Int J Pharm 36:199–203
McElnay JC, Elliott DS, Cartwright-Shamoon J, D'Arcy PF (1988) Stability of methotrexate and vinblastine in burette administration sets. Int J Pharm 47:239–247
McGinity JW, Lach JL (1976) In vitro adsorption of various pharmaceuticals to montmorillonite. J Pharm Sci 65:896–902

McMillan M (1989) Clinical pharmacokinetics of cyclosporin. Pharmacol Ther 42:135–156

Moorhatch P, Chiou WL (1974) Interactions between drugs and plastic intravenous fluid bags. II. Leaching of chemicals from bags containing various solvent media. Am J Hosp Pharm 31:149–152

Morgan DJ, McCormick Y, Cosolo W, Roller L, Zalcberg J (1992) Prolonged release of morphine alkaloid from a lipophilic suppository base in vitro and in vivo. Int J Clin Pharmacol Ther 30:576–581

Newerly K, Berson SA (1957) Lack of specificity of insulin I-131 binding by isolated rat diaphragm. Proc Soc Exp Biol Med 94:751–755

Parker WA, McCara ME (1980) Compatibility of diazepam with intravenous fluid containers and administration sets. Am J Hosp Pharm 37:496–500

Pearson SD, Trissel LA (1993) Leaching of diethylhexyl phthalate from polyvinyl chloride containers by selected drugs and formulation components. Am J Hosp Pharm 50:1405–1409

Petty C, Cunningham NL (1974) Insulin adsorption by glass infusion bottles, polyvinyl chloride infusion containers and intravenous tubing. Anesthesiology 40:400–404

Riley SA, Flegg PJ, Mandal BK (1986) Contact lens staining due to sulphasalazine. Lancet I:972

Roberts MS, Kowaluk EA, Polack AE (1991) Prediction of solute sorption by polyvinyl chloride plastic infusion bags. J Pharm Sci 80:449–455

Salomies HEM, Heinonen RM, Toppila MAI (1994) Sorptive loss of diazepam, nitroglycerin and warfarin sodium to polypropylene-lined infusion bags (Softbags). Int J Pharm 110:197–201

Sautou V, Chopineau I, Gremeau R, Chevrier R, Bruneaux F (1994) Compatibility with medical plastics and stability of continuously and simultaneously infused isosorbide dinitrate and heparin. Int J Pharm 107:111–119

Schildt B, Ahlgren T, Bergham L, Wendt Y (1978) Adsorption of insulin by infusion materials. Acta Anaesth Scand 22:556–562

Shaw LM, Bowers LD, Demers L, Freeman DJ, Moyer T, Sarghvi A, Seltman H, Venkataramanan R (1987) Critical issues of cyclosporine monitoring: report of the Task Force on cyclosporine monitoring. Clin Chem 36:1841–1846

Sugar J (1974) Adenochrome pigmentation of hydrophilic lenses. Arch Ophthalmol 91:11–12

Taormina D, Abdallah HY, Venkataramanan R, Logue L, Burckart GJ, Ptachcinski RJ, Todo S, Fung JJ, Starzl TE (1992) Stability and sorption of FK 506 in 5% dextrose and 0.9% sodium chloride injection in glass, polyvinyl chloride and polyolefin containers. Am J Hosp Pharm 49:119–122

Trissel LA, Xu Q, Kwan J, Martinez JF (1994) Compatibility of paclitaxel injection vehicle with intravenous administration and extension sets. Am J Hosp Pharm 51:2804–2810

Twardowski ZJ, Nolph KD, McGary TJ, Moore HL, Collin P, Ausman RK, Slimack WS (1983a) Insulin binding to plastic bags: a methodologic study. Am J Hosp Pharm 40:575–579

Twardowski ZJ, Nolph KD, McGary TJ, Moore HL (1983b) Nature of insulin binding to plastic bags. Am J Hosp Pharm 40:579–582

Twardowski ZJ, Nolph KD, McGary TJ, Moore HL (1983c) Influence of temperature and time on insulin adsorption to plastic bags. Am J Hosp Pharm 40:583–586

Tyrer JH, Eadie MJ, Sutherland JM, Hooper WD (1970) Outbreak of anticonvulsant intoxication in an Australian city. Br Med J 2:271–273

Venkataramanan R, Burckart GJ, Ptachcinski RJ, Blaha R, Logue LW, Bahnson A, Giam C-S, Brady JE (1986) Leaching of diethylhexyl phthalate from polyvinyl chloride bags into intravenous cyclosporine solution. Am J Hosp Pharm 43:2800–2802

Weinberger M, Hendeles L, Bighley L (1978) The relation of product formulation to absorption of oral theophylline. N Engl J Med 299:852–857

Wideroe TE, Smeby LC, Berg KL, Jorstad S, Svartas TM (1983) Intraperitoneal (^{125}I) insulin absorption during intermittent and continuous peritoneal dialysis. Kidney Int 23:22–28
Yahya AM, McElnay JC, D'Arcy PF (1985) Binding of chloroquine to glass. Int J Pharm 25:217–223
Yahya AM, McElnay JC, D'Arcy PF (1986) Investigation of chloroquine binding to plastic materials. Int J Pharm 34:137–143
Yuen PH, Denman SL, Sokoloski TD, Burkman AM (1979) Loss of nitroglycerin from aqueous solution into plastic intravenous delivery systems. J Pharm Sci 68:1163–1166

CHAPTER 10

Age and Genetic Factors in Drug Interactions

J.C. McElnay and P.F. D'Arcy

A. Introduction

It has long been known that specific patient factors can influence the course of therapy and the adverse drug reactions and interactions that may follow (Wallace and Watanabe 1977; Ouslander 1981; Braverman 1982; Greenblatt et al. 1982; Shaw 1982; D'Arcy and McElnay 1983; Royal College of Physicians 1984; Nolan and O'Malley 1988a,b; Denham 1990; Williams and Lowenthal 1992). Of these factors, age and genetic influences have been pinpointed as significant contributors to problems with drug therapy.

This chapter deals with these two influences in turn: Sect. B, "Age", and Sect. C, "Genetic Factors", and it will highlight some of their more important influences on drug action. Where possible, it will indicate the mechanisms that underline these influences.

B. Age

It has long been established that elderly patients use more medicaments than younger age groups (Landahl 1987; Williams and Lowenthal 1992; Sloan 1992; Stewart and Cooper 1994) and thus have a greater risk of drug-drug interactions occurring (Kellaway and McCrae 1973; Lawson and Jick 1976; Levy et al. 1980; Williamson and Chopin 1980; Simons et al. 1992; Schenker and Bay 1994; Stanton et al. 1994; Stewart and Cooper 1994) (Fig. 1). It must be understood, however, that there is virtually no direct evidence in the literature that age per se can cause drug-drug interactions. Its role is more in enhancing the effects of drug-drug interactions when they occur. Age thus tends to exert a quantitative influence on the interaction but does not alter its qualitative spectrum.

Senescence is frequently evoked to explain the unwanted sequelae to therapy, undoubtedly rightly – provided senescence is regarded as being accompanied, for example, by physiological changes due to age (Lamy 1991), by small body mass, poor renal function, and impaired function of other organs, notably the liver. Table 1 shows some of the key changes that occur as a result of physiological ageing. In elderly patients the reserve capacity of many organs may be considerably reduced, and because of this erosion there is a narrowing

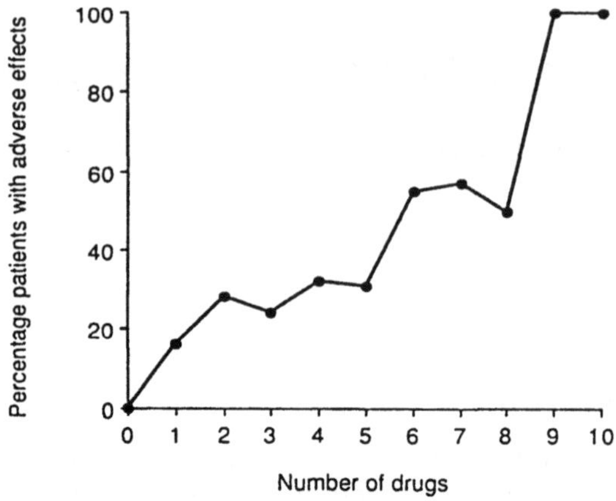

Fig. 1. Relationship between percentage of patients with ADR and the number of drugs prescribed. (Denham 1990)

of the safety margin between the therapeutic and toxic dose of many drugs. As a result of this the elderly, as a group, get rather more than their fair share of drug-induced disease and the complications of drug-drug interactions. This especially applies to psychogeriatric patients in whom the nature of drugs, the number used concomitantly, and the long-term use presents peculiar hazards.

Table 2 shows, as an example, the drug-drug interactions identified by Gosney and Tallis (1984) in their survey of interacting drugs in elderly patients admitted to hospital. Nine of these combinations were potentially life-threatening; 51 combinations were likely to lead to potentially serious side effects, and 27 were significant in as much as they could lead to suboptimal treatment. Other accounts of common drug-drug interactions are given throughout this volume and will not be repeated here. It is sufficient to emphasise that qualitatively the elderly appear to suffer the same adverse effects as younger age groups but they suffer them at an enhanced level.

Many adverse effects of drugs or drug combinations that may simply be a nuisance to younger patients are much intensified in the elderly due, for example, to their decreased resistance to the adverse consequences. Indeed, the adverse effects of drug treatments may convert the elderly patient from a functional sentinent human being into a chair-fast incontinent wreck.

In their simplest form, drug-drug interactions cause either an enhanced pharmacological or toxicological effect of one or more of the combination drugs, or conversely a reduced therapeutic effect of one or more of the components. A number of investigations have shown that age is not an independent risk factor (Nolan and O'Malley 1989; Carbonin et al. 1991; Gurwitz and Avorn 1991; Chrischilles et al. 1992). However, regardless of whether the elderly are more sensitive to the adverse effects of medication, they have more disease, consume more medications and are overrepresented in nearly every study of adverse drug reactions (Schneider et al. 1992).

Table 1. Some key changes as a result of physiological aging. (COLLIER et al. 1993)

System	Key changes
Cardiovascular system	Decrease in cardiac output
	Decrease in stroke volume
	Loss of elasticity of blood vessels
	Decrease in recovery rate to restore basal heart rate following exercise
	Pre-disposition to blood hypercoagulability
	Decrease in deposition of fibrin in blood vessel walls
	Decrease in peripheral perfusion
	Decrease in organ blood flow
Central nervous system	Changes in receptor and target organ sensitivity
	Decrease in cerebral perfusion
	Slowing of cerebral processes
Respiratory system	Decrease in the elasticity of lung tissue
	Lessening of ciliary activity
	Decrease in dead space in lungs
	Decrease in forced expiratory volume
	Decrease in peak flow
	Wasting of muscles in diaphragm
Endocrine system	Decrease in glucose tolerance
	Decrease in peripheral responsiveness to insulin
	Decrease in thyroid functional activity but increased tissue sensitivity to thyroid hormones
	Decrease in normal thyroxine turnover
	Reduction of plasma binding of thyroid hormones
	Decrease in basal adrenal activity
	Reduction of oestrogen level
	Decrease in the production of progesterone after the menopause
	Decrease in free androgens in men
	Altered hormonal control of bone homeostasis
	Compromised immunological systems
Gastrointestinal system	Decrease in acid secretion
	Decrease in gastric emptying rate
	Decrease in gut motility
	Decrease in peristaltic activity
	Decrease in gastric blood flow
	Decrease in absorption of calcium and iron from small intestine
Hepatic system	Reduction in hepatic blood flow
	Decreased metabolic capacity
Genitourinary system	
Kidney	Decrease in glomerular filtration rate
	Decrease in functioning renal glomeruli
	Decrease in renal blood flow
	Decrease in excretory and reabsorptive capacities of renal tubules
Bladder	Loss of supporting elastic tissue
	Detrusor muscle hypertrophy
Reproductive tract	Decrease in vaginal secretions
	Increase in pH of vaginal secretions

Table 2. Drug interactions. (Gosney and Tallis 1984)

Drug(s)		Nature of clinically most important interaction	Comment	No. cases
Frusemide	Aminoglycosides	Increased ototoxicity and nephrotoxicity	Potentially serious, probably avoidable	24
Frusemide	Cephalosporins	Increased nephrotoxicity	Potentially serious, probably avoidable	9
Warfarin	Phenylbutazone	Increased risk of haemorrhage	Potentially dangerous, avoidable	1
Warfarin	Aspirin	Increased risk of haemorrhage	Potentially dangerous, avoidable	3
Heparin	Aspirin	Increased risk of haemorrhage	Potentially dangerous, avoidable	1
Verapamil	Propranolol (within 6h)	Risk of asystole	Potentially dangerous, avoidability uncertain	2
Frusemide	Prednisolone (with potassium supplementation)	Antagonism	Significant interaction, avoidability uncertain	16
Frusemide	Prednisolone (without potassium supplementation)	Increased risk of potassium loss as well as antagonism (four of these had $K^+ < 2.8$ mmol/l)	Potentially serious, risk could have been reduced by potassium supplementation	14
Loop diuretics	Indomethacin	Antagonises diuretic effect	Potentially serious, could have been avoided by using alternative analgesic	4
Loop diuretics	Other non-steroidal anti-inflammatory	May antagonise diuretic effect	Possibly significant, avoidability uncertain	15

Carbenoxolone	Spironolactone drugs	Inhibition of ulcer healing, interference with treatment of cardiac failure	Potentially significant, should have been avoided	1
Phenylbutazone	Chlorpromazine	Increased risk of leucopenia	Potentially dangerous, avoidable	2
Iron	Antacid	Interference with action/absorption	Possibly significant, avoidable	2
Cimetidine	Antacid	Interference with action/absorption	Possibly significant, avoidable	20
Digoxin	Antacid	Interference with action/absorption	Possibly significant, avoidable	5
Tetracycline	Antacid	Interference with action/absorption	Significant, avoidable	3
Tetracycline	Iron	Interference with action/absorption	Significant, avoidable	2
Warfarin	Co-trimoxazole	Interference with control, increased risk of bleeding	Significant, could have been avoided by using trimethroprim alone	5
Phenytoin	Co-trimoxazole	Potentiation	Possibly significant, could have been avoided by using trimethoprim alone	1
Phenylbutazone	Thyroxine	Interference with thyroid function test	Possibly significant, avoidable	1
Tolbutamide	Co-trimoxazole	Potentiation	Possibly significant, could have been avoided by using trimethoprim alone	2
			Total	133

I. Pharmacokinetics in the Elderly

There are striking examples of age-related alterations of both the kinetics and dynamics of drugs. Every drug used in the elderly should have its pharmacology and toxicology specifically evaluated in the elderly, yet few elderly patients, if any, participate in clinical trials on new drug substances. One does not question that it is desirable, but it must be appreciated that the ethical situation is not at all clear. In a paper given to the Medico Pharmaceutical Forum, Griffin (1986a, cited in Griffin 1989) stated that "on a personal basis I believe that it is unethical to conduct studies in the elderly, where the elderly person is being selected for the study by virtue of some presumed abnormality of metabolism associated with the process of ageing. Such data should be obtained in the course of clinical trials primarily to evaluate efficacy".

Pharmacokinetic effects of medication may be increased or decreased in the elderly patient. Posner and Rolan (1994) have categorised age-related differences in kinetics between the elderly and young as being primarily due to: diminished renal function, altered proportions of body fat and water, reduced cardiac output and some degree of altered hepatic metabolism. Altered kinetics in the elderly are discussed below under the four main kinetic processes of absorption, distribution, metabolism and elimination.

1. Absorption

The extent of drug absorption is generally little affected by age. It is true that physiological changes occur with ageing that can influence drug absorption, for example, decreased gastric acid secretion, decreased pancreatic lipase activity, decreased gastric emptying and gastrointestinal motility, and decreased gastric blood flow. However, although delays in absorption have been described for some substances, they are generally of little clinical importance since the extent of absorption remains largely unchanged. The absorption of levodopa is an exception. It has been demonstrated, for example, that there can be a threefold increase in its absorption in elderly patients probably due to an age-related reduction in gastric dopamine decarboxylase and consequently greater availability of levodopa for absorption into the systemic circulation (Evans et al. 1980). This increase in absorption is probably one of the main reasons for the substantially lower dose of levodopa required for elderly patients (Broe and Caird 1973) and for levodopa being one of the commoner causes of adverse drug reactions (ADRs) in elderly patients.

2. Distribution

Several blood flow changes occur with advancing years. Firstly there is a decrease in cardiac output of 1%/year after the age of 25 years. Secondly, in older adults there is a decreased perfusion of limbs, liver and mesentery of up to 45%. Thirdly, in elderly people there is a decreased blood flow to the heart

(30%) and brain (15%). Although little data are available, these changes in blood flow could obviously influence the rate of drug distribution to various tissues. In the elderly, however, it is the changes in body composition, including the concentration of plasma albumin, which have the greatest impact on drug distribution.

In the elderly, lean body mass is reduced, fat mass is increased and intracellular fluid volume is decreased. The implications of these changes for drug kinetics will depend largely on the degree of lipid and aqueous solubility of the drug in question and its degree of binding to plasma albumin. The volume of distribution of lipid-soluble drugs, e.g., diazepam and lignocaine, will obviously be increased while the opposite effect will be seen with water-soluble drugs, e.g., gentamicin and nadolol. These effects will tend towards lower plasma concentrations of lipid-soluble drugs and higher concentrations of water-soluble drugs. Lipid-soluble drugs, as well as distributing into body fat, can concentrate in the brain, which is high in lipoid tissue. There is also some evidence to suggest that the blood-brain barrier is less intact in the elderly, thus allowing drugs to distribute into the brain in increased concentrations. Changes in drug distribution are of particular significance when the therapeutic ratio of the drug is low as in the case of digoxin, where the 30% reduction in its volume of distribution in the elderly is probably related to reduction in lean body mass. This necessitates a reduction in the loading dose of digoxin in elderly patients.

For drugs which are highly bound to plasma proteins, the decreased concentrations of albumin that occur, particularly in sick or malnourished elderly patients, can also have marked effects on drug distribution and therefore drug plasma levels. As a general rule, drugs which are highly bound to plasma proteins distribute very little into organs since bound drug cannot cross the capillary membranes, due to the large size of the protein molecules. Warfarin is a good example of this effect – it is highly bound to plasma albumin (>99%) and is largely retained in the plasma compartment. In the elderly, however, since albumin concentrations are decreased, less drug is bound and more drug is free to distribute throughout the body and to be eliminated by the liver and kidney. This redistribution leads to decreased total concentrations (free + bound) of drug in the plasma, but, in the absence of effects on drug elimination, one would expect the free, pharmacologically active drug concentrations to remain relatively similar. In the elderly, however, drug elimination is often decreased and therefore the free drug concentrations in plasma may become elevated with consequent increases in pharmacological effects and toxicity. When drugs are bound to albumin there is also a greater risk of displacement due to interaction between drugs competing for the same bonding sites if the serum albumin levels are reduced (see Chap. 4). In contrast to albumin, plasma globulin concentrations tend to increase with increasing age. This effect is relatively unimportant since few drugs are extensively bound to these proteins.

Some drugs are taken up actively into various tissues, particularly if they resemble a naturally occurring nutrient. This method of tissue uptake is utilised by some of the antitumour drugs, for example, the uptake of melphalan, an amino acid derivative, into tumour cells has been shown to involve active transport. One might expect the active uptake of drugs into tissues to be decreased in the elderly due to declining metabolic function; however, good data are not available to support this hypothesis.

In summary, the main effects of ageing on drug distribution include a more widespread distribution of lipid-soluble drugs, more limited distribution of water-soluble drugs and a more extensive distribution of drugs bound to plasma albumin. A summary of the effects of ageing on drug distribution is presented in Table 3 (Ritschel 1980).

Table 3. Change of drug parameters in the aged. (Ritschel 1980)

Drug	Biological half-life	Volume of distribution	Total clearance
Acetaminophen (paracetamol)	↑	U	↓
Acetanilide	↑	↓	U
Aminopyrine	↑	NI	NI
Amitriptyline	↑	↓	↓
Amobarbital	↑	U	↓
Ampicillin	↑	U	↓
Antipyrine (phenazone)	↑	↓	↓
Aspirin	NI	↑	↓
Carbenicillin	↑	↑	NI
Carbenoxolone	↑	↓	↓
Cefamandole	↑	↑	U
Cefazoline	↑	U	↓
Cefradin	↑	U	↓
Chlordiazepoxide	↑	↑	↓
Chlormethiazole	↑	↑	↓
Chlorthalidone	↑	U	↓
Cimetidine	↑	↓	↓
Cyclophosphamide	↑	↑	NI
Desipramine	↑	NI	NI
Desmethyldiazepam	↑	↑	↓
Diazepam	↑	↑	U or ↓
Digoxin	↑	↓	↓
Dihydrostreptomycin	↑	NI	NI
Doxycycline	↑	U	↓
Flurbiprofen	U	U	U
Gentamicin	↑	↓	↓
Imipramine	↑	NI	NI
Indomethacin	↑	NI	NI
Indoprofen	U	U	U
Kanamycin	↑	NI	↓
Levomepromazine	↑	U	↓
Lidocaine (lignocaine)	↑	↑	U
Lithium	NI	NI	↓
Lorazepam	U or ↑	↓	U or ↓

Table 3. *Continued*

Drug	Biological half-life	Volume of distribution	Total clearance
Methotrexate	↑	↓	↓
Metoprolol	↑	NI	NI
Morphine	↑	NI	NI
Netilmicin	↑	U	↓
Nitrazepam	↑		U
Nortriptyline	↑	U	↓
Oxazepam	↑	↑	NI
Penicillin G	↑	NI	NI
Phenobarbital	↑	NI	NI
Phenylbutazone	↑	U	↓
Phenytoin	NI	NI	↑
Procaine penicillin	↑	NI	NI
Practolol	↑	↓	NI
Propicillin	↑	↓	NI
Propranolol	↑	↓	↓
Protriptyline	↑	↓	↓
Quinidine	↑	↓	↓
Spironolactone	↑	NI	NI
Sulbenicillin	↑	↑	NI
Sulfamethizole (sulphasomidine)	↑	U or ↓	↓
Tetracylline	↑	NI	NI
Theophylline	↑	U or ↓	↓
Thioridazine	↑	NI	NI
Tobramycin	↑	↓	↓
Tolbutamide	↑	↓	↓
Warfarin	↑	U	↓

3. Metabolism

Since the early animal experiments of CONNEY et al. (1956, 1957) and CONNEY (1967) it has often been observed that the rate of oxidative metabolism of various substrates can differ markedly depending on a number of factors of which age is important. However, direct measurement of enzyme levels in liver biopsy samples has shown few age-related alterations in the concentration of several drug-metabolising enzymes in man (WOODHOUSE et al. 1984). Human studies have suggested that the changes observed in clearance of oxidatively metabolised drugs are more likely due to physiological changes in elderly patients than to changes in their cytochrome P450 systems. There are few, if any, detectable changes related to ageing in cytochrome P450 protein or mRNA (SCHMUCKER and KONTAK 1990). Age-related reduction in overall liver mass and absolute liver blood flow are most likely the important factors in diminishing drug metabolism (CALLOWAY et al. 1965; SKAUNIC et al. 1978). Age may affect hepatic oxidative processes to a greater extent than conjugative metabolic processes (see Table 4). Although an individual's liver function is difficult to assess, problems are more likely in patients with ascites, jaundice and encephalopathy.

Table 4. Studies on the relation of age to the clearance of drugs by hepatic biotransformation. (GREENBLATT et al. 1982)

Drug or metabolite	Initial pathway of biotransformation[a]
Evidence suggests age-related reduction in clearance	
Antipyrine[b] (phenazone)	Oxidation (OH, DA)
Diazepam[b]	Oxidation (DA)
Chlordiazepoxide	Oxidation (DA)
Desmethyldiazepam[b]	Oxidation (OH)
Desalkylflurazepam[b]	Oxidation (OH)
Clobazam[b]	Oxidation (DA)
Alprazolam[b]	Oxidation (OH)
Quinidine	Oxidation (OH)
Theophylline	Oxidation
Propranolol	Oxidation (OH)
Nortriptyline	Oxidation (H)
Small or negligible age-related change in clearance	
Oxazepam	Glucuronidation
Lorazepam	Glucuronidation
Temazepam	Glucuronidation
Warfarin	Oxidation (OH)
Lidocaine (lignocaine)	Oxidation (DA)
Nitrazepam	Nitroreduction
Flunitrazepam	Oxidation (DA), nitroreduction
Isoniazid	Acetylation
Ethanol	Oxidation (alcohol dehydrogenase)
Metroprolol	Oxidation
Digitoxin	Oxidation
Prazosin	Oxidation
Data conflicting or not definitive	
Meperidine (pethidine)	Oxidation (DA)
Phenylbutazone	Oxidation (OH)
Phenytoin	Oxidation (OH)
Imipramine	Oxidation (OH, DA)
Amitriptyline	Oxidation (OH, DA)
Acetaminophen (paracetamol)	Glucuronidation, sulfation
Amobarbital	Oxidation (OH)

[a] OH denotes hydroxylation and DA dealkylation.
[b] Evidence suggests that the age-related reduction in clearance is greater in men than in women.

4. Excretion

The two main routes by which drugs are eliminated from the body are metabolism by liver enzymes, and excretion by the kidneys. The chemical changes that result from liver metabolism generally (but not always) result in molecules that are less active pharmacologically and are less lipid soluble and/or more easily ionised. This causes them to be more readily excreted by the kidneys.

In the geriatric patient, drug overdose is particularly likely to occur if the drug remains active in the body until it is excreted by the kidneys. This is because renal function (glomerular filtration and tubular excretion) diminish with increasing age even in the absence of clinically detectable disease. A reduction of about 30% in the glomerular filtration rate and tubular function has been demonstrated in otherwise normal patients over 65 years of age when compared with young adults. Indeed at 90 years of age the functional capacity of the "normal" kidney may only be half what it was at 30 years of age (AGATE 1963). Diminished renal function due to age may be made worse by dehydration, congestive heart failure urinary retention and diabetic nephropathy, all of which are more frequent in the elderly patient.

It has been suggested that the most important single kinetic cause of adverse reactions to drugs in the elderly is impaired renal elimination (CAIRD 1985). This affects drugs eliminated by glomerular filtration (e.g., digoxin, cimetidine and procainamide) and those eliminated by tubular excretion (e.g., penicillins and aminoglycoside antibiotics). At all ages, the renal clearance of digoxin is related linearly to creatinine clearance. The mean digoxin clearance in elderly patients treated with the drug has been shown to be only 40 ml/min with values as low as 10 ml/min as compared with the average normal glomerular filtration in age-comparable healthy elderly patients of 80–90 ml/min (ROBERTS and CAIRD 1976).

II. Pharmacodynamics in the Elderly

Pharmacodynamic interactions between drugs may produce similar or antagonistic pharmacological effects or side effects. They may be due to competition at receptor sites, or occur between drugs acting on the same physiological system. A large number of drug interactions are of no clinical significance and it is really only interactions involving those drugs with a narrow therapeutic index (e.g., lithium, anticoagulants, antiepileptics) and those which require careful control of dosage (e.g., antihypertensives, antidiabetics) that are of major clinical importance.

The pharmacodynamic effect of drugs can be influenced by age as illustrated in the following examples:

The elderly show an increased hypotensive response to angiotensin-converting enzyme (ACE) inhibitors; they show a reduced responsiveness to propranolol. Data are conflicting for the calcium antagonists. The inotropic effect of theophylline is increased with age, but its bronchodilator effect is decreased. The anticoagulant effect of warfarin is increased in elderly patients due to the greater fragility of the hepatic synthesis of clotting factors (SHEPHARD et al. 1977). Warfarin is a drug which is notorious for its involvement in drug interactions and therefore this increased warfarin sensitivity could predispose the elderly patient to greater adverse sequelae should an interaction take place.

Feely and Coakley (1990) emphasised in their review on altered pharmacodynamics in the elderly that the importance of age-related changes in drugs sensitivity is increasingly appreciated. The type, intensity and duration of drug action can be affected, ranging from therapeutic failure to major drug toxicity. Alterations in physiologic and homeostatic systems, including the autonomic nervous system, baroreceptors, thermoregulation and balance, have been described. The increased sensitivity of elderly patients to the postural hypotensive effects of medication has been well documented. These effects can result in falls and dizzy spells (Gribbin et al. 1971; Baker and Harvey 1985; Tinetti et al. 1988), which can not only result in poor quality of life in affected individuals but also can be catastrophic in relation to broken bones, which often result from falls. Ramsay (1981) showed that there was a weak correlation between age and postural fall in diastolic pressure both before and after treatment with methyldopa. He was careful, however, to comment that this correlation might not be with age per se but with the severity of the hypertension.

The elderly patient is less susceptible to developing tachycardia in response to equivalent concentrations of isoprenaline (Vesta et al. 1979; Lakatta 1979). β-Adrenoceptor antagonists are less effective at comparable blood levels in elderly patients compared with the young (Scott and Reid 1982); these effects are likely due to decreased numbers of available β-receptors.

Phenothiazines lower body temperature and are more likely to induce hypothermia in the old than in the young. Similarly, elderly patients are more likely to develop parkinsonian effects with phenothiazines than younger patients (Exon-Smith 1964; Jones and Meade 1964; Collins et al. 1980).

Drugs affecting the CNS produce a relatively greater response in the elderly than in the young for a given plasma concentration (Williams and Lowenthal 1992). The elderly are particularly vulnerable to the side effects of major tranquillisers. Antidepressant agents are particularly likely to cause postural hypotension, urinary retention and sedation. Falls leading to hip fractures have been associated with psychoactive drugs and antidepressants. Nonsteroidal anti-inflammatory agents (NSAIDs) have an increased risk of causing hyperkalaemia or renal failure or death from gastrointestinal haemorrhage. They also cause sodium and fluid retention and so can reduce the effects of drugs used to treat hypertension or congestive heart failure. This latter pharmacodynamic interaction is often ignored.

With diuretics, the elderly are more susceptible to fluid and electrolyte disorders, including volume depletion, hypokalaemia, hyponatraemia and hypomagnesaemia. Adverse reactions to lignocaine are frequent in the elderly (Williams and Lowenthal 1992). The increased effect of some benzodiazepines is well known but it is uncertain whether this is due to receptor alterations, or to the balance of neurotransmitters in the ageing brain, or perhaps to better penetration of the blood-brain barrier.

All of the drugs mentioned in this (pharmacodynamics) and the preceding section (pharmacokinetics) are commonly susceptible to interaction with other drugs, and if interactions occur, and if they potentiate the activity of one of the component drugs, then they will almost certainly present a greater hazard than if they occurred in a younger age group. The results of the interaction will be qualitatively similar in the different age groups but more pronounced in its effects in the elderly.

III. Inappropriate, Unnecessary and Interacting Medication

In a paper entitled "Inappropriate Prescribing in the Elderly", ADAMS et al. (1987) investigated the medication of 1094 patients who were pensioners attending the Accident and Emergency Department of a United Kingdom District General Hospital. Of these 871 (79.6%) were taking at least one prescribed medication. The most commonly consumed medications as a percentage of all subjects in the study were diuretics 32%, analgesics and non-steroidal anti-inflammatory agents 25.2%, hypnotics and sedatives 18.5%, bronchodilators 14.5% and nitrites 10%. Of all the prescriptions written, 4.8% were identified as having been written for conditions for which they were contraindicated according to the British National Formulary. In addition, 356 interacting combinations were identified in the 2353 prescriptions, affecting 216 (19.7%) of patients taking medication. Therefore in this study at least 24.5% of prescriptions written for elderly patients were inappropriate.

LINDLEY et al. (1992) carried out a similar study. They examined 416 successive admissions of elderly patients to a United Kingdom teaching hospital. Interacting drug combinations and drugs with relative contraindications were common, but not as important in producing ADRs as drugs with absolute contraindications or unnecessary drugs. Forty-eight patients (11.5% of admissions) were taking a total of 51 drugs with absolute contraindications (3.8% of prescriptions).

One hundred and seventy-five drugs were discontinued on or shortly after admission in 113 (27%) patients because they were deemed to be unnecessary. One hundred and three patients (27%) of those on medication experienced 151 ADRs of which 75 (49.7%) were due to drugs with absolute contraindications and/or that were unnecessary, a significantly higher rate of ADRs than observed for all prescriptions. Of 26 (6.3%) admissions attributed to ADRs, 13 (50%) were due to inappropriate prescriptions. The admission rate per prescription was significantly higher for inappropriate than for appropriate drugs. The authors concluded that much drug-related morbidity in the elderly population was due to inappropriate prescribing.

Some idea of the relative frequency of drug interactions among elderly patients has been given by more specific publications:

GOSNEY and TALLIS (1984) from Liverpool, United Kingdom, surveyed, retrospectively, the prescription of contraindicated and interacting drugs in elderly patients admitted to hospital. Contraindicated or adversely interacting drugs were identified in 200 (3.2%) of 6160 prescriptions. One hundred and thirty-six (23.7%) patients were affected. The most common interactions (potential or actual) were frusemide with aminoglycosides (increased ototoxicity); cimetidine with antacids (interference with action/absorption); frusemide with prednisolone (potassium loss); loop diuretics with NSAIDs (antagonism of diuretic effect), and frusemide with cephalosporins (increased nephrotoxicity).

Nolan and O'Malley (1989) evaluated the risk of potential drug interactions in 11 private nursing homes in Dublin. The percentage of patients prescribed potentially interacting combinations increased greatly with the number of medications taken.

Potential drug-drug interactions were also studied in an ambulatory population in Florida (Hale et al. 1989). Ten major interaction classifications were studied. It was found that 40% of patients taking quinidine were also taking a digitalis glycoside, and nearly one-third of patients taking warfarin were also taking a drug with an interacting potential.

Schneider et al. (1992) investigated adverse drug reactions in an elderly outpatient population attending an interdisciplinary geriatric clinic and a medical clinic in Cleveland, Ohio, United States. The sample size of the study was 463 patients, of whom 332 attended the medical clinic and 131 attended the geriatric clinic. Potential drug interactions were identified in the records of 143 (31%) subjects.

Doucet et al. (1993), from Rouen and Bois Guillaume, studied drug interactions in French patients over 65 years old. Of 513 elderly patients admitted to hospital, the principal drug interactions occurred with diuretics and benzodiazepines. Of this population 63% had one or more drug interactions leading to an adverse effect (124 patients) and to hospitalisation in half these cases.

Stanton et al. (1994) studied drug-related admissions to an Australian hospital. The median age of the patients was 67 years; 4.4% of admissions were due to drug interactions. Schenker and Bay (1994) working in a Veterans Medical Administration Center in Texas, United States, investigated drug disposition and hepatotoxicity in the elderly and concluded that drug-drug interactions and concurrent derangements accompanying advanced age made a significant contribution to adverse drug effects.

Apart from these group studies, there have been a large number of single patient reports in which elderly patients suffered drug interactions. For example, Kerr (1993) in the United States described an elderly woman who developed gastrointestinal bleeding probably potentiated by the interaction of fluconazole with warfarin. Scarfe and Israel (1994) also in the United States, reported the case of an elderly woman who experienced a significant increase in INR after levamisole and fluorouracil were added to an established regimen of warfarin.

These studies confirm the obvious, that as the number of medications in a patient's regimen increases, the potential for interacting combinations also increases. Multiple drug therapy and differences in pharmacokinetics and pharmacodynamics are thus factors that predispose the elderly to adverse reactions to drugs, e.g., to digoxin, to diuretics and to psychotropic agents (Nolan and O'Malley 1989). Therefore, if elderly patients also receive inappropriate medication (commonly digoxin, diuretics and psychotropic agents), or interacting combinations, and if they are subject to age-related alterations of drug dynamics and kinetics, then it is quite obvious that drug-drug interactions that potentiate the activity of one of the components might well present a particular and peculiar hazard.

IV. Logistics: Age and Medicine Consumption

A World Health Organisation report on health care in the elderly (World Health Organization 1981) concluded that elderly patients consume a lot of drugs, that polypharmacy seems to be the rule in acute hospital settings and in institutions, and that psychoactive drugs appear to be used where most of the mentally disabled are found. One may question whether the position has changed in the 14 years that has elapsed since this report was published.

Throughout the past 40 years, Sweden, Norway and the United Kingdom have recorded the highest ratio of elderly people (men aged over 65 years, women over 60 years) to population among the industrialised countries (Organisation for Economic Cooperation and Development, OECD). The projected figures for the percentage changes in United Kingdom elderly population for the years 1990–2001 (and 2001–2011) are 2.4% (5.0%) over the age of 65 years, 11.0% (1.5%) over the age of 75 years, and 34.6% (12.8%) over the age of 85 years. United Nations has calculated for OECD countries that the figure for elderly persons (aged 65 and over) expressed as a percentage of the total population will be at the year 2000: 15.2% for the United Kingdom, 12.8% for the United States and an average of 13.9% for the OECD countries (Office of Health Economics 1992).

Medication prescribed for the elderly in the National Health Service rose by 29% from 98 million items in 1978 to 143 million by 1988. The total number of prescription items was 350 million in 1988 for the whole population; thus the elderly's share of the total advanced from 32% to 41% during the same period. On average the elderly had between them 16.3 prescription items per annum in 1988, as compared with 12.2 items in 1978. In sharp contrast, the number of prescriptions given to those of working age fell by 10% from 6.1 items per person to 5.5 items between 1978 and 1988 (Office of Health Economics 1989). In the United States, it seems that geriatric patients consume 33% of prescription drugs (Sloan 1992).

V. Long-Term Treatments

Most of the drugs involved in clinically relevant interactions are those on which patients are carefully stabilised for relatively long periods (e.g., oral anticoagulants, anticonvulsants, oral antidiabetics, cardiovascular drugs, psychotropic agents).

Past experience has clearly shown that the elderly often receive these types of medications, and it is the elderly, drug-stabilised patient who will be at special risk of any changes in therapy or environment that will influence the potency or bioavailability of normal medication. It is under these circumstances that drug-drug interactions may become dangerous to the elderly patient (see, for example, Table 2). In order to make drug therapy safer in elderly patients, two main strategies may be employed: Firstly, improve patient compliance with medication instructions. Secondly, improve patient prescribing and monitoring via avoiding unnecessary treatments, reviewing drug therapy regularly, simplifying drug regimens and monitoring for adverse reactions and interactions.

C. Genetic Factors

The term "pharmacogenetic disorder" was originally coined by Vogel (1959) and was originally limited to hereditary disorders revealed solely by the use of drugs. Its meaning has now been enlarged to embrace all genetic contributions to the considerable variation that exists in the interaction between man and the pharmacological agents that he uses. The term can therefore be taken to cover the adverse drug reactions and interactions that occur when particular drugs are given to a patient with a genetically determined dysfunction of an organ or body system, for example, renal polycystic disease.

As with the effect of age on the incidence or severity of drug-drug interactions, there is no reason to suggest that genetic influences cause drug interactions per se. What is clear, however, is that patients who suffer the effects of pharmacogenetic disorders may well show a predisposition to specific drug-drug interactions, or may show enhanced effects when such interactions occur.

The risk of drug toxicity usually arises from enzyme deficiency states. Examples of these have been well described by Bennett (1993). For example:

Hepatic porphyrias, where deficient conversion of porphyrins to haem exposes affected individuals to risk from drugs which have the common property of increasing the activity of delta-aminolaevulinic acid synthetase, the rate-limiting enzyme of porphyrin synthesis, in their livers. Disorders of porphyrin metabolism have been comprehensively reviewed by Fletcher and Griffin (1986) and they have presented a figure of the overall scheme of porphyrin synthesis and a list of those drugs which have been recognised as being associated as precipitating agents of porphyria. The figure (Fig. 2) and list of drugs (Table 5) are reprinted here.

Malignant hyperpyrexia is a rare pharmacogenetic disease (rigid and non-rigid types) occurring both in man and in the Landrace strain of pigs. In susceptible individuals any potent inhalation anaesthetic or any skeletal muscle relaxant can cause fever, rigidity, hyperventilation, cyanosis, hypoxia,

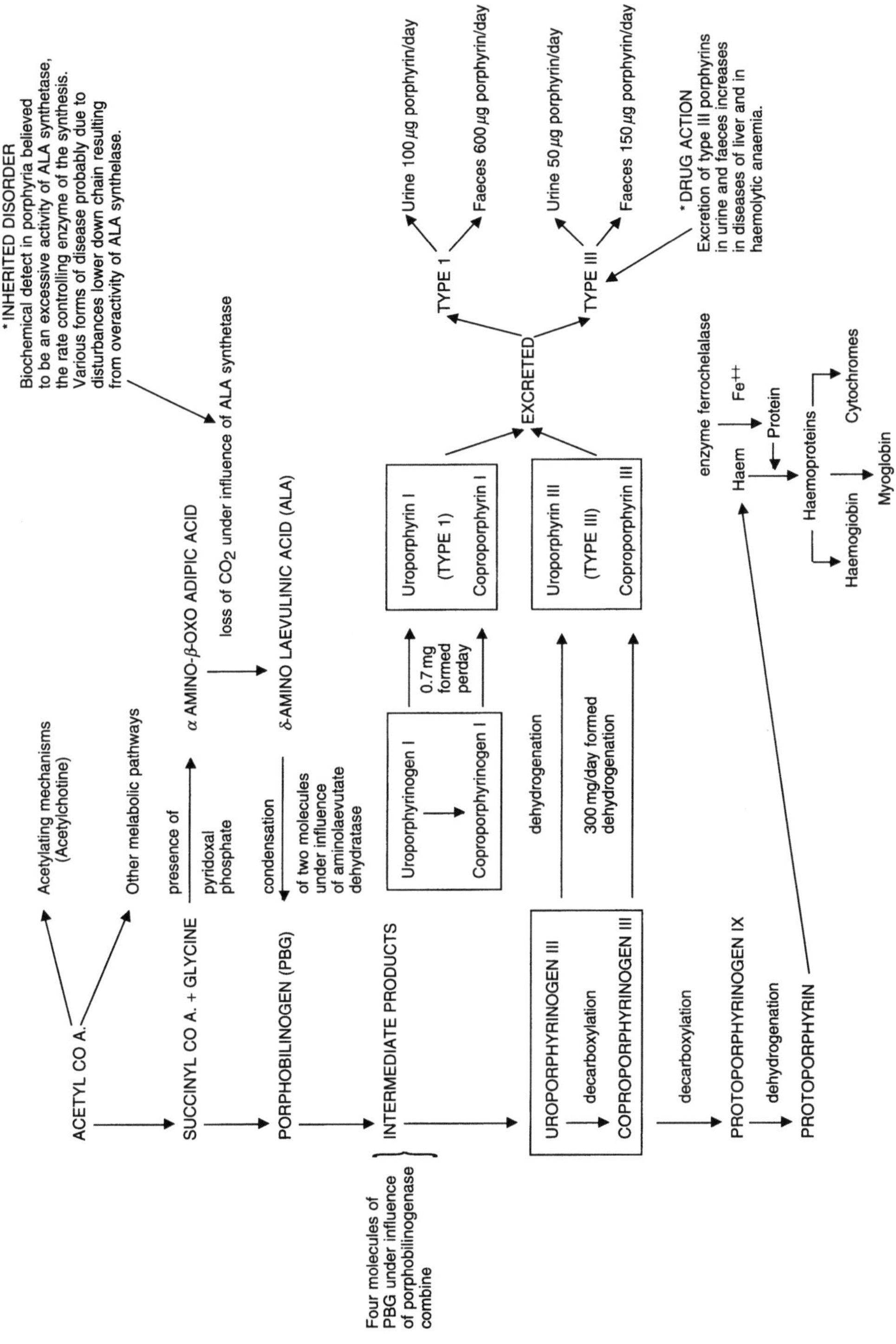

Fig. 2. Overall scheme of porphyrin synthesis. (FLETCHER and GRIFFIN 1986)

Table 5. Drugs classified as to their association with porphyria. (FLETCHER and GRIFFIN 1986)

Potentially porphyrogenic drugs	Drugs believed not to precipitate porphyria	Potentially porphyrogenic drugs	Drugs believed not to precipitate porphyria
Alcohol	Acetazolamide	Flufenamic acid	Heparin
Alphaxalone	Adrenaline	Flunitrazepam	*Hydrallazine*
Aluminium	Alclofenac	Fluroxine	Hyoscine
2-Allyloxy-3	*Amitriptyline*	Frusemide	
methylbenzamide	Aspirin		Ibuprofen
Amidopyrine (A)	Atropine	Glutethimide	*Imipramine*
Aminoglutethemide		Gold preparations	*Indomethacin*
Amitriptyline	B vitamins (except	Griseofulvin (A)	Insulin
Amphetamines	pyridoxine)		
Androgens (A, C)	Bethanidine	Halothane	Ketamine
Apronalide	Biguanides	Hydantoins	Ketoprofen
Avapyrazone	Bromides	(phenytoin,	
Azapropazone	Bumetanide	ethotoin,	Labetolol
	Bupivacaine	mephentoin)	Lithium
Barbiturates (A, C)	Buprenorphine	Hydrallazine	Lorazepam
Bemegride		Hydrochlorthiazide	
Busulphan	Cephalexin	Hyoscine *N*-butyl	Mandelamine
	Cephalosporins	bromide	Mecamylamine
Carbromal	Chloral hydrate		Meclozine
Carbamazepine	*Chloramphenicol*	Imipramine	*Mefenamic acid*
Chlorambucil	Chlormethiazole	Indomethacin	Mersalyl
Chloramphenicol	Chlorpheniramine		Metformin
Chlordiazepoxide	Chlorpromazine	Isoniazid	Methadone
(A)	*Chloroquine*	Isopropylmeprobamate	Methenamine
Chlormezanone	Chlorthiazides	Lignocaine	mandelate
Chloroform	Clobazam		Methylphenidate
Chloroquine (C)	Clofibrate	Mefenamic acid	Morphine
Chlormethiazole	*Clonazepam*	Mephenazine	
Chlorpropamide	Codeine	Meprobamate	Naproxen
(A, C)	Colchicine	Mercury compounds	Neostigmine
Cimetidine	Corticosteroids	Methoxyflurane	*Nitrofurantoin*
Clonazepam	Cyclizine	Methsuximide	Nitrous oxide
Clonidine	Cyclopropane	Methyldopa	Nortriptyline
Cocaine		Methyprylone	
Colistin	Dexamethasone	Metoclopramide	*Oxazepam*
Cyclophosphamide	Diamorphine	Metyrapone	Oxpentifylline
	Diazepam	Metronidazole	
Danazol	Diazoxide		Paracetamol
Dapsone	Digitalis compounds	Nalidixic acid	*Paraldehyde*
Diazepam	Diphenhydramine	Nikethamide	Penicillamine
Dichloralphenazone	Dicoumarol	Novobiocin	Penicillins
(A)	anticoagulants	Nitrazepam	*Pethidine*
Diclofenac	Disopyramide	Nitrofurantoin	Phenformin
Diethyhexyl	Domperidone		Pheniramine
phthalate	Droperidol	Oral contraceptives	Phenothiazines
Diethylpropion		Oxazolidinediones	(e.g.,
Dimenhydrinate	EDTA	(paramethadione	chlorpromazine)
	Ether (diethyl)	and trimethadione)	Phenoperidine
Enflurane		Oestrogens (A, C)	*Phenylbutazone*
Ergot preparations	Fentanyl	Oxazepam	Prednisolone
Erythromycin	Flurbiprofen		Procaine
Ethchlorvynol	Fusidic acid	Pancuronium	Prilocaine
Ethinamate		Paraldehyde	Promethazine
Ethosuximide	Gentamicin	Pargyline	Propantheline
Etomidate	Glipizide	Pentazocine	bromide
Eucalyptol	Guanethidine	Pentylenetetrazol	*Propanidid*

Table 5. *Continued*

Potentially porphyrogenic drugs	Drugs believed not to precipitate porphyria	Potentially porphyrogenic drugs	Drugs believed not to precipitate porphyria
Pethidine	Primaquine	Spironolactone	Tubocurarine
Phenazone	Propoxyphene	Steroids	*Sodium valproate*
Phenelzine	Propranolol	Streptomycin	
Phenoxybenzamine	Prostigmine	Succinimides	Vitamin C
Phensuximide	*Pyrimethamine*	(ethosuximide,	
Phenylbutazone		methsuximide,	
Phenylhydrazine	Quinine	phensuximide)	
Primidone		Sulphonal	
Probenecid	Reserpine	Sulphonamides (A)	
Progestogens (A, C)	Resorcinol	Sulphonylureas	
Propanidid	*Rifampicin*	Sulthiame	
Pyrazinamide			
Pyrazolones	Streptomycin	Tetracyclines	
(antipyrine,	Succinylcholine	Theophylline	
isopropylantipyrine,		Tolazamide	
dipyrone, sodium	*Tetracyclines*	Tolbutamide (A, C)	
phenyl dimethyl	Tetraethylammonium	Tranylcypromine	
pyrazolone)	bromide	Trional	
Pyridoxine	Thiouracils	Troxidone	
Pyrimethamine	Tricyclic		
	antidepressants	Valproic acid	
Ranitidine	(amitriptyline)	Viloxazine	
Rifampicin	Trifluoperazine		
	Thiazides	Xylocaine	
Sodium valproate	Tripelennamine		

A, acute porphyria; C, cutaneous porphyria.
Substances marked in italics have been variously described as causing porphyria or being safe to administer to patients with porphyria.

respiratory and metabolic acidosis, and hyperphosphataemia with a raised glucose level. Initial hyperkalaemia and hypercalcaemia are followed by hypokalaemia and hypocalcaemia. The mortality of the condition is high (60%–70%) (BRITT and KALOW 1968; BRITT et al. 1969; FURNISS 1970; ISAACS and BARLOW 1970; DENBOROUGH et al. 1970, 1973; CARPOPRESSO et al. 1975; SUTHERLAND and CARTER 1975).

Glucose-6-phosphate Dehydrogenase Deficiency. The life-span of the erythrocyte in a normal subject varies between 110 and 125 days. In haemolytic anaemias the red cells are destroyed more rapidly and the average life-span of the erythrocyte is correspondingly shorter. It has been known for some years that, when certain ordinary, otherwise harmless oxidant drugs are administered to some individuals in ordinary therapeutic doses, this is accompanied by rapid red-cell destruction and the development of acute haemolytic anaemia (Table 6). The cause of this can be an inherited deficiency of glucose-6-phosphate dehydrogenase (G6PD). GRIFFIN (1986b) has reviewed the types of G6PD deficiency [e.g., African type (A), Mediterranean type, the variants Canton and Union] and has also discussed the occurrence of other types of haemolytic anaemias not limited to patients with G6PD deficiency.

Table 6. Adverse reactions due to poor microsomal oxidation. (Smith and Shah 1984, personal communication; Smith and Idle 1981)

Adverse reaction	Drugs concerned
Lactic acidosis	Metformin
Agranulocytosis	Carbimazole Phenylbutazone Chlorpromazine Nortriptyline Imipramine Thioridazine Captopril
Neuropathy or sensory disturbances	Phenytoin
Hepatic adenoma	Oral contraceptives
Cerebellar signs	Phenytoin
Vitamin-D-deficiency-like state	Phenytoin Phenobarbitone
Folate-deficiency-like state	Phenytoin Phenobarbitone
Syncope	Prazosin
Malignant ventricular arrhythmias	Mexiletine Disopyramide
Bradycardia	Propranolol Metoprolol

Defective Carbon Oxidation (Oxidative Genetic Polymorphism). Mono-oxidation represents the single most important step in hepatic metabolism of drugs taking place in the P450 system. Heterogeneity of this oxidative system has been well described (see Chap. 5). The first to be documented related to the genetic polymorphism of the hydroxylation of the antihypertensive agent debrisoquine and this has been used extensively as a tool to further elucidate the nature of genetic polymorphism.

Affected individuals exhibit adverse responses to standard doses of drugs whose inactivation involves oxidation of their carbon centres. In general 90% of Caucasian populations are extensive metabolisers and 10% poor metabolisers. The latter are virtually unable to oxidise and so the unchanged drug is liable to accumulate with adverse effect. Table 6 summarises the published work of Smith and Idle (1981) and the communicated results of Smith and Shah (1984, personal communication). The table shows those adverse drug reactions that are considered to have a metabolic basis and that are attributable to interindividual differences in oxidative status. Bennett (1993) has added additional information: increased β-blockade with bufuralol and timolol, postural hypotension with nortriptyline, and prolonged cardiovascular action with nifedipine.

Acetylator Status. Many drugs are acetylated during their metabolism, for example, the hypotensive hydralazine, the antitubercular isoniazid, the monoamine oxidase inhibitor (MAOI) phenelzine, and the sulphonamides. Subjects vary in their speed of acetylation. This variation is genetically linked and this acetylation polymorphism is an excellent example of a pharmacogenetic phenomenon that is clinically relevant. Slow acetylators respond adversely to standard doses of hydralazine and procainamide (antinuclear antibodies in plasma, and some proceed to systemic lupus erythematosus) and dapsone (haemolytic anaemia). On the other hand, fast acetylators often respond less favourably to treatment because they have lower blood levels of a drug and suffer from inadequate dosage. They also appear to be at a greater risk from treatment with isoniazid, which may cause acute hepatocellular necrosis due to the formation of an active metabolite (BENNETT 1993).

Abnormalities of Plasma Pseudocholinesterase. Variation in the duration of the neuromuscular blocking action of suxamethonium is probably the classical example of this particular category. Suxamethonium is a useful and potent neuromuscular blocking agent of the depolarising type with a very brief dura-

Table 7. Drugs reported to induce haemolysis in subjects with G6PD deficiency

Drug	Haemolysis	
	Black subjects	Caucasian subjects
Acetanilide	+++++	
Dapsone	++	+++
Furazolidone	++	
Nitrofural	++++	
Nitrofurantoin	++	++
Sulphanilamide	+++	
Sulphapyridine	+++	+++
Sulphacetamide	++	
Salazosulphapyridine	+++	
Sulphamethoxypyridazine	++	
Thiazosulphone	++	
Quinidine		++
Primaquine	+++	+++
Pamaquine	++++	
Pentaquine	+++	
Quinocide	+++	++
Naphthalene	+++	+++
Neoarsphenamine	++	
Phenylhydrazine	+++	
Toluidine blue	++++	
Trinitrotoluene		+++

tion of effect. This is because suxamethonium is rapidly hydrolysed and inactivated by pseudocholinesterase in the serum.

There have been reports, since suxmethonium was introduced in 1949 by Bovet et al., of prolonged apnoea and muscular pain and stiffness in some patients (Bourne et al. 1952; Evans et al. 1952; Lehmann and Ryan 1956; Kaufman et al. 1960; Telfer et al. 1964). Investigation showed that these adverse effects were associated with low plasma pseudocholinesterase levels. A familial incidence of abnormal sensitivity to suxamethonium was demonstrated in patients who were shown to have an atypical pseudocholinesterase activity due to genetic abnormalities. Four allelic genes seem to control the inheritance of pseudocholinesterase; one normal, two atypical and one silent allelic gene. They form pseudocholinesterases with varying activity. Such patients do not have any other recognised abnormality and they usually present as cases of prolonged apnoea after a single injection of suxamethonium.

This atypical enzyme hydrolyses suxamethonium at a much slower rate than does the normal type of serum cholinesterase. Consequently, the apnoea that the drug induces is excessively prolonged. Fortunately, Kalow et al. (Kalow and Genest 1957; Kalow and Staron 1957) were able to develop a technique to detect the presence of this atypical enzyme.

Raised Intraocular Pressure and Glaucoma with Topical Glucocorticoids. Repeated topical application of glucocorticoids to the eye is followed by an increase in intraocular pressure (Armaly 1968; Schwartz et al. 1972). The extent to which this happens depends on the age of the subject and his/her genetic make-up. The European population studied showed a trimodel distribution: 66% with low, 29% with intermediate, and 5% with high-pressure changes in response to the local application. Familial studies have suggested that the response is influenced by two alleles P^L and P^H, with the three groups represented by the genotypes P^LP^L, P^LP^H, and P^HP^H, respectively. The rise in intraocular pressure is totally reversible by withdrawal of the drug.

The risk of developing open-angle hypertensive glaucoma spontaneously is greatly increased in patients with P^HP^H genetic make-up. Although the other genotypes are found in cases of glaucoma, the P^LP^L is much smaller than could be expected for the normal population.

D. Comment

This chapter is not able to show that age or genetic make-up is directly responsible for drug-drug interactions. The examples of therapeutic hazard that have been given in this text are largely those due to adverse drug reactions. It has been shown, however, that both age and genetic factors can seriously affect drug therapy either by reducing its effect or alternatively enhancing its effect, including increased toxicity.

It has been shown in many studies that the elderly as a group have to take more medicines than younger age groups because of the disease conditions that occur as age progresses. It is also a well-established fact that the more medicines that are taken together the greater the possibility of drug-drug interactions occurring. It therefore follows that the elderly are likely to suffer the effects of more drug-drug interactions than younger age groups and that, due to decreased resilience in the elderly, such interactions will probably have greater effect on their continued health than similar interactions would have on younger patients.

Genetic abnormalities can increase the toxicity of drug therapy at normal therapeutic dosage levels. This may therefore increase the possibility of hazard from specific drug-drug interactions. Medical opinion does not generally favour multiple-drug therapy, but on occasion it is necessary. The price that may be required for using multiple drugs in elderly patients and in those persons with an abnormal genetic make-up may be the occurrence of relatively more and more serious drug-drug interactions.

References

Adams KRH, Al-Hamouz S, Edmund E, Tallis RC, Vellodi C, Lye M (1987) Inappropriate prescribing in the elderly. J R Coll Physicians Lond 21:39–41
Agate J (1963) The practice of geriatrics. Heinemann, London, pp 217–222
Armaly MF (1968) Genetic factors related to glaucoma. Ann NY Acad Sci 151:861–875
Baker SP, Harvey AH (1985) Fall injuries in the elderly. Clin Geriatr Med 1:501–512
Bennett PN (1993) Evaluation of toxicity in human subjects. In: Ballantyne B, Marrs T, Turner P (eds) General and applied toxicology, vol 1. Macmillan, Basingstoke, pp 367–384
Bourne JG, Collier HOJ, Somers GF (1952) Succinylcholine (succinoylcholine) muscle-relaxant of short action. Lancet I:1225–1229
Bovet D, Depierre F, Courvoisier S, de Lestrange Y (1949) Synthetic curarizing agents. II. Phenolic ethers with quaternary ammonium groups. The action of tris (diethylamino-ethoxy) benzene triiodoethylate (2559F). Arch Int Pharmacodyn 80:172–188
Braverman AM (1982) Therapeutic considerations in prescribing for elderly patients. Eur J Rheumatol Inflamm 5:298–293
Britt BA, Kalow W (1968) Hyperrigidity and hyperthermia associated with anaesthesia. Ann NY Acad Sci 151:947–958
Britt BA, Locher WG, Kalow W (1969) Hereditary aspects of malignant hyperthermia. Can Anaesth Soc J 16:89–98
Broe GA, Caird FI (1973) Levodopa for parkinsonism in elderly and demented patients. Med J Aust 1:630–635
Caird FI (1985) Towards rational therapy in old age. The F.E. Williams Lecture 1985. J Roy Coll Physicians Lond 19:235–239
Calloway NO, Foley CF, Lagerbloom P (1965) Uncertainties in geriatric data. II. Organ size. J Am Geriatr Soc 13:20–28
Carbonin P, Pahor M, Bernabei R et al. (1991) Is age an independent risk factor of adverse drug reactions in hospitalized medical patients? J Am Geriatr Soc 39:1093–1099
Carpopresso PR, Gittleman MA, Reilly DJ, Paterson LJ (1975) Malignant hyperthermia associated with enflurane anaesthesia. Arch Surg 110:1491–1493
Chrischilles EA, Segar ET, Wallace RB (1992) Self-reported adverse drug reactions and related resource use. A study of community-dwelling persons of 65 years of age and older. Ann Intern Med 117:634–640

Collier P, Maguire T, Morrow N, McCollum A, McElnay J, Scott E, Scott M, Singleton M (1993) Care for the elderly. A self-study course for pharmacists. Northern Ireland Centre for Postgraduate Pharmaceutical Education and Training of Pharmacists, and the Pharmacy Practice Group, School of Pharmacy, The Queen's University of Belfast
Collins KJ, Exton-Smith AN, James MH, Oliver JD (1980) Functional changes in automatic nervous responses with ageing. Age Ageing 9:17–24
Conney AH (1967) Pharmacological implications of microsomal enzyme induction. Pharmacol Rev 19:317–366
Conney AH, Miller EC, Miller JA (1956) The metabolism of methylated aminoazo dyes. Evidence for the induction of enzyme synthesis in the rat by 3-methylcholanthrene. Cancer Res 16:450–459
Conney AH, Miller EC, Miller JA (1957) Substrate-induced synthesis and other properties of benzpyrene hydroxylase in rat liver. J Biol Chem 228:703–706
D'Arcy PF, McElnay JC (1983) Adverse drug reactions and the elderly patient. Adv Drug React Ac Pois Rev 2:67–101
Denborough MA, Ebeling P, Kimng JO, Zapf P (1970) Myopathy and malignant pyrexia. Lancet I:1138–1140
Denborough MA, Dennett X, Anderson RMcD (1973) Central core disease and malignant hyperpyrexia. Br Med J 1:272–273
Denham MJ (1990) Adverse drug reactions. Brit Med Bull 46:53–62
Doucet J, Chassagne P, Pauty MD, Breton T, Drieux S, Hemet C, Denis P, Menard JF, Bercoff E (1993) Drug-interactions in patients over 65 – effects and their types – prospective-study of 517 cases. Rev Med Interne 14:439
Evans FT, Gray PWS, Lehmann H, Silk E (1952) Sensitivity to succinylcholine in relation to serum cholinesterase. Lancet I:1229–1230
Evans MA, Triggs EJ, Broe GA, Saines N (1980) Systemic activity of orally administered L-dopa in the elderly Parkinsonian patient. Eur J Clin Pharmacol 17:215–222
Exon-Smith AN (1964) Accidental hypothermia in the elderly. Br Med J 2:1255–1258
Feely J, Coakley D (1990) Altered pharmacodynamics in the elderly. Clin Geriatr Med 6:269–283
Fletcher AP, Griffin JP (1986) Disorders of porphyrin metabolism. In: D'Arcy PF, Griffin JP (eds) Iatrogenic disease, 3rd edn. Oxford Medical, Oxford, pp 67–81
Furniss P (1970) Hyperpyrexia during anaesthesia. Br Med J 4:745
Gosney M, Tallis R (1984) Prescription of contraindicated and interacting drugs in elderly patients admitted to hospital. Lancet II:564–567
Greenblatt DJ, Sellers EM, Shader RI (1982) Drug disposition in old age. N Engl J Med 306:1081–1088
Gribbin B, Pickering TG, Sleight P, Peto R (1971) Effect of age and high blood pressure on baroreflex sensitivity in man. Circulation 29:424–431
Griffin JP (1986a) Lecture given to the Medico Pharmaceutical Forum, October 1986, cited in Griffin JP (1989) Drugs and the elderly. In: Griffin JP (ed) Medicines: regulation research and risk. Greystone Books, The Queen's University of Belfast, p 203
Griffin JP (1986b) Pharmacogenetic and iatrogenic disease. In: D'Arcy PF, Griffin JP (eds) Iatrogenic diseases, 3rd edn. Oxford Medical, Oxford, pp 59–66
Griffin JP (1992) Drugs and the elderly. In: Griffin JP (ed) Medicines: regulation research and risk. Greystone Books, The Queen's University of Belfast, pp 373–387
Gurwitz JH, Avorn J (1991) The ambiguous relation between aging and adverse drug reactions. Ann Intern Med 114:956–966
Hale WE, Monks RG, Stewart RB (1989) Drug use in a geriatric population. J Am Geriatr Soc 27:374–377
Isaacs H, Barlow MB (1970) Malignant hyperpyrexia during anaesthesia: possible association with subclinical myopathy. Br Med J 1:275–277
Jones IH, Meade TW (1964) Hypothermia following chlorpromazine therapy in myxoedematous patients. Gerontol Clin (Basel) 6:252–256

Kalow W (1956) Familial incidence of low pseudocholinesterase level. Lancet II:576–577
Kalow W (1959) In: Wolstenholme GE, O'Conner MC (eds) Ciba Foundation Symposium on the biochemistry of human genetics. Churchill-Livingstone, London, p 39
Kalow W, Genest K (1957) A method for the detection of atypical forms of human serum cholinesterase. Determination of dibucaine number. Can J Biochem Physiol 35:339–346
Kalow W, Staron N (1957) On distribution and inheritance of atypical forms of human serum cholinesterase, as indicated by dibucaine numbers. Can J Biochem Physiol 35:1305–1320
Kaufman L, Lehmann H, Silk E (1960) Suxamethonium apnoea in an infant: expression of familial pseudocholinesterase deficiency in three generations. Br Med J 1:166–167
Kellaway GSM, McCrae E (1973) Intensive monitoring for adverse drug effects in patients discharged from acute medical wards. N Z Med J 78:525–528
Kerr HD (1993) Potentiation of warfarin by fluconazole. Am J Med Sci 305:164–165
Lakatta EG (1979) Alterations in the cardiovascular system that occur in advanced age. Fed Proc 38:163–167
Lamy PP (1991) Physiological changes due to age. Pharmacodynamic changes of drug action and implications for therapy. Drugs Aging 1:385–404
Landahl S (1987) Drug treatment in 70-82-year-old persons. A longitudinal study. Acta Med Scand 221:179–184
Lawson DH, Jick H (1976) Drug prescribing in hospital: an international comparison. Am J Public Health 66:644–648
Lehmann H, Ryan E (1956) Familial incidence of low pseudocholinesterase level. Lancet II:124
Levy M, Kewitz H, Altivein W, Hellebrand J, Eliakin S (1980) Hospital admissions due to adverse drug reactions: a comparative study from Jerusalem and Berlin. Eur J Clin Pharmacol 17:25–31
Lindley CM, Tully MP, Paramsothy V, Tallis RC (1992) Inappropriate medication is a major cause of adverse drug reactions in elderly patients. Age Ageing 21:294–300
Nolan L, O'Malley K (1988a) Prescribing for the elderly. I. Sensitivity of the elderly to adverse drug reactions. J Am Geriatr Soc 36:142–149
Nolan L, O'Malley K (1988b) Prescribing for the elderly. II. Prescribing patterns: differences due to age. J Am Geriatr Soc 36:245–254
Nolan L, O'Malley K (1989) The need for a more rational approach to drug prescribing for elderly people in nursing homes. Age Ageing 18:52–56
Office of Health Economics (1992) Compendium of health statistics, 8th edn. Office of Health Economics, London
Ouslander JG (1981) Drug therapy in the elderly. Ann Intern Med 95:711–722
Posner J, Rolan PE (1994) Clinical pharmacokinetics. In: Griffin JP, O'Grady J, Wells FO (eds) The textbook of pharmaceutical medicine, 2nd edn. Greystone Books, The Queen's University of Belfast, pp 333–351
Ramsay L (1981) The use of methyldopa in the elderly. J R Coll Physicians Lond 15:239–244
Ritschel WA (1980) Disposition of drugs in geriatric patients. Pharm Int 1:226–230
Roberts MA, Caird FI (1976) Steady-state kinetics of digoxin in the elderly. Age Ageing 5:214
Royal College of Physicians (1984) Report on medication for the elderly. J Roy Coll Physicians Lond 18:7–17
Scarfe MA, Israel MK (1994) Possible drug-interaction between warfarin and combination of levamisole and florouracil. Ann Pharmacother 28:464–467
Schenker S, Bay M (1994) Drug disposition and hepatotoxicity in the elderly. J Clin Gastroenterol 18:232–237
Schmucker WD, Kontak JR (1990) Adverse drug reactions causing hospital admission in an elderly population: experience with a decision algorithm. J Am Board Fam Pract 3:105–109

Schneider JK, Mion LC, Frengley JD (1992) Adverse drug reactions in an elderly outpatient population. Am J Hosp Pharm 49:90–96
Schwartz JT, Reuling FH, Feinlieb M, Garrison RJ, Collie DJ (1972) Twin heritability study of the effect of corticosteroids on intraocular pressure. J Med Genet 9:137–143
Scott PJ, Reid JL (1982) The effect of age on the response of human isolated arteries to noradrenaline. Br J Clin Pharmacol 13:237–256
Shaw PG (1982) Common pitfalls in geriatric drug prescribing. Drugs 23:324–328
Shephard AM, Hewick DS, Moreland TA, Stevenson IH (1977) Age as a determinant of sensitivity to warfarin. Br J Clin Pharmacol 4:315–320
Simons LA, Tett S, Simons J, Lauchlan R, McCallum J, Friedlander Y, Powell I (1992) Multiple medication use in the elderly – use of prescription and nonprescription drugs in an Australian community setting. Med J Aust 157:242
Skaunic V, Hulek P, Martinkova K (1978) In: Kitani K (ed) Liver and ageing. Elsevier/North Holland, Amsterdam, pp 115–130
Sloan RW (1992) Principles of drug therapy in geriatric patients. Am Fam Physician 45:2709–2718
Smith RL, Idle JR (1981) Genetic polymorphism in drug oxidation. In: Davis M, Tredger JM, Williams E (eds) Drug reactions and the liver. Pitman, Bath, pp 95–104
Stanton LA, Peterson GM, Rumble RH, Cooper GM, Polack AE (1994) Drug-related admissions to an Australian hospital. J Clin Pharmacol Ther 19:341–347
Stewart RB, Cooper JW (1994) Polypharmacy in the aged. Practical solutions. Drugs Aging 4:449–461
Sutherland FS, Carter JR (1975) Malignant hyperpyrexia during enflurane anaesthesia. J Tenn Med Assoc 68:785–786
Telfer ABM, Macdonald DJZ, Dinwoodie AJ (1964) Familial sensitivity to suxamethonium due to atypical pseudocholinesterase. Br Med J 1:153–156
Tinetti ME, Speechley M, Ginter SF (1988) Risk factors for falls among elderly persons living in the community. N Engl J Med 319:1701–1707
Vesta RE, Wood AJ, Shand DG (1979) Reduced beta-adrenoreceptor sensitivity in the elderly. Clin Pharmacol Ther 26:181–186
Vogel F (1959) Modern problems of human genetics. Ergeb Inn Med Kinderheilkd 12:52–125
Wallace DE, Watanabe AS (1977) Drug effects in geriatric patients. Drug Intell Clin Pharm 11:597–603
Williams L, Lowenthal DT (1992) Drug therapy in the elderly. South Med J 85:127–131
Williamson J, Chopin JM (1980) Adverse reactions to prescribed drugs in the elderly: a multicentre investigation. Age Aging 9:73–80
Woodhouse KW, Mutch E, Williams FM, Rawlins MD, James OF (1984) The effect of age on pathways in drug metabolism in the human liver. Age Ageing 13:328–334
World Health Organization (1981) Health care in the elderly: report of the technical group on the use of medicaments by the elderly. Drugs 22:279

CHAPTER 11

Drugs Causing Interference with Laboratory Tests

S. YOSSELSON-SUPERSTINE

A. Introduction

Laboratory tests, along with the patient's history and the physical examination, often provide the main key to accurate diagnosis. In some cases their abnormality is the only clue to diagnosis. Results of laboratory tests are also a guide to rational therapy. They can reflect the effectiveness of the therapeutic agent employed as well as indicate the appearance of adverse reactions. No proper diagnosis and therapy can be provided without accurate and reliable laboratory tests. Thus the sensitivity and the specificity of the test method should be among the most important factors, more than cost and ease of performance, in choosing and adopting a method or equipment. Specificity is defined as the proportion of the true negatives that are correctly identified by the test (ALTMAN and BLAND 1994) and is affected inversely by the number of false-positive measurements as shown in the following formula (GADDIS and GADDIS 1990):

$$\text{Specificity (\%)} = \frac{\text{True negatives}}{\text{True negatives} + \text{false positives}} \times 100$$

Many factors can cause a laboratory error or false-positive results. They include human error, equipment or environmental changes, chemicals added to specimens and the presence of endogenous substances. Another important factor which has an effect is the presence of medications in the fluid tested or influencing the functions in the body; this is then reflected in the test result. This last factor – medications – is often overlooked. The mechanism of their interference is the subject of this review. It is not the purpose of this chapter to survey the literature and summarize the drug interferences which have been documented with the various laboratory tests, nor is it the aim to establish the clinical significance of drug-test interference. Only examples of important interferences representing different mechanisms will be discussed in more detail. The design of a study for the determination of a clinically significant drug-test interaction resulting from methodological interference will also be described. The reader is referred to excellent compilations of published studies on this topic (SALWAY 1990; YOUNG 1990) and to critical evaluations of some of the relevant literature (YOSSELSON-SUPERSTINE 1986, 1989).

B. Mechanisms of Drug-Test Interactions

In general, drug interferences can be classified into two major groups: (1) pharmacological interferences, also known as biological or in vivo interferences and (2) methodological interferences, also known as analytical or in vitro interferences.

I. Pharmacological Interferences

Pharmacological interferences are by far the most frequent type of interference. They affect the result of the test by virtue of the activity of the drug or its metabolites in the human body, regardless of the method employed in the test. A pharmacological effect of a drug on a laboratory test is easy to detect when the change in the test value is expected and wanted, but much more difficult to interpret when it is an unexpected, toxicological affect of the drug. The following examples demonstrate these effects:

1. Effect on Glucose Determination

When a hypoglycemic agent is administered to a diabetic patient, we expect the results of the blood or urine glucose tests to be affected in a downward direction. The sulfonylureas cause hypoglycemia by stimulating release of insulin from pancreatic β cells and by increasing the sensitivity of peripheral tissue to insulin (KAHN and SCHECHTER 1990). Therefore it is not surprising that glipizide, a sulfonylurea agent, reduces glucose levels when given in a therapeutic dose. However, when the pharmacological effect, detected in the test, is not the reason for prescribing the drug, but rather an adverse reaction to its administration, it is much more difficult to detect and interpret, especially when the reaction is not expected or necessarily seen in every patient or in each course of therapy. Thus thiazide diuretics prescribed, for instance, to treat hypertension, could elevate blood glucose from weeks to years after therapy is started (JOSEPH and SCHUNA 1990) and the prevalence could be as high as 30%. The adverse effect on glucose tolerance could be a result of several possible mechanisms. One possibility is that the effect is not due to the thiazide molecule itself, but to hypokalemia induced by the drug. Another possibility is that thiazide compounds directly inhibit insulin secretion from the β cells (BRASS 1984).

2. Effect on Uric Acid Determination

Another example is the pharmacological interference of drugs with uric acid determination. When allopurinol is administered to a hyperuricemic patient in a therapeutic dose it is expected to reduce uric acid levels. Both allopurinol and its primary metabolite, alloxanthine, are inhibitors of xanthine oxidase – an enzyme catalyzing oxidation of xanthines to form uric acid (INSEL 1990). On the other hand, thiazides and loop diuretics cause uric acid elevation as a side effect in as many as half the patients receiving them (ANON 1980). The prevalence of asymptomatic hyperuricemia could even rise to 65%–75% in elderly patients being treated with diuretic, while the development of symptomatic gout is commonly seen in only 1%–2% of these individuals. Uric acid retention begins soon after diuretic therapy is started, is dose dependent and produces an average increase in serum uric acid concentration of 1.2–1.5mg/dl (WADE et al. 1989). The precise mechanism is unclear. Several mechanisms have been suggested, among them increased proximal tubular renal reabsorption, decreased tubular secretion and increased post-secretory reabsorption of uric acid (MAY et al. 1992). The increase in uric acid will be demonstrated by all laboratory methods for uric acid measurement, regardless of the technique employed, further proof that the interference is pharmacological or toxicological in its nature.

3. Intramuscular Injections and Muscle Enzyme Determination

Another type of drug-related interference in vivo is interference resulting from the mode of drug administration, rather than the chemical moiety of the drug. This is seen with the injection of medications intramuscularly, which result in the elevation of muscle enzymes, such as creatine kinase (CK) or aspartate aminotransferase (AST), and the possible erroneous attribution of the rise to cardiac disease, mainly acute myocardial infarction, or to liver disease. The rise in CK, for instance, could be of the order of two to six times the normal concentration and lasting up to 48h postinjection (SEIFERT et al. 1992).

4. Effect of Drug-Drug Interactions

Drug-drug interactions are another mode of drug effect in vivo on a laboratory test or an analysis employed to monitor the effect of another drug, or to determine its blood concentrations. For example, there will be a dramatic increase in prothrombin time values when phenylbutazone is added to warfarin therapy as a result of combined effects of displacement of warfarin from plasma protein-binding sites as well as inhibition of hepatic warfarin metabolism. This is a pharmacokinetic interaction affecting the measurement of prothrombin time (HANSTEN 1992). The administration of quinidine can result in an increase in serum digoxin concentration of 0.5ng/ml and more (ANON 1994). The mechanism is variable and involves a quinidine-induced 30%–40% reduction in the apparent volume of distribution of digoxin and a reduction of binding of digoxin in the cardiac muscle (SCHENCK-GUSTAFSSON et al. 1981). Renal and nonrenal clearances of digoxin are also reduced by 30%–40% (ANON 1994).

There is a very high variability in the outcome of this interaction as a result of the numerous affecting factors, such as the sequence of administration, duration of therapy, dose of drugs and other drugs taken by the patient. This is why it remained undetected for half a century (HANSTEN 1992), and when first described was even thought to be an interference with the digoxin assay methodology, which, if true, would not have necessitated an adjustment in the digoxin dosage. However, this possibility was excluded by measuring digoxin in plasma of patients on quinidine alone and finding no measurable concentrations and by looking for an effect of quinidine on digoxin levels while adding both drugs to plasma in vitro (EJVINSSON 1978).

II. Methodological Interferences

A drug or its metabolites can cause analytical error by interfering in the various stages of the analysis, for instance, by competing with or imitating the analyte participating in a chemical, immunological or enzymatic reaction; by changing the chemical or physical conditions of the reaction environment; or by falsely increasing the readings at the end of the analysis, such as color or fluorescence measurements. It is important to study and document these interferences and especially to research their clinical significance, so we can choose a different method for a specific patient whose drug therapy cannot be interrupted. There are alternatives to many of the laboratory tests available today.

1. Colorimetric Interferences

Most of the laboratory tests in chemical pathology still have a colorimetric component, which could be subject to the effect of foreign chromagens. The following are examples of such interferences:

a) Uric Acid Analysis

The colorimetric method for the determination of uric acid in serum is an example of a method in which drugs can react with the reagent and falsely add to the reading of uric acid, if no modifications are made to eliminate reducing substances. This method is still used in the SMAC Automated System (TECHNICON SMAC 1976), and is based on the quantitative method of FOLIN and DENIS (1912) as modified by MUSSER and ORTIGOZA (1966). In this method phosphotungstate reagent is reduced by uric acid to phosphotungstite – a blue-colored complex. Sodium tungstate is used as an alkalizing agent and hydroxylamine is added to intensify the color. Drugs can interfere here if they also reduce phosphotungstate. A survey of the literature (YOSSELSON-SUPERSTINE et al. 1980; YOSSELSON-SUPERSTINE 1986) found that the most significant interferences were those with drugs such as acetaminophen, aspirin, ascorbic acid and levodopa, which are commonly used and whose effect on uric acid determination has been studied in sera of volunteers or patients and not just in a solution in vitro.

Acetaminophen increased serum uric acid by 17% when a high dose of 2g was administered to 11 subjects (SINGH et al. 1972). In urine, uric acid increased in concentration by 100%. No increase in serum uric acid concentration was noticed when the method of HENRY et al. (1957) was employed. This method used carbonate-phosphotungstate and eliminated the use of cyanide in the reaction used previously in order to reduce the turbidity of the final colored solution, which made colorimetric measurement difficult. In addition to modifying the test to eliminate acetaminophen interference, SINGH et al. were also able to demonstrate a log-linear relationship between the concentration of the drug in aqueous solution and the concentration of the apparent uric acid, further suggesting that the interference was with the method of uric acid determination and not merely a pharmacological effect of the drug on uric acid. CARAWAY (1969) demonstrated that ascorbic acid in a concentration of 10 μg/ml, which could be found in the blood after the consumption of 2 g of the vitamin, increased serum uric acid concentration by 1.34 mg/dl in pooled sera and in sera of 21 patients, when determined by a carbonate-phosphotungstate method. This interference is not significant in the more sensitive cyanide methods but is of importance in methods where cyanide is replaced by sodium silicate or carbonate (ALPER and SEITCHIK 1957). Ascorbic acid was eliminated by mild alkaline treatment prior to adding phosphotungstic acid (CARAWAY 1963). Incubation for 10 min after adding sodium carbonate effectively destroys the ascorbic acid present in serum and urine in physiological concentrations.

High concentrations of uricase-resistant chromagens have been reported in the serum of gouty patients receiving high maintenance doses of salicylate (GRAYZEL et al. 1961). These metabolites have not been identified but were not found to be gentisic acid – a salicylate metabolite interfering with uric acid colorimetric assay in urine but not in plasma (CARAWAY 1969). The plasma levels of salicylates in GRAYZEL's study were above 13 mg/100 ml. It should be noted that at these levels there is also a pharmacological interference, a uricosuric effect of salicylates, resulting in a lowering of uric acid levels. This can affect the 7%–119% rise in uric acid as measured by a colorimetric method (YÜ and GUTMAN 1959; GRAYZEL et al. 1961). On the other hand, the rise in uric acid levels at salicylate doses of less than 2 g/day is again caused by a pharmacological effect of the drug leading to urate retention. Unconjugated salicylate in the renal tubule may act by blocking a postulated tubular secretion (YÜ and GUTMAN 1959). The distinction between the two types of interferences – pharmacological and methodological – was possible by measuring uric acid twice, once by the colorimetric assay and once by an enzymatic spectrophotometric assay (GRAYZEL et al. 1961). These interferences are of great clinical significance since they can lead to an erroneous diagnosis of gout in rheumatoid arthritic patients receiving various doses of aspirin containing medications.

CAWEIN and HEWINS (1969) have demonstrated a 20% increase in serum uric acid in 18 patients who received 3–7 g levodopa daily. This elevation was absent when the uricase method was used. However, a pharmacological effect cannot be ruled out since an elevation in serum uric acid, measured by the uricase method, in two patients (AL-HUJAJ and SCHONTHAL 1971, 1972) and the appearance of symptoms of gout subsequent to the initiation of levodopa treatment, were also reported (HONDA and GINDIN 1972). It should also be noted that no methodological interference was found with therapeutic levels of α-methyldopa – a chemically related substance, and uric acid determination (SMALL et al. 1976). Combined enzymatic-colorimetric or spectrophotometric methods for the determination of true uric acid, based on the destruction of true uric acid by converting it to allantoin by uricase, and on the difference in color development between total chromagens and nonurate chromagens, should be employed whenever a drug interference with uric acid determination is suspected (CARAWAY and MARABLE 1966; MARTINEK 1970; SIGMA 1977; KODAK EKTACHEM 1992a).

b) Creatinine Analysis

Creatinine measurement is one of the most valuable laboratory tools in clinical practice. Not only does it provide an excellent estimation of the patient's kidney function, but it also serves as a guide to dosage adjustments of drugs whose elimination is mainly by the renal route. Two major techniques are used today for the analysis of creatnine: colorimetric and enzymatic. The colorimetric method is based on Jaffé's reaction, which was developed more than a century ago and is still the most common assay for the determination of creatinine either in serum or in urine (JAFFÉ 1886). It is employed in many reagent kits and instruments, such as Beakman ASTRA-8, Technicon SMAC and Du Pont ACA. In this reaction there is development of a red color when creatinine reacts with picric acid in an alkali environment. The colored end product absorbs light between 490 and 520nm and is measured by spectrophotometer. The Jaffé reaction is efficient and inexpensive, but is influenced not only by environmental conditions, and the amount of the chemicals used, but also by noncreatinine chromagens, which could account for 20% of the total measured creatinine in serum (NARAYANAN and APPLETON 1980). These chromagens could be physiologic substances such as glucose or protein (DATTA et al. 1986), as well as medications such as some of the cephalosporin (GRÖTSCH and HAJDU 1987) and penicillin antibiotics (KROLL et al. 1984a), lactulose (BRUNS 1988), high-dose furosemide infusion (MURPHY et al. 1989), the today rarely used methyldopa (MADDOCKS et al. 1973; NANJI and WHITLOW 1984), acetohexamide (ROACH et al. 1985) and phenacemide (RICHARDS 1980). Of the many cephalosporins studied, only cefoxitin and cephalothin produced a significant interference at therapeutic concentrations of 100mg/l (GREEN et al. 1990), as well as cafazolin, which caused a false increase of 10–20μmol/l (0.1–0.2mg%) of creatinine for every 20mg/l cefazolin (NANJI et al. 1987).

The many publications on the interferences of these medications, and especially cephalosporins, with the Jaffé analysis were evaluated very extensively in a recent review (DUCHARME et al. 1993). The mechanism of interference appears to be the presence of a carbonyl group on the offending compound that alters the absorptivity of the complex (KROLL et al. 1987). Some of the modifications suggested over the years, such as changing the pH or incubation temperature (LETELLIER and DESJARLAIS 1985a), usage of kaolin and Lloyd's reagent (Fullers' earth) (HAECKEL 1981; BJERVE et al. 1988) and the addition of a dialysis step could increase the specificity of the method. Thus, for instance, a dialysis step diminished interference by cefoxitin using the SMAC analytical system (KROLL et al. 1984b).

Timing of absorbance reading or usage of single-point kinetic methods will further reduce drug interferences. The rate of color formation is different for the cephalosporins and for creatinine. It reaches equilibrium in a shorter period (8min) for creatinine (GROTSCH and HAJDU 1987).

The effect of timing of reading was also demonstrated in a study in which creatinine was measured in the Beckman ASTRA 8, which performs two-point kinetic reading, and the Technicon SMAC II, which makes a delayed measurement. The ASTRA measurement exaggerated the apparent creatinine contribution from cephalothin (HICKMAN and MATHER 1988). However, usage of the enzymatic assay (TOFFALETTI et al. 1983) is a better solution when an unexplained high concentration of creatinine is measured in a patient with an intact kidney, such as demonstrated many times by a normal blood urea nitrogen (BUN) reading.

If creatinine is not determined by an enzymatic method in a patient receiving a cephalosporin antibiotic, it is best that samples for creatinine analysis be obtained at least 4–6h after infusion of the drug. The interference is maximal when the sample is drawn within 30min after starting the infusion (NANJI et al. 1987).

It is also recommended that creatinine be estimated by an enzymatic method in patients on lactulose therapy since increases of 3mg% and 6.5mg% in creatinine, using the alkaline picrate assay, by the kinetic method (Beckman ASTRA) and by the continuous flow method (Technicon SMAC), respectively, have been detected when a lactulose concentration of 100g/l was present in the solution. The drug caused no problems with the enzymatic method (BRUNS 1988). Likewise the enzymatic method is preferred when an unusually low creatinine level is found, such as in cases of high-dose furosemide consumption (MURPHY et al. 1989). In this case there is a possibility of a metabolite blocking the colorimetric reaction since the parent drug did not cause an interference in vitro. This error is of great clinical significance because it could mask diminished renal function in patients receiving high doses of furosemide (1–2g/dose).

c) Discoloration of Urine and Feces

Unusual color and appearance of urine or feces could provide useful diagnostic information and should not be overlooked. For instance, red, pink or brown urine suggests hematuria, and orange or dark yellow urine suggests the presence of urobilin. Bleeding from the upper gastrointestinal tract may cause the stool to be black.

Drugs excreted in urine or feces can change the normal color and appearance of these body wastes and obscure the elementary gross examination of these specimens. Some also have the potential to affect tests performed on urine or feces, especially when they are based on colorimetric methods.

Therefore, before proceeding with other diagnostic measures, a detailed history of dietary intake of food and drugs should be obtained from the patient. For instance, orally administered iron can cause false-positive reactions for fecal occult blood when tested by the Hemoccult and Hematest methods (LIFTON and KREISER 1982). Hemoccult uses guaic as a reagent and

Table 1. Drugs which may discolor urine

Drug/drug class	Color produced
Acetanilide	Yellow to red
Aloe	Yellow-pink to red-brown in alkaline urine
Aminopyrine	Red
Aminosalicylic acid	Discoloration; red in hypochlorite solution
Amitriptyline	Blue-green
Anisindione	Pink or red to red-brown in acidic urine; orange in alkaline urine
Antipyrine	Red-brown
Azuresin	Blue or green
Cascara	Yellow-brown in acidic urine; yellow-pink in alkaline urine, turning black on standing
Chloroquine	Rust yellow to brown
Chlorzoxazone	Orange or purplish red
Cimetidine (injection)	Green
Clofazimine	Red to brownish black
Danthron	Pink to red or red-brown in alkaline urine
Daunonrubicin	Red
Deferoxamine mesylate	Reddish
Dimethylsulfoxide (DMSO)	Reddish, due to hemoglobinuria
Diphenadione	Orange in alkaline urine
Doxorubicin	Red
Emodin	Pink to red or red-brown in alkaline urine
Ethoxazena	Orange to orange-brown
Ferrous salts	Black
Furazolidone	Rust yellow to brown
Idarubicin	Red
Indigotindisulfonate	Blue or green
Indomethacin	Green due to biliverdinemia
Iron sorbitex	Black
Levodopa	Dark on standing in hypochlorite solution
Methocarbamol	Dark to brown, black or green on standing
Methyldopa	Dark on standing in hypochlorite solution
Methylene blue	Blue or green
Metronidazole	Dark
Mitoxantrone	Dark blue or green

Table 1. *Continued*

Drug/drug class	Color produced
Nitrofurantoin	Rust yellow to brown
Pamaquine	Rust yellow to brown
Phenacetin	Dark brown to black on standing
Phenazopyridine	Orange to orange-red
Phenindione	Orange-red in alkaline urine
Phenolphthalein	Pink to purplish red in alkaline urine
Phenolsulfonphthalein (PSP)	Pink to red in alkaline urine
Phenothiazines	Pink to red or red-brown
Phensuximide	Pink to red or red-brown
Phenytoin	Pink to red or red-brown
Primaquine	Rust yellow to brown
Promethazine (injection)	Green
Propofol (injection)	Green
Quinacrine	Deep yellow in acidic urine
Quinine	Brown to black
Resorcinol	Dark green
Riboflavin	Yellow fluorescence
Rifampin	Bright red-orange
Santonin	Yellow in acidic urine, pink in alkaline urine
Senna	Yellow-brown in acidic urine; pink to red in alkaline urine; brown on standing
Sulfasalazine	Orange-yellow in alkaline urine
Sulfonamides, antibacterial	Rust yellow to brown
Thiazolsulfone	Pink or red
Tolonium	Blue-green
Triamterene	Pale blue florescence
Warfarin	Orange

Table 2. Drugs which may discolor feces

Drug/drug class	Color produced
Antacids, aluminum hydroxide types	Whitish or speckling
Antibiotics, oral	Greenish gray
Anticoagulants, all	Pink to red or black (if bleeding)
Bismuth-containing preparations	Greenish black
Charcoal	Black
Clofazimine	Red to brownish black
Danthron	Brownish staining of rectal mucosa
Dithiazanine	Green to blue
Ferrous salts	Black
Heparin	Pink to red or black (if bleeding)
Indocyanine green	Green
Indomethacin	Green due to biliverdinemia
Nonsteroidal anti-inflammatory drugs	Pink to red or black (if bleeding)
Phenazopyridine	Orange-red
Pyrvinium pamoate	Red
Rifampin	Red-orange
Salicylates, especially aspirin	Pink to red or black
Senna	Yellow

Hematest uses orthatolidone as a reagent. The reaction between these reagents and a substance acting as a peroxidase (hemoglobin) causes a blue-colored reaction in the product tested. Other potential interactions with the test causing false-positive results are with ascorbic acid, aspirin, povidone-iodine and medications that irritate the gastroinetstinal tract, such as corticosteroids and nonsteroidal anti-inflammatory agents (ENGLE et al. 1988).

A comprehensive listing of drugs which may discolor feces and urine is given in Tables 1 and 2 (ANON 1993).

2. Effect on pH of the Assay Environment

The pH of a solution is one of the major conditions for a specific chemical reaction to be carried out. Many of the colorimetric reactions in laboratory tests are pH dependent. Urine is one of the biological fluids whose pH is most affected by diet and drugs. Acid urine may be produced by a diet high in meat protein and in some fruits such as cranberries and by drug and chemicals, such as ammonium chloride, methionine, methenamine mandelate, ascorbic acid or acid phosphate. Alkaline urine may be induced by a diet high in certain fruits and vegetables, especially citrus fruits and drugs such as sodium bicarbonate, potassium citrate and acetazolamide (BRADLEY et al. 1979).

The following test is an example of a test affected by the change in urine pH:

a) Urine Protein Reagent Strip Test

The colorimetric reagent strip test is based on the ability of proteins to alter the color of some acid-base indicators without altering the pH. When an indicator such as tetrabromophenol blue is buffered at pH 3 it is yellow in solutions without protein, but in the presence of protein the color changes to green, and then to blue with increasing protein concentrations.

The Albustix reagent strip is a protein test strip that contains a single test area. This area consists of a small square of absorbant paper impregnated with a buffered solution of tetrabromophenol blue. Uristix, *N*-Uristix, Combistix, Hema-Combistix, Labstix, Bililabstix, Multistix and *N*-Multistix and *N*-Multistix-SG reagent strips are multideterminant reagent strips, each containing an area for protein determination along with test areas for other urinary constituents. Protein is determined simply by dipping the strip into well-mixed uncentrifuged urine, and immediately comparing the resultant color with the chart provided on the reagent strip bottle. The results are reported as negative (yellow color), trace, or one "plus" to four "plus". Trace readings may detect protein in a concentration of 5–20 mg/dl. "Plus" readings are approximately equivalent to protein concentrations of 30, 100, 300 and over 2000/dl, respectively, and are reliable indicators of increasingly severe proteinuria. Albumin reacts with the indicator more strongly than do the other proteins.

Today there is automated and semi-automated instrumentation using reflector photometers that could read the reagent strips, allowing for elimination of the variable that has commonly caused erroneous and confused results when color has been interpreted by the human eye (Anon 1982).

Highly buffered, alkaline urines may give false-positive results when the buffer systems in the reagent area are overcome and an actual shift in pH of the buffers occurs. Addition of sodium carbonate (43 g/l urine standard) caused a +4 reading by the Albustix method (Gyure 1977).

Therapeutic doses of the commonly used antacids in over the counter medications do not elevate urine pH by more than one unit (Gibaldi et al. 1974). Since normal pH varies from 4.6 to 8, it is unlikely that they affect urine protein determination by the reagent strip method unless, at least theoretically, there is an abuse of antacid products.

Therapeutic doses of acetazolamide (375–1000 mg/day) have also not been found to elevate urine pH to 10 or above and did not falsely elevate urine protein levels (Yosselson-Superstine and Sinai 1986).

3. Interference with Chromatographic Methods

Drugs can interfere with various chromatographic methods. They, or their metabolites, can confuse spots of analytes on thin layers or solvent chromatography or they can elute with the tested analyte using high-pressure liquid chromatography, causing an erroneous increase in its concentration. The following are examples of such interferences:

a) Urine Amino Acid Screening

Urine screening for amino acid metabolic disorders can be performed effectively by a semiquantitative chromatographic technique. First the amino acids are separated by electrophoresis at acid pH, followed by solvent chromatography for separation in the second dimension. Alternatively, thin-layer chromatography can be used. Staining is performed by different reagents. Most alpha amino acids and compounds containing primary or secondary amino groups attached to an aliphatic carbon atom react with Ninhydrin (triketohydrindene hydrate) reagent to form purple colors. Isatin reagent is particularly useful for locating proline and hydroxyproline. Proline, which is yellow with Ninhydrin reagent, reacts with isatin to form a blue color, which is much more visible. Other amino acids also react to form unstable colors, varying from blue to red-purple.

Urea and citrulline react with Ehrlich's reagent to form a yellow color. This reagent is also useful for the location of tryptophan and indoles. Histidine, carnosine and other imidazole derivatives react with Pauly's reagent to form reddish colors. Hydroxyproline also reacts to form an orange-brown color. Arginine and other monosubstituted guanidine derivatives react with Sakaguchi's reagent to form an orange color.

Most sulfur-containing amino acids react with platinic-iodide reagent to create white spots on a light pink background. With fast blue reagent, methylmalonic acid (MMA) appears as a deep-purple spot and ethylmalonic acid (EMA) appears as a dark-blue/gray spot above MMA. Phenylpyruvic acid and the branched-chain keto acids appear as bright-purple spots above EMA. Homogentisic acid appears as a light-brown spot and lactic acid appears as a bleached spot.

Several drugs or their metabolites can react with the mentioned reagents and obscure the chromatographic results if they chromatograph near amino acids or give false-positive results. A careful history of medications should, therefore, be obtained when an unusual amino acid pattern is found in the urine. In addition, authentic amino acids should be chromatographed with the urine to show dissimilarity and a repeat urine amino acid screening should be done after the patient has been off all medications for at least 3 or 4 days. Drugs implicated in interfering with this test are ampicillin and other synthetic penicillins, which appear as several purple and brown ill-defined spots to the right of phenylalanine. Kanamycin antibiotic is excreted as a compound which can react with Ninhydrin and which migrates faster than lysine. Carbenicillin causes an ill-defined purple area at the right upper corner of the chromatogram, in addition to a purple spot to the right of phenylalanine and one to the right of glutamine. Cephalexin gives a Ninhydrin-purple spot next to phenylalanine. Leukemic patients receiving folate antimetabolite therapy (e.g., methotrexate) and patients with vitamin B_{12} or folic acid deficiency may excrete 4(5)-amino-5(4)-imidazole carboxamide (AIC), which has a mobility similar to that of β-aminoisobutyric acid, β-alanine and δ-aminolevulinic acid and reacts with Ninhydrin to form a yellow-brown color. *N*-Acetylcysteine, a mucolytic agent and an agent also used today to treat acetaminophen poisoning, is excreted both in its oxidized form (*N*-acetylcysteine) and as a mixed disulfide of *N*-acetylcysteine and cysteine. Both derivatives react with the cyanide-nitroprusside reagent and with Ninhydrin reagent (Shih et al. 1991).

b) High-Pressure Liquid Chromatography Drug Assays

High-pressure liquid chromatography (HPLC) assays are possible on all therapeutically monitored drugs except lithium and digoxin. They can be developed easily with no need for preparation of commercial reagent kits; however, they are more costly and not without the problem of drug interference (Yosselson-Superstine 1984, 1989; Bottorff and Stewart 1986). For instance, in one of the HPLC methods which was developed to measure serum procainamide, as well as its major active metabolite *N*-acetylprocainamide (NAPA), sulfathiazole was found to interfere, when added in vitro, with the estimation of NAPA (Shukur et al. 1977). In another HPLC method (Dutcher and Strong 1977), a presumed metabolite of quinidine in therapeutic concentrations (>3 μg/ml) caused the overestimation of NAPA by up to 0.2 μg/ml. Aminophylline, paracetamol (acetaminophen), codeine and caffeine were also reported to be elevated by one of these methods (Stearns 1981); however, except for caffeine, which was injected in a concentration of 1.5 μg/ml (peak plasma concentration after a dose of 100 mg given as coffee), this happened with toxic concentrations of these drugs. The interference of caffeine was resolved in another HPLC procedure by changing the buffer concentration (COYLE et al. 1987).

In another similar assay method (GANNON and PHILLIPS 1982), metronidazole HCl in a dose of 500 mg given intravenously every 6 h coeluted with procainamide and raised the baseline of the procainamide peak, thus causing an erroneous increase in its blood concentration.

Of all the drugs commonly monitored in clinical practice, theophylline is the one for which the largest list of reports has been published about assay methodologies and clinically significant drug interferences. Most of these interferences were reported with the use of HPLC methods, which are still used (KELSEY et al. 1987) but have been replaced in many laboratories by immunoassays. Among the interfering drugs are many anti-infective agents such as many of the cephalosporins, ampicillin, methicillin, sulfonamides, metronidazole, chloramphenicol, analgesics such as salicylic acid and acetaminophen and many other drugs. The interference is dependent on the assay conditions, such as type of mobile phase, pH adjustments and incorporation of an ion-pair reagent. For a detailed list and discussion refer to YOSSELSON-SUPERSTINE (1989). Because of their frequent coadministration it should be emphasized that most of the interferences caused by cephalosporins are clinically significant especially when the patient suffers from renal disease, in which case an immunoassay method should be preferred (GANNON and LEVY 1984).

4. Interference with Enzymatic Reactions

Many of the laboratory tests used today include an enzymatic reaction step. These reactions could be a part of a simple colorimetric method, as well as a more sophisticated immunoassay. Drugs can interfere with these reactions by activation of the reaction, by inhibition of the reaction, or by simply reacting as the substrate and thus affecting the results of the analysis. The following examples demonstrate these effects:

a) Creatinine Analysis

The enzymatic method for creatinine determination, which is used in the automated drug-slide system, is based on the enzymatic hydrolysis of creatinine by creatinine iminohydrolase and the production of ammonia and *N*-methylhydantoin. The ammonia reacts with bromophenol blue to form a blue chromagen, the amount of which is measured spectrophotometrically. Its concentration thus reflects the concentration of creatinine (TOFFALETTI et al. 1983).

The 4-amino group of flucytosine, a systemic antifungal drug, can be converted to ammonia by the same enzymatic system and cause false elevation of creatinine. Therapeutic plasma levels of flucytosine are between 400 and 775 μmol/l. When the flucytosine level is at the upper limit, the serum creatinine level may be falsely elevated by a factor of 14, whereas at subtherapeutic flucytosine levels, such as 75 μmol/l, the serum creatinine level may be falsely elevated by a factor of 2.5 (HERRINGTON et al. 1984; MITCHELL 1984; SOUNEY and MARIANI 1985; O'NEILL et al. 1987). This interaction could be of much clinical significance since flucytosine is frequently administered together with amphotericin B and renal function is evaluated constantly along the treatment period to detect any possible nephrotoxic reaction to the drug therapy. The Ektachem 700 system of the second generation and above uses a single-slide enzymatic method. It is more specific and does not cause false elevation with flucytosine (COUTTIE et al. 1987). In this method the creatine formed is converted to sarcosine and urea by creatine amidinohydrolase. The sarcosine, in the presence of sarcosine oxidase, is oxidized to glycine, formaldehyde and hydrogen peroxide. The final step involves the peroxidase-catalyzed oxidation of triarylimidazole leuko dye to produce a blue dye (KODAK EKTACHEM 1992b).

This improved method did not completely eliminate the interference by another drug, lidocaine (GOSNEY et al. 1987; SPINLER et al. 1989), and patients on long-term lidocaine therapy may show an increase in creatinine concentration of up to 1.0 mg/dl (88 μmol/l). Most patients receiving intravenous lidocaine will show an increase in creatinine concentration of 0.3 mg/dl (26 μmol/l) or less (KODAK EKTACHEM 1992b). The interfering substance was found to be *N*-ethylglycine (NEG), an inactive metabolite of lidocaine (SENA et al. 1988). NEG is similar in structure to sarcosine (*N*-methylglycine) and can thus serve as an additional substrate for sarcosine oxidase, producing false elevation of creatinine. Because this metabolite is found in both serum and urine the interference is noticed in creatinine determination in both urine and serum. Until a further improvement in the single-slide method becomes available, serum creatinine should be analyzed by the picric acid method in patients on lidocaine therapy.

b) Alkaline Phosphatase Analysis

In the determination of alkaline phosphatase, *p*-nitrophenyl phosphate, used as the substrate, is hydrolyzed to *p*-nitrophenol at alkaline pH. *p*-Nitrophenol, which is yellow, is read by a spectrophotometer. Alkaline phosphatase is the catalyzing enzyme in this reaction and its concentration is correlated with the concentration of the end product. Mg^{2+} is an enzyme activator in this reaction. A drug metabolite such as cysteine (a metabolite of *N*-acetylcysteine) can chelate with magnesium ion, causing a decrease in alkaline phosphatase activity (LETELLIER and DESJARLAIS 1985b). Theophylline is another drug that can affect this test result by directly inhibiting alkaline phosphatase activity. It can cause a negative bias of up to 7% (KODAK EKTACHEM 1992c). Two further drugs, methotrexate and nitrofurantoin, in five times therapeutic concentrations, can falsely elevate alkaline phosphate levels. However, the interference is spectral and not enzymatic in its nature (LETELLIER and DESJARLAIS 1985b).

5. Interference with Immunoassays

Immunoassays are gaining more and more acceptance as the method of choice in the biochemical and in the clinical laboratory, for instance, in disposition studies for biotechnology products or in therapeutic drug monitoring (TDM) (BOTTORFF and STEWART 1986). Various types of immunoassays are employed, the most frequently used being the homogeneous enzyme multiplied immunotechnique (EMIT), enzyme-linked immunosorbent assay (ELISA), radioimmunoassay (RIA) and fluorescence immunoassay (FIA). All of these assays involve antigen-antibody reactions, so one important mechanism of drug interference with these methods is that involving nonspecificity of antibodies. This is demonstrated in the following assay:

a) Digoxin Assay

Digoxin is one of the most important drugs where the therapy is guided by its concentrations in the patient's blood. All the immunoassays, radioimmunoassays, enzyme immunoassays and fluoroimmunoassays used for measurement of the drug suffer from the cross-reactivity of digoxin metabolites and from endogenous substances known as digoxin-like immunoreactive substances (DLIs). These substances accumulate in the blood, especially in neonates, during the third trimester of pregnancy, renal failure and liver disease (MORRIS et al. 1990). The most important exogenous substance to interfere with digoxin assays is the diuretic spironolactone – a synthetic steroid. Canrenone and its 20-hydroxy derivative, both metabolites of spironolactone, have apparently more cross-reactivity with digoxin than the parent drug (HUFFMAN 1974; SILBER et al. 1979; MORRIS et al. 1987). The wide range of falsely increased levels of digoxin ranging from no effect to an increase of 4.0ng/ml after the administration of therapeutic doses of spironolactone (25–200mg/day) can be attributed to the difference in the specificity of the digoxin antiserum used in the different RIA kits (MORRIS et al. 1987).

Endogenous steroids are known to cross-react with digoxin. An apparent digoxin concentration of 0.5mg/l has also been reported in the serum of a patient receiving no digoxin, 30min after the administration of 6-methylprednisolone (GAULT et al. 1985). The interference with steroids was noted with both radioimmunoassay and with fluorescence polarization immunoassays (SOLDIN et al. 1984). Fab fragments infused into patients with digoxin toxicity pose another problem with digoxin radioimmunoassays, as well as with fluorescence excitation transfer immunoassays (NATOWICZ and SHAW 1991). Digoxin serum levels after Fab treatment could be as much as tenfold greater than pretreatment values, when the solid-phase methods are used. The interference could be eliminated when polyethylene glycol and charcoal methods are used (GIBB et al. 1983). It is best not to use digoxin blood levels as a guideline for continuation of the antidote therapy and to base the decision on the clinical presentation of the patient, although the newer techniques for measuring unbound digoxin could also be helpful (BANNER et al. 1992). It is interesting to note that unconventional medicines could also participate in drug-test interactions. This is important to remember when unusual results are obtained since such medicines often do not appear on the patient's drug list. Such an interference was noted with kyushin – a Chinese medicine that is very popular in Japan, where it is obtainable without prescription. This medicine was responsible for digoxin levels of more than 2.5 ng/ml in a patient who was on a daily dose of 0.25 mg digoxin and did not exhibit any signs of digoxin toxicity. The main components of the drug have chemical structures similar to those of digoxin and digitalis-like cardiotonic actions. One tablet of kyushin had a digoxin-like immunoreactivity equivalent to 1.9 μg (TDX analyzer, Abbot Laboratories), 1.5 μg (Du Pont Aca V analyzer) and 72 μg digoxin (Enzymun-Test, Boehringer Mannheim). The different equivalencies are attributed to differences in cross-reactivity of the antibody used in the immunoassays.

Two healthy volunteers took a typical dose of kyushin – two tablets three times a day, and digoxin-like immunoreactivity reached almost 0.4 ng/ml in half a day. The authors concluded that digoxin serum levels should be interpreted very carefully in patients taking Chinese medicine (FUSHIMI et al. 1989).

C. Design of a Study for Evaluation of Analytical Interference

The Expert Panel on Drug Effects in Clinical Chemistry (EPDECC) of the International Federation of Clinical Chemistry (IFCC) suggested guidelines for evaluation of analytical interference (GALTEAU and SIEST 1984). The first step in the guidelines is to accumulate the relevant information (e.g., physicochemical and pharmacokinetic characteristics) about the drug, its metabolites and formulations. Drugs to be selected for evaluation are those most frequently used, those specific for the disease monitored by the laboratory test studied or those suspected of interfering because of their physicochemical characteristics.

The second step is to select the laboratory method to be studied and choose a reference method of analysis. Once we have the drugs and the methods, we are ready to start with in vitro studies on biological specimens, serum, plasma or urine obtained from healthy subjects not on medication. We prepare tenfold concentrated solutions of the drug and add to the serum, plasma or urine pool. A control sample of equivalent volumes of the solvent is also prepared and added to the biological specimens. The concentration studied is the one which is ten times the highest therapeutic concentration reported in the literature or, if unknown, the concentration calculated from ten times the therapeutic daily dose diluted in 5 or 15 l, depending on the volume of distribution of the drug. All the drug and control samples should be tested in duplicate.

If an interference is established, we follow with studies to measure the magnitude of the analytical interference. At least five different concentrations are used with two of them in the therapeutic range. A dose-effect curve is obtained and the slope calculated.

Further steps used in the various studies for the detection and evaluation of a drug-laboratory method interaction are the repetition of the study using many different assays after establishing the basic difference in the results obtained by them, that is when no interfering substance is present. The studies should then be carried out on volunteers' or patients' biological fluids after the consumption of the potential interfering drug in various dosages or concentrations.

D. Conclusions

In the present chapter, an attempt has been made to shed light on a very important phenomenon, the effect of drugs on laboratory tests in clinical practice. Laboratory tests are an indispensable tool for appropriate diagnosis and therapy, and we need to be able to rely upon them. We must strive to make them highly specific but, when this is not possible, the alternative is to become familiar with the list of drug-test interferences. Knowledge of the mechanisms of the interferences will assist us greatly in remembering and in understanding them.

When we come across a suspected drug-test interference, there are certain practical steps that we should follow. This is well presented in the suggested algorithm by TROUB (1992). When the laboratory result is inconsistent with other laboratory results or the clinical picture, the possibility of transcription error should be ruled out and, if no such error is found, the test should be repeated (with the new and original specimens). If the doubtful results are consistent, a pharmacological or dietary effect should be ruled out. The literature should then be consulted for past documentation of possible analytical interference. If this is found, either the offending drug should be discontinued and the test repeated, or, if this is not possible, a different method of analysis should be employed.

Whenever we come across a newly suspected drug-laboratory test interference, it should be made known to the medical community, so that others will be able to perform further studies to evaluate the nature of the interference and suggest ways to eliminate it. The documentation provided by such studies will assist in minimizing possible errors in diagnosis, as well as in treatment.

References

Al Hujaj M, Schonthal H (1971) Hyperuricemia and levodopa. N Engl J Med 285:859–860

Al-Hujaj M, Schonthal H (1972) Gout and levodopa. N Engl J Med 286:376

Alper C, Seitchik J (1957) Comparison of the Archibald-Ken and Stransky colorimetric procedure and the Praetorius enzymatic procedure for the determination of uric acid. Clin Chem 3:95–101

Altman DG, Bland JM (1974) Diagnostic tests 1: Sensitivity and specificity. Br Med J 308:1552

Anon (1980) Diuretics, hyperuricemia and tienilic acid. Lancet II:681–682

Anon (1982) Proteins in urine. In: Modern urine chemistry, Ames Division, Miles Laboratories, Elkhart, Indiana

Anon (1993) Drug-induced discoloration of feces and urine. In: Knoben E, Anderson PO (eds) Handbook of clinical drug data. Drug Intelligence Publications, Hamilton, Illinois, p 22

Anon (1994) Digoxin-quinidine. In: Tatro DS, Olin BR, Hebel SK (eds) Drug interaction facts. Facts and comparisons, St. Louis, p 290

Banner JW, Bach P, Burke B, Freestone S, Gooch WM (1992) Influence of assay methods on serum concentrations of digoxin during Fab fragment treatment. Clin Toxicol 30:259–267

Bjerve KS, Egens J, Lampinen LM, Masson P (1988) Evaluation of several creatinine methods in search of a suitable secondary reference method: report from the subcommittee on reference method for creatinine. Nordic Society for Clinical Chemistry. Scand J Clin Lab Invest 48:365–373

Bottorff MB, Stewart CF (1986) Analytical techniques and quality control. In: Taylor WJ, Diers Caviness MH (eds) A textbook for the clinical application of therapeutic drug monitoring. Abbott Laboratories, Texas, pp 51–57

Bradley M, Schuman GB, Ward PCJ (1979) Examination of urine. In: Henry JB (ed) Clinical diagnosis and management by laboratory methods. WB Saunders Company, Philadelphia, pp 559–634

Brass EP (1984) Effects of hypertensive drugs on endocrine function. Drugs 27:447–458

Bruns DE (1988) Lactulose interferes in the alkaline picrate assay for creatinine. Clin Chem 34:2592–2593

Caraway WT (1963) Uric acid. Stand Meth Clin Chem 4:239–247

Caraway WT (1969) Non-urate chromagens in body fluids. Clin Chem 24:54–57

Caraway WT, Marable H (1966) Comparison of carbonate and uricase-carbonate methods for the determination of uric acid in serum. Clin Chem 12:18–24

Cawein MJ, Hewins J (1969) False rise in serum uric acid after L-dopa. N Engl J Med 281:1489–1490

Couttie K, Earle J, Loakley J (1987) Evaluation of single-slide creatinine method on the Kodak Ektachem 700 shows positive interference from lidocaine metabolites. Clin Chem 33:1674

Coyle JD, Mackichan JJ, Boudoulas H, Lima JJ (1987) Reversed-phase liquid chromatography method for measurement of procainamide and three metabolites in serum and urine: percent of dose excreted as deethyl metabolites. J Pharm Sci 76:402–405

Datta P, Graham GA, Schoen I (1986) Interference by IgG paraproteins in the Jaffe method for creatinine determination. Am J Clin Pathol 85:463–468

Ducharme MP, Smythe M, Strohs G (1993) Drug induced alterations in serum creatinine concentrations. Ann Pharmacother 27:622–633

Dutcher JS, Strong JM (1977) Determination of plasma procainamide and N-acetylprocainamide concentration by high pressure liquid chromatography. Clin Chem 23:1318–1320

Ejvinsson G (1978) Effect of quinidine on plasma concentrations of digoxin. Br Med J 1:279–280

Engle JP, Donnelly AJ, Lewis RK (1988) Fecal occult blood tests: potential drug interactions. Consult Pharm 3:356–358
Folin O, Denis W (1912) A new colorimetric method for the determination of uric acid in blood. J Biol Chem 13:469–475
Fushimi R, Tachi J, Amino N, Miyai K (1989) Chinese medicine interfering with digoxin immunoassays. Lancet I:339
Gaddis GM, Gaddis ML (1990) Introduction to biostatistics: part 3. Sensitivity, specificity, predictive value, and hypothesis testing. Ann Emerg Med 19:591–597
Galteau MM, Siest G (1984) Drug effects in clinical chemistry. Part 2. Guidelines for evaluation of analytical interference. J Clin Chem Clin Biochem 22:275–279
Gannon RH, Levy RM (1984) Interference of third-generation cephalosporins with theophylline assay by high-performance liquid chromatography. Am J Hosp Pharm 41:1185–1186
Gannon RH, Phillips LR (1982) Metronidazole interference with procainamide HPLC assay. Am J Hosp Pharm 39:1966–1967
Gault MH, Longerich L, Dawe M, Vasder SC (1985) Combined liquid chromatography/radioimmunoassay with improved specificity for serum digoxin. Clin Chem 31:1272–1277
Gibaldi M, Grundhofer B, Levy G (1974) Effect of antacids on pH of urine. Clin Pharmacol Ther 16:520–525
Gibb I, Adams PC, Parnham AJ, Jenning K (1983) Plasma digoxin: assay anomalies in Fab-treated patients. Br J Clin Pharmacol 16:445–447
Gosney K, Adachi-Kirkland J, Schiller HS (1987) Evaluation of lidocaine interference in the Kodak Ektachem 700 Analyzer single-slide method for creatinine. Clin Chem 33:2311
Grayzel AL, Liddle L, Seegmiller JE (1961) Diagnostic significance of hyperuricemia in arthritis. N Engl J Med 265:763–768
Green AJE, Halloran SP, Mould GP (1990) Interference by newer cephalosporins in current methods for measuring creatinine. Clin Chem 36:2139–2140
Grötsch H, Hajdu P (1987) Interference by the new antibiotic cefpirome and other cephalosporins in clinical laboratory tests, with special regard to the "Jaffe" reactions. J Clin Chem Clin Biochem 25:49–52
Gyure WL (1977) Comparison of several methods for semiquantitative determination of urinary protein. Clin Chem 23:876–879
Hansten PD (1992) Drug interactions. In: Koda-Kimble MA, Young LY (eds) Applied therapeutics, the clinical use of drugs. Applied Therapeutics, Vancouver, Washington, pp 2-1–2-12
Haeckel R (1981) Assay of creatinine in serum with use of Fuller's earth to remove interferents. Clin Chem 27:179–183
Henry RJ, Sobel C, Kim J (1957) A modified carbonate-phosphotungstate method for the determination of uric acid and comparison with the spectrophotometric uricase method. Am J Clin Pathol 28:152–160
Herrington D, Drusano GL, Smalls U, Standiford HC (1984) False elevation in serum creatinine levels. JAMA 252:2962
Hickman PE, Mather P (1988) Variation among instruments in interference by cephalosporin in the Jaffe reaction for creatinine. Clin Chem 34:215–216
Honda H, Gindin RA (1972) Gout while receiving levodopa for parkinsonism. JAMA 219:55–57
Huffman DH (1974) The effect of spinonolactone and canrenone on digoxin radioimmunoassay. Res Commun Chem Pathol Pharmacol 9:787
Insel PA (1990) Analgesic-antipyretics and antiinflammatory agents: drugs employed in the treatment of rheumatoid arthritis and gout. In: Gilman AG, Rall TW, Nies AS, Taylor P (eds) Goodman and Gilman's The pharmacological basis of therapeutics. Pergamon Press, New York, pp 638–681
Jaffé M (1886) Über den Niederschlag welchen Pikrinsaeure in normalen Harn erzeugt und über eine neue Reaktion des Kreatinins. Z Physiol Chem 10:391–400

Joseph JC, Schuna AA (1990) Management of hypertension in the diabetic patient. Clin Pharm 9:864–873
Kahn CR, Schechter Y (1990) Insulin, oral hypoglycemic agents, and the pharmacology of the endocrine pancreas. In: Gilman AG, Rall TW, Nies AS, Taylor P (eds) Goodman and Gilman's The pharmacological basis of therapeutics. Pergamon Press, New York, pp 1463–1495
Kelsey HC, Ball MJ, Kay JDS (1987) Interference by acetazolamide in theophylline assay depends on the method. Lancet II:403
Kodak Ektachem (1992a) Test methodology notebook. URIC. Eastman Kodak Company, Rochester, New York
Kodak Ektachem (1992b) Test methodology notebook. CREA. Eastman Kodak Company, Rochester, New York
Kodak Ektachem (1992c) Test methodology notebook. ALKP. Eastman Kodak Company, Rochester, New York
Kroll MH, Hagengruber C, Elin RJ (1984a) Reaction of picrate with creatinine and cepha-antibiotics. Clin Chem 30:1664–1666
Kroll MH, Hagengruber C, Elin RJ (1984b) Effect of dialysis on interference by cefoxitin with determination of creatinine. Clin Chem 30:1386–1388
Kroll MH, Roach NA, Poe B, Elin J (1987) Mechanism of interference with the Jaffe reaction for creatinine. Clin Chem 33:1129–1132
Letellier G, Desjarlais F (1985a) Analytical interference of drugs in clinical chemistry II – The interference of three cephalosporins with the determination of serum creatinine concentration by the Jaffe reaction. Clin Biochem 18:352–356
Letellier G, Desjarlais F (1985b) Analytical interference of drugs in clinical chemistry: 1 study of twenty drugs on seven different instruments. Clin Biochem 18:345–351
Lifton LJ, Kreiser J (1982) False positive stool occult blood tests caused by iron preparations: a controlled study and a review of the literature. Gastroenterology 83:860–863
Maddocks J, Hann S, Hospkins M, Coles GA (1973) Effect of methyldopa on creatinine estimation. Lancet I:157
Martinek RG (1970) Review of methods for determining uric acid in biologic fluids. J Am Med Technol 32:233–241
May DB, Young LY, Wiser TH (1992) Essential hypertenstion. In: Koda-Kimble MA, Young LY (eds) Applied therapeutics, the clinical use of drugs. Applied Therapeutics, Vancouver, pp 7-1–7-32
Mitchell EK (1984) Flucytosine and false elevation of serum creatinine level. Ann Intern Med 101:278
Morris RG, Lognado PY, Lehmann DR, Fremin DB, Glistak ML, Bunret RB (1987) Spironolactone as a source of interference in commercial digoxin immunoassay. Ther Drug Monit 9:208–211
Morris RG, Frewin DB, Saccoia NC, Goldsworthy WL, Jeffries WS, McPhee AJ (1990) Interference from digoxin-like immunoreactive substance(s) in commercial digoxin kit assay methods. Eur J Clin Pharmacol 39:359–363
Murphy J, Hurt TL, Griswold WR, Peterson BM, Rodarte A, Krous HF, Reznik VM, Mendoza SA (1989) Interference with creatinine concentration measurement by high dose furosemide infusion. Crit Care Med 17:889–890
Musser WA, Ortigoza C (1966) Automated determination of uric acid by the hydroxylamine method. Tech Bull Reg Med Tech 30:21–25
Nanji AA, Whitlow KJ (1984) Spurious increase in serum creatinine associated with intravenous methyldopa therapy. Drug Intell Clin Pharm 18:896–897
Nanji AA, Poon R, Hinherg I (1987) Interference by cephalosporins with creatinine measurement by desk-top analysers. Eur J Clin Pharmacol 33:427–429
Narayanan S, Appleton HD (1980) Creatinine: a review. Clin Chem 26:1119–1126
Natowicz M, Shaw L (1991) Digoxin assay anomalies due to digoxin specific Fab immunotherapy. Drug Intell Clin Pharm 25:739–741
O'Neill MB, Hill JG, Arbus GS (1987) False elevation of the serum creatinine level associated with the use of flucytosine. Can Med Assoc J 137:133–134

Richards RK (1980) Structure-activity relationships in the effects of phenacemide analogs on serum creatinine and anticonvulsant activity. Arch Int Pharmacodyn 244:107–112

Roach NA, Kroll MH, Elin RJ (1985) Interference by sulfonylurea drugs with the Jaffe method for creatinine. Clin Chim Acta 151:301–305

Salway JG (1990) Drug-test interactions handbook. Chapman and Hall Medical, London

Schenck-Gustafsson K, Jogestrand T, Nordlander R, Dahlquist R (1981) Effect of quinidine on digoxin concentration in skeletal muscle and serum in patients with atrial fibrillation. N Engl J Med 305:209–211

Seifert CF, Bradberry JC, Resman-Targoff BH (1992) Clinical laboratory tests and interpretation. In: Herfindal ET, Gourley DR, Hart LL (eds) Clinical pharmacy and therapeutics. Williams & Wilkins, Baltimore, pp 61–81

Sena SF, Syed D, Romeo R, Krzymowski GA, McComb RB (1980) Lidocaine metabolite and creatinine measurements in the Ektachem 700: steps to minimize its impact on patient care. Clin Chem 34:2144–2148

Shih VE, Mandell R, Sheinhait I (1991) General metabolic screening tests. In: Hommes FA (ed) Techniques in diagnostic human biochemical genetics. Willey-Liss, New York, pp 61–81

Shukur LR, Powers JL, Marques RA, Winter ME, Sadee W (1977) Measurement of procainamide and N-acetylprocainamide in serum by high-performance liquid chromatography. Clin Chem 23:636–638

Sigma Chemical Co (1977) The enzymatic-calorimetric determination of uric acid in serum or urine. Technical Bulletin, No. 680, St. Louis

Silber B, Sheiner LB, Powers JL, Winter ME, Sadee W (1979) Spironolactone-associated digoxin assay interference. Clin Chem 25:48–50

Singh HP, Hebert MA, Gault MH (1972) Effect of some drugs on clinical laboratory values as determined by the Technicon SMA 12/60. Clin Chem 18:137–144

Small RE, Fredy HR, Small BJ (1976) Alpha-methyldopa interference with the phosphotungstate uric acid test. Am J Hosp Pharm 33:556-560

Soldin SJ, Papanastasiou-Diamandi A, Heyes J, Lingwood C, Olley P (1984) Are immunoassays for digoxin reliable? Clin Biochem 17:317–320

Souney PF, Mariani G (1985) Effect of various concentrations of flucytosine on the accuracy of serum creatinine determinations. Am J Hosp Pharm 42:621–622

Spinler SA, Anderson BD, Kindwall KE (1989) Lidocaine interference with Ektachem Analyzer determinations of serum creatinine concentration. Clin Pharm 8:659–663

Stearns FM (1981) Determination of procainamide and N-acetylprocainamide by high-performance liquid chromatography. Clin Chem 27:2064–2067

Technicon Instrument Corporation (1976) Technicon SMAC high-speed computer-controlled biochemical analyzer, product labelling. Technicon Instrument Corporation, New York

Toffaletti J, Blosser N, Hall T, Smith S, Tompkins D (1983) An automated drug-slide enzymatic method evaluated for measurement of creatinine in serum. Clin Chem 29:684–687

Troub SL (1992) Evaluating potential drug interferences with laboratory tests. In: Troub SL (ed) Basic skills in interpreting laboratory data. American Society of Hospital Pharmacists, Bethesda, pp 11–21

Wade WE, Waite WW, Cobb HH (1989) Hyperuricemia associated with thiazide diuretic therapy in an ambulatory geriatric hypertensive population. Consult Pharm 4:94–98

Yosselson-Superstine S (1984) Drug interferences with plasma assays in therapeutic drug monitoring. Clin Pharmacokinet 9:67–87

Yosselson-Superstine S (1986) Drugs causing interference with laboratory tests. In: D'Arcy PF, Griffin JP (eds) Iatrogenic diseases, 3rd edn. Oxford University Press, Oxford, pp 954–969

Yosselson-Superstine S (1989) Drug interferences in therapeutic drug monitoring. Giorn It Chim Clin 14:23–34

Yosselson-Superstine S, Sinai Y (1986) Drug interference with urine protein determination. J Clin Chem Clin Biochem 24:103–106
Yosselson-Superstine S, Granit D, Superstine E (1980) Drug interference with the phosphotungstate uric acid test. Am J Hosp Pharm 37:1458–1462
Young DS (1990) Effects of drugs on clinical laboratory tests. AACC, Washington
Yü TF, Gutman AB (1959) Study of the paradoxical effects of salicylate in low, intermediate and high dosage on the renal mechanisms of excretion of urate in man. J Clin Invest 38:1298–1315

CHAPTER 12

Drug Interactions with Herbal and Other Non-orthodox Remedies

P.A.G.M. De Smet and P.F. D'Arcy

A. Introduction

Medicines derived from plants formed the majority of the earlier materia medica because chemically synthesised compounds were then not available. Many of these herbs have stood the test of time and critical clinical assessment and have found their way into the pharmacopoeias of orthodox medicines sometimes as the isolated and chemically standardised active ingredient. Such drugs as cocaine, colchicine, coumarin anticoagulants, digoxin, ephedrine, morphine, quinine and quinidine, reserpine, tubocurarine, sennosides, and the ergot and vinca alkaloids entered orthodox medicinal use by this route.

There are many types of herbal remedies, ranging from self-made teas prepared from self-collected herbs to officially registered drug products which have passed through the same rigid registration procedure as synthetic medicines (see Table 1).

It is rather difficult to classify a particular herb in its most appropriate category. The same botanical product which is a registered medicine in one country may be a dietary supplement or recreational herb in the next, and in one country the same herb may be available as an official medicine, as a health food preparation and as a raw ingredient (De Smet 1993a). Yet it is important to keep these different categories in mind, when the risks of herbal remedies are discussed, because the nature and magnitude of these risks can vary considerably with the specific type of product. For instance, a consumer of a high-quality registered herbal medicine should not have to worry about the correct identity of the ingredients, whereas this should be a primary concern for an individual who goes out into the field to collect his own herbal materials.

The majority of preparations used in herbal or non-orthodox medicines are a mixture of herbal ingredients. Herbs having diverse actions may be incorporated into one concoction. Also herbal products are, in some circumstances, a mixture of herbal remedies with other ingredients of non-herbal origin (e.g., arsenic or lead) or undeclared Western drugs (e.g., prednisolone, non-steroidal anti-inflammatory/antirheumatic agents and paracetamol) as found recently by Karunanithy and Sumita (1991) in traditional Chinese antirheumatic medicines. Readers are also referred to the reviews of D'Arcy (1991, 1993) and to a comprehensive review by De Smet (1992a) on adultera-

Table 1. Different types of herbal remedies

Raw materials for self-preparation
– By self-collection
– Through commercial channels
(Semi)finished non-medicinal products
– Dietary supplements
– Health foods
– Recreational herbs, etc,
Registered medicines
– By special procedure
– By regular procedure

tion and contamination of herbal products with toxic metals and synthetic drug substances and other drugs used in non-orthodox medicine.

It must be appreciated that the quality control exerted over most herbal preparations and other non-orthodox remedies which are not registered as medicines is often poor and more likely to be non-existent and that most preparations are not standardised for potency in biological test systems. As a consequence their potency may vary considerably from sample to sample.

Some of the interactions indicated in this review are based on a firm pharmacological basis but their clinical relevance still needs to be established. Others have been presented in anecdotal reports with lack of data essential to establish a firm cause-effect relationship. In conformity with the objective of this volume, mechanisms of the interaction are given where this is possible; that it is not always possible is acknowledged since often the available data do not permit even an assumption of the cause.

This approach in the context of this chapter is necessary if warnings are to be given about likely interactions since the sparse reports of drug interactions between non-orthodox products and Western medicines neither confirms their safety in use nor indeed suggests that the incidence of such interactions is low. The simple fact is that most interactions of this type will not be recognised as such by the self-medicated patient and will not be reported to an orthodox medical practitioner and therefore will not appear in the medical or pharmaceutical literature.

When such original reports of reactions and interactions, as are available, are carefully analysed, it becomes obvious that many of the cases where herbal products have been associated with actual human poisoning were not in fact caused by the herbs alleged to be in the product, but resulted from substitution or contamination of the declared ingredients, intentionally or by accident, with a more toxic botanical, a poisonous metal, or a potent non-herbal drug substance (De Smet 1992a).

Although herbal medicines are by far the largest component of non-orthodox remedies they do not have exclusive claims; it must be made clear that there are various types of other alternative treatments ranging from preparations of animal origin, minerals, vitamins and amino acids. Many of these are capable of interfering with orthodox medicines. Interactions between minerals and vitamins and orthodox medicines have already been excellently reviewed in standard drug interaction texts (GRIFFIN et al. 1988; STOCKLEY 1991; HANSTEN et al. 1993) and therefore this present chapter will largely confine itself to interactions involving herbals, animal origin preparations and amino acids. It is accepted that firm evidence for many of the interactions is slight but it is thought better at this stage of knowledge to include them and give warnings rather than neglect them to the possible detriment of the medicated patient.

For convenience, and to illustrate the points that various types of alternative treatments can interfere with orthodox drugs, the subject will be discussed under the following headings: animal agents; amino acids; vitamins and minerals; dietary fads; and, in much more detail, herbal medicines (DE SMET 1992b).

B. Animal Agents

There are few parts of animals that have not been used at one time or another for their medicinal properties. For example, crude thyroid hormones may occur as ingredients of slimming preparations and/or adulterants of herbal products (DE SMET 1992a).

I. Fish Oil

Fish oil can upset coagulation control by increasing the bleeding time and reducing platelet aggregation. It is difficult therefore to exclude the possibility of untoward consequences in stabilised anticoagulated patients if this oil is taken concomitantly with anticoagulant medication (DE SMET 1989).

II. Chinese Toad Venom

A traditional component of the Chinese medicine "Kyushin" is the venom of the Chinese toad (*Bufo bufo gargarizans*). The venom contains bufalin and cinobufaginal, which are chemically similar to digoxin. The venom may therefore interfere with digoxin immunoassays; the constituents can react with digoxin antibodies and create the false impression of high plasma digoxin levels (FUSHIMI et al. 1989, 1990; LIN et al. 1989; KWAN et al. 1992; PANESAR 1992).

C. Amino Acids

I. L-Tryptophan

L-Tryptophan (LT) is used in the orthodox treatment of depressed patients not responsive to other therapy and also as a sedative-hypnotic. In non-orthodox treatments, LT is a common health food supplement. It is capable of interacting with other serotonergic drugs, and combinations of LT with monoamine oxidase inhibitor (MAOI) or fluoxetine have repeatedly induced a so-called serotonin syndrome which is characterised by neurobehavioural problems such as myoclonus, shivering, diaphoresis, hyperthermia, hyperreflexia, ocular oscillations, ataxia and/or hypomanic changes (DE SMET 1991).

A marked and rapid deterioration has been observed in parkinsonian patients when 100mg pyridoxine (vitamin B_6)/day was added to a daily regimen of 8–9g LT. This deleterious effect was not produced by pyrodoxine or LT alone (DE SMET 1991). Pyridoxine also reduces the effect of levodopa in the treatment of parkinsonism (DUVOISIN et al. 1969). Pyridoxine is a codecarboxylase and by facilitating the decorboxylation of levodopa it reduces its blood level. Another reported interaction is the significant reduction of blood concentrations of levodopa when LT is given concomitantly (DE SMET 1991).

D. Vitamins

Vitamin and mineral supplements are a common combination and are readily obtainable as over-the-counter medicines; they may be regarded by consumers as food supplements rather than medication and the possibility of their interaction with prescribed medicines may not be appreciated. Many drugs are known to evoke vitamin deficiency states, for example isoniazid, hydralazine and penicillamine are antagonistic to vitamin B_6; anticonvulsants cause folate and vitamin D deficiency and have been implicated in osteomalacia (DENT et al. 1970); and vitamin B_{12} deficiency may occur after prolonged treatment with the antidiabetic metformin (CALLAGHAN et al. 1980; ADAMS et al. 1983).

They can influence the effects of various orthodox drugs and sometimes this may be used intentionally as an advantage (see SOKOL et al. 1991), but more often unwanted effects occur. For example health foods and food supplements containing appreciable quantities of vitamin K can reduce the effect of oral anticoagulants (HEALD and POLLER 1974; STOCKLEY 1991). Cases of impaired anticoagulation have been recorded after concomitant intake of "Gon" (a non-orthodox remedy containing vitamin K_4), enteral nutrition, and excessive amounts of green vegetables (UDALL and KROCK 1968; HEALD and POLLER 1974; HOGAN 1983; KEMPIN 1983; WALKER 1984; WATSON et al. 1984).

Vitamin B_6 can nullify the beneficial effects of levodopa in parkinsonism (COTZIAS 1969; CALNE and SANDLER 1970). A further example is that folic acid may occasionally decrease serum phenytoin to a clinically significant degree (HANSTEN et al. 1993). Vitamins may also influence the activity of other non-orthodox remedies, for example the absorption of selenium from sodium selenite can be drastically reduced by megadoses of vitamin C (ROBINSON et al. 1985), and the use of large doses of vitamin C (up to 5g/day), taken prophylactically against the common cold, may alter urinary pH sufficiently to influence the renal elimination and therefore the clinical response of alkaline drugs (MEDICAL LETTER 1971).

The amount of vitamin consumed is dependent on the specific product and its specific dose recommendations. While many multivitamin preparations provide daily amounts below or just above their Recommended Daily Allowance (RDA) values, there are also products which provide megadoses of one or more vitamins, This latter category entails a risk of toxic effects (EVANS and LACEY 1986; BROWN and GREENWOOD 1987; HELSING 1992) as well as drug interactions. In general, however, with the majority of multivitamin/mineral supplements, it is doubtful whether the amount of vitamin contained in the product plays any great part in interactions with orthodox drugs.

E. Minerals

It is normally the mineral salt component of the combination as well as the mineral content of antacids and some laxatives that is responsible for the majority of interactions. For example, the absorption of tetracyclines is reduced by the formation of insoluble chelates in the gut in the presence of milky and other foods containing bivalent and trivalent metal ions (Ca, Mg, Al, Fe) (KUNIN and FINLAND 1961; BRAYBROOKS et al. 1975; CHIN and LACH 1975), and serum levels of oral tetracyclines may be decreased by more than 50% if they are taken with ferrous sulphate or milk (NEUVONEN 1976).

Iron supplements are among the most frequently prescribed drugs (LA PIANA SIMONSEN 1989) and are taken in over-the-counter remedies by many other people. CAMPBELL and HASINOFF (1991) have listed some 19 orally administered drugs, including antimicrobials, captopril, levodopa, salicylic acid and thyroxine, which have functional groups in a configuration that will bind iron to form an iron-drug complex. The formation of a stable iron-drug complex reduces the extent of drug absorption but does not appear to reduce the rate of drug absorption. Iron can also catalyse oxidation and reduction interactions. Zinc also can produce various interactions with orthodox drugs (BARRY et al. 1991).

Materials with a high calcium content may reduce the absorption of the bisphosphonates. Vitamins with mineral supplements such as iron, calcium supplements, laxatives containing magnesium, or antacids containing calcium or aluminium all dramatically reduce the biological availability of the bisphosphonates (FLEISCH 1987; FELS et al. 1989).

F. Dietary Fads

Dietary fads may lead to an excessive intake of certain nutrients, which in turn could result in interference with orthodox medication. Certain diets, including vegetable-rich weight-reducing diets, are rich in vitamin K and if taken in sufficient amounts will interfere with warfarin treatment (HEALD and POLLER

1974; Hogan 1983; Patriaca et al. 1983; Walker 1984; Watson et al. 1984). For example, a patient maintained on warfarin 8 mg daily required an increase in dosage to 13 mg/day when placed on enteral nutrition rich in vitamin K; the warfarin dosage returned to 8 mg/day when the food supplement was stopped. Problems have also been experienced with anticoagulated patients consuming large amounts of broccoli (Kempin 1983; Stockley 1984).

Beside the risk of pharmacodynamic problems, there is also the possibility that the pharmacokinetic fate of orthodox drugs could be affected by a dietary change. For instance, dietary fibres may affect absorption (see below), whereas certain other foodstuffs may exert an influence on drug biotransformations in humans (Alvares 1984; Anderson 1988). Also certain foods and food supplements can affect the urinary excretion of drugs. A balanced diet with adequate protein provides an acid urine, whereas a low-protein diet or a strict vegetarian diet may give an alkaline urine. This opens up the possibility that diet-induced changes in urinary pH affect the rates of excretion of weakly acidic or weakly basic drugs (Wesley-Hadzija 1971; Pierpaoli 1972). Acid urine favours the ionisation of alkaline drugs (and vice versa), so reabsorption is reduced and renal excretion is increased. Acid fruit juices and squashes may therefore reduce the efficacy of the antimalarials quinine and chloroquine (P.F. D'Arcy, personal observation).

Grapefruit juice, but not orange juice, greatly augments the bioavailability of the antihypertensive, calcium-antagonists felodipine, nifedipine and nitrendipine (Baily et al. 1989, 1991; Soons et al. 1991; Edgar et al. 1992); similar results are apparent on the clearance of caffeine (Fuhr et al. 1993). These affects are due to the inhibition of cytochrome P450 (isoforms CYP1A2 and CYP3A4) enzymes by several flavonoid glycosides present in grapefruit juice, of which the bitter principle naringin (4′,5,7-trihydroxyflavanone 7-rhamno-glucoside) is the most abundant (Fuhr et al. 1993).

Interestingly, grapefruit juice has also recently been found to inhibit the 7-hydroxylation of coumarin, as measured by its effect on urinary excretion of 7-hydroxycoumarin in healthy volunteers (Merkel et al. 1994). This implies that grapefruit juice may interfere with the pharmacokinetics of coumarin-yielding medicinal herbs, such as *Melilotus officinalis* (sweet clover), *Asperula odorata* (sweet woodruff), *Dipteryx odorata* (tonka bean) and *Anthoxanthum odoratum* (sweet vernal grass) (Hoppe 1975; Lewis and Elvin-Lewis 1977; Teuscher and Lindequist 1980). A comprehensive review on the significance of interactions between grapefruit juice and drugs was recently presented by Bailey et al. (1994).

G. Herbal Drugs

This category includes crude herbal remedies as well as plant-derived drugs; discussion will be divided into the following categories:

1. Effects of herbal drugs on orthodox drug pharmacokinetics: effects on absorption
2. Effects of herbal drugs on orthodox drug pharmacokinetics: effects on elimination
3. Effects of orthodox drugs on herbal drug pharmacokinetics
4. Pharmacodynamic interactions between herbal drugs and orthodox drugs
5. Multiple or unclarified interactions between herbal drugs and orthodox drugs
6. Interactions between different herbal drugs

I. Effects of Herbal Drugs on Orthodox Drug Pharmacokinetics: Effects on Absorption

1. Dietary Fibres

BROWN et al. (1977) showed that digoxin bioavailability is decreased by almost 20% when given with a high-fibre meal. In vitro studies by FLOYD et al. (1977) supported this and indicated that up to 45% of the digoxin may be sequestered in or bound to the bran. Later work by REISSELL and MANNINEN (1982) showed that the effect of fibre was small but that an interactive effect on absorption of digoxin might occur if fibre and digoxin were ingested simultaneously.

ROSSANDER (1987) suggested, on the basis of in vitro studies, that dietary fibres might interfere with the bioavailability of iron in the diet. Confirmatory work in healthy volunteers showed that bran had a marked inhibitory effect on the absorption of dietary iron but not cellulose or pectins. ROE et al. (1988) showed that dietary fibre in combined wheat bran and psyllium biscuits or biscuits containing psyllium alone reduced the absorption of riboflavin by 6.4% and 5.7%, respectively. No effect of the wheat bran supplement alone was detected.

The possibility that drug interactions with dietary fibre or other bulk-forming herbal drugs may interfere with the gastrointestinal absorption of orthodox drugs has become of more concern and there have been a number of pivotal publications. For example, PERLMAN (1990) has reported a case which raises the possibility of an interaction between lithium salts and ispaghula husk (psyllium hydrophilic mucilloid), a bulk-forming laxative.

The patient had a long history of psychiatric problems and was treated with lithium carbonate and then lithium citrate. She also received ispaghula husk (one teaspoonful in water twice daily). Despite increasing her lithium dosage, her blood lithium concentrations fell to below acceptable levels. Ispaghula husk was discontinued and her blood lithium concentrations increased.

RICHTER et al. (1991) have reported that concomitant fibre intake (pectin 15 mg/day or oat bran 50–100 g/day) may decrease the absorption of the lipid-lowering agent lovastatin. Their patients showed greatly increased low-density lipoprotein (LDL) cholesterol levels when these agents were introduced into a lipid-lowering diet plus lovastatin (80 mg/day) regimen. LDL cholesterol levels normalised when the pectin or oat bran was stopped.

Stewart (1992) has reported that excessive dietary fibre may reduce the efficacy of tricyclic antidepressants. Three of her patients with recurrent major depression, who had been successfully treated with intermittent tricyclics in the past, became refractory to therapy after beginning to ingest a high-fibre diet. Serum antidepressant levels were lower than those previously achieved.

2. Guar Gum

Todd et al. (1990) have reviewed the pharmacological properties of guar gum. Administration of up to 30 g/day guar gum to diabetic patients does not seem to alter mineral or electrolyte balance during long-term therapy (Behall et al. 1989; McIvor et al. 1985; Uusitupa et al. 1989), although plasma levels of fat-soluble vitamins (A and E) have tended to decrease during extended treatment, but not to a clinically important extent (Uusitupa et al. 1989).

Because of its effect on prolonging gastric retention and because drugs will diffuse more slowly out of viscous matrices than from solutions, guar gum may affect the absorption of certain concomitantly administered drugs. Often only the rate of absorption is affected [e.g., bumetanide, digoxin, paracetamol (acetaminophen)] and such interactions are of minor clinical significance. However, the absorption of certain other drugs (e.g., metformin (Gin et al. 1989), phenoxymethylpenicillin (Huupponen et al. 1985) and some formulations of glibenclamide (Neugebauer et al. 1983) may be reduced by a clinically significant degree.

Guar gum is a component of many slimming preparations and it might be advisable that women taking oral contraceptives take additional contraceptive precautions whilst taking guar gum (Anonymous 1987). The same warning could be extended to women taking low-dose oestrogen, combined oral contraceptives which are on on any slimming diet that contains bulk-forming agents. Theoretically, they could be at risk of contraceptive failure if the absorption of the contraceptive is compromised. However, we know of no clinical study that has assessed the clinical importance of this theoretical notion, nor any report that has linked slimming diets with contraceptive failure.

3. Tannins

From the 1960s to the present time, it has been variously suggested and refuted that among the botanical constituents that might also interfere with the absorption of some orthodox drugs are tannins in tea or coffee (Carrera et al. 1973) and phytates (Davies 1982). For example, Carrera et al. (1973) showed in rat studies that the absorption of vitamin B_{12} was reduced proportionately to the oral dose of tannic acid administered. This interaction, it was claimed, was due to the formation of a non-absorbable complex between tannic acid, the vitamin and glycoproteins in the gut.

There has also been much discussion about the possibility of tea and coffee comsumption weakening the effect of antipsychotic medication. These discussions had their foundation in a report by Lever and Hague in 1964 that phenothiazines can be precipitated in vitro by many fruit juices, milk, tea and coffee. The mixtures so produced are rendered unpalatable and, since some of the precipitates do not redissolve in hydrochloric acid, it was suggested that they might be poorly absorbed in vivo. On this basis it has been variously suggested that tea and coffee might antagonise the efficacy of antipsychotic drugs, especially neuroleptic drugs (Hirsch 1979; Kulhanek et al. 1979).

Support for these views was given by the results of experiments in rats which suggested that tea and coffee alter the pharmacokinetics of neuroleptic drugs (Kulhaneck and Linde 1981). Additional support was given by Lasswell et al. (1984), who investigated the in vitro interaction of neuroleptics and tricyclic antidepressants with tea, coffee and gallotannic acid and they showed that there was significant precipitation with a variety of agents including several phenothiazines, amitriptyline, haloperidol, imipramine and loxapine. The strong complex formed between these drugs and tannins was thought to be the basis of the interaction of these drugs with tea and coffee. They did not, however, produce any evidence of their own to suggest that such interactions were of any clinical significance. However, in a controlled study in 16 female patients in a psychiatric hospital, the effect of tea and coffee drinking was investigated on steady-state blood levels and clinical efficacy of chlorpromazine, haloperidol, fluphenazine and trifluoperazine (Bowen et al. 1981). Withdrawal of these beverages did not increase the bioavailability of the drugs studied, nor did they affect the individual variation in the plasma levels of these drugs. The conclusion was that limitation of tea and coffee intake in medicated psychiatric patients could not be justified (Bowen et al. 1981).

It must be remembered, however, that apart from a tannin-based interaction of caffeinated beverages there is the pharmacodynamic possibility that there are psychotropic effects of caffeine in psychiatric patients drinking up to 10 cups of coffee per day; these effects of caffeine cannot be neglected (De Freitas and Schwartz 1979).

Ordinary tea is sometimes reputed to affect oral iron absorption and some researchers have assessed the effects of other herbal teas. Hesseling et al. (1979) found that freshly prepared rooibos tea (*Aspalathus linearis*) did not affect iron absorption in contrast to ordinary tea. More recently, studies by El-Shobaki et al. (1990) examined the influence of various herbal teas on iron absorption; they found that ordinary tea inhibited iron absorption, whereas anise, mint, caraway, cumin, tilia and liquorice promoted the absorption of iron. It should be noted, however, that other publications dispute that ordinary tea affects the absorption of pharmacologic doses of oral iron preparations (Koren et al. 1982).

II. Effects of Herbal Drugs on Orthodox Drug Pharmacokinetics: Effects on Elimination

1. Cola Nut

An extensive study by Fraser et al. (1976) of factors affecting antipyrine metabolism in West African villagers showed that cola nut consumption inhibited antipyrine metabolism and prolonged the antipyrine half-life by 3.5 h. It was suggested that unidentified constituents of cola nuts competed with antipyrine for oxidation by the microsomal enzyme system. There were three other predictors of oxidative capacity: sex, haemoglobin in women and height in men.

Vesell et al. (1979) failed to show any effect of cola nuts, chewed for either 14 or 28 consecutive days, on antipyrine disposition in Caucasian males. Genetic factors may therefore be of importance and although many questions remain unanswered, the results of the antipyrine studies are interesting enough to suggest that this type of work be reopened since further research in this direction and particularly with other herbal remedies would seem to be warranted.

2. Eucalyptus Species

A number of studies have demonstrated that eucalyptus leaves, the oil and the major active principle, eucalyptol, can all induce microsomal enzyme activity in both in vitro and in vivo tests (Jori et al. 1969; Seawright et al. 1972). However, there have not been any recorded interactions between eucalyptus and orthodox drugs at the clinical level.

3. Grapefruit Juice

There is evidence that grapefruit juice augments the bioavailability of the antihypertensives felodipine, nifedipine and nitrendipine, and the clearance of caffeine. These interactions have already been mentioned previously under "Dietary Fads".

4. Herbal Smoking Preparations

Herbal smoking preparations are sufficiently common in the alternative drug market (Siegel 1976) to suggest the possibility that they, like tobacco smoking, may contain polycyclic hydrocarbons in their smoke and that they might also affect the metabolism of orthodox drugs. However, we know of no work done in this field.

5. Kampo Medicines

Oriental Kampo medicines often contain glycyrrhizin and these have been reported to influence prednisolone pharmacokinetics. Homma et al. (1992a) showed that three Kampo remedies, which also contained Saiko (*Bupleuri radix*), and were commonly co-administered with prednisolone in the treatment of asthma, nephrotic syndrome and collagen diseases, had a variable and different effect on prednisolone pharmacokinetics in healthy subjects.

Only one of the three remedies had a steroid-sparing effect due to decreased 11β-hydroxysteroid dehydrogenase (11-beta-HSD) activity. The other two either increased or did not change the activity of this enzyme. Interestingly, the authors attributed the enzyme inhibitory effect not to glycyrrhizin, which was present in the three Kampos, but to magnolol, which was only contained in the Kampo with a positive effect on prednisolone metabolism.

Follow-up work by HOMMA et al. (1994) was aimed at confirming the inhibitor of prednisolone metabolism contained in Saiboku-To. In in vitro experiments they studied the effects of 10 herbal constituents on 11-beta-HSD and showed that five herbal extracts had inhibitory activity: *Glycyrrhiza glabra* > *Perillae frutescens* > *Zizyphus vulgaris* > *Magnolia officinalis* > *Scuttellaria baicalensis*. Seven chemical constituents which were identified as the major urinary products of Saiboku-To in humans were studied; magnolol derived from *M. officinalis* showed the most potent inhibition of the enzyme and although this was less than that of glycyrrhizin, the non-competitive inhibition mechanism was different from a known competitive mechanism. These results suggested that magnolol might contribute to the inhibitory effects of Saiboku-To on prednisolone metabolism through inhibition of 11-beta-HSD.

Studies on ofloxacin in healthy volunteers by HASEGAWA et al. (1994) found no significant effects of the Kampo medicines Sho-saiko-To (TJ-9), Rikkunshi-To (TJ-43) and Sairei-To (TJ-114) on any estimated bioavailability parameter.

6. Liquorice

Liquorice, the dried rhizome and roots of *Glycyrrhiza glabra*, has long been a popular and traditional ingredient of both herbal and orthodox medicines. It is also found in confectionary, soft drinks, chewing tobacco and chewing gum. There is evidence to suggest that liquorice may interfere with the pharmacokinetics of orthodox medicines. For example, CHEN et al. (1990) showed that an intravenous infusion of the active principle, glycyrrhizin (GL), increased the plasma concentrations of prednisolone and influenced the pharmacokinetics of prednisolone in man. Interestingly, the main component in plasma is different when GL is administered by different routes. After intravenous administration the main component in plasma is glycyrrhizin (KATO et al. 1984; CHEN et al. 1990), while after oral dosage the component is glycyrrhetinic acid (GA), a metabolic derivative of GL (NAKADA et al. 1986; SHIMADA et al. 1989).

In vitro, the inhibiting effect of GA on the metabolism of corticosteroids by 5α-,5β-reductase and 11β-dehydrogenase is stronger than that of GL. It is not surprising therefore that in later studies CHEN et al. (1991) showed that oral administration of GL also modified the pharmacokinetics of both total and free prednisolone. The area under the curve (AUC) of prednisolone was significantly increased, the total plasma clearance was significantly reduced, and the mean residence time was significantly prolonged. However, the volume of distribution showed no evident change.

These results suggested that the oral administration of GL increases the plasma prednisolone concentrations and influences its pharmacokinetics by inhibiting its metabolism, but not affecting its distribution. It has been suggested that this combination would be advantageous in the treatment of rheumatoid conditions. The basis of the interaction has been suggested as inhibition of the metabolism of prednisolone by microsomal enzymes; inhibition of urine clearance and interference with plasma protein binding were both discounted. There is direct evidence that GL and GA can inhibit corticosteroid 5α- and 5β-reductase and 11β-dehydrogenase activities in rat liver and kidney in vitro and in vivo (Tamura et al. 1979; Monder et al. 1989).

III. Effects of Orthodox Drugs on Herbal Drug Pharmacokinetics

Not only may herbal drugs affect the pharmacokinetics of orthodox drugs, but in reverse orthodox drugs may influence the pharmacokinetics of herbal drugs. Some illustrative examples are as follows:

1. Caffeine-Containing Herbs

Certain antibacterial 4-quinolones and fluoroquinolones (ciprofloxacin, enoxacin, pipemidic acid and temafloxacin) inhibit the hepatic metabolism of caffeine, increase its elimination half-life and decrease its clearance (Carbo et al. 1989; Harder et al. 1989; Healy et al. 1989; Mahr et al. 1992). Users of caffeine-containing beverages and herbals should therefore be advised that they have an increased risk of adverse effects (e.g., tremor, tachycardia, insomnia, CNS excitation) when they are taking such quinolones. The most important herbal remedies which contain substantial amounts of caffeine are derived from *Cola*, *Ilex* and *Paullinia* species (De Smet 1989).

2. Sparteine-Containing Herb

Sparteine is a quinolizidine alkaloid from *Cytisus scoparius* which was recently found in a herbal slimming remedy on the United Kingdom market. Substantial doses of this preparation in slow metabolisers could be expected to be associated with many adverse reactions including circulatory collapse (Galloway et al. 1992). The antiarrhythmic agent quinidine is a potent inhibitor of the oxidative metabolism of sparteine (Schellens et al. 1991) and a similar effect has been observed with haloperidol (Gram et al. 1989) and with moclobemide (Gram et al. 1993).

In addition, Crewe et al. (1992) conducted in vitro experiments in which several antidepressants (tricyclic or selective serotonin reuptake inhibitors) inhibited human liver microsomal P4502D6 (CYP2D6) activity, which resulted in a reduced oxidative conversion of sparteine to dehydrosparteine.

3. *Teucrium chamaedrys*

LOEPER et al. (1994) evaluated the molecular mechanism of the hepatotoxicity which had been observed in numerous users of a French preparation containing *Teucrium chamaedrys*. A dose-dependent increase in serum alanine aminotransferase (ALT) activity in mice was observed after the intragastric administration of the lyophilisate of a tea prepared from blooming aerial parts. Hepatotoxicity could also be produced by intragastric administration of 0.125 g/kg of an enriched fraction, which contained the same level of furano neo-clerodane diterpenoids as 1.25 g/kg of the lyophilisate. The increase in serum ALT activity could be enhanced by inducers of the 3A family (troleandomycin). Toxicity was also increased by pretreatment with phorone (a depletor of hepatic glutathione), whereas it could be attenuated by inducers of microsomal epoxide hydrolase (such as clofibrate). These findings suggest that the hepatotoxicity of *T. chamaedrys* resides in one or more reactive metabolites of its furanoditerpenoids and that orthodox drugs may influence their formation.

IV. Pharmacodynamic Interactions Between Herbal Drugs and Orthodox Drugs

It is not possible in a chapter of this size to be comprehensive in relating all the pharmacodynamic interactions involving herbal remedies, it is only possible to be illustrative. For convenience therefore this present section has been divided into two categories:

Herbal drugs with well-known constituents, which are also used in orthodox medicine. Many of these are already well known and appreciated and therefore will be presented in brief detail in tabular form (Table 2). It should be noted that familiar botanical constituents with potent pharmacological effects may sometimes occur in very unfamiliar plants. For example, the traditional Chinese medicine Zangqie, which is derived from *Anisodus tanguticus*, yields the same toxic tropane alkaloids as *Atropa belladonna* and *Datura* species (CHANG and BUT 1987).

Herbal drugs with less well-known constituents, which will be discussed in some more detail here.

Herbal drugs, which do not have generally known potent constituents but may nevertheless be capable of pharmacodynamic interactions with orthodox drugs, include the following:

1. Betel Nut (*Areca catechu*)

Two chronic schizophrenic patients who were maintained on depot neuroleptics developed serious extrapyramidal symptoms after a period of

Table 2. Herbal drugs with well-known constituents and their interactions. (De Smet 1992b; D'Arcy 1993)

Herbal drug	Interactions
Adonis vernalis *Convallaria majalis* *Digitalis* species *Nerium oleander* *Strophanthus* species *Urginea maritima* *Xysmalobium undulatum*	Due to the presence of cardioactive glycosides digitalis-like effect and potentiation of toxicity are possible. Although *Crataegus* extracts do not contain digitalis-like glycosides, potentiation of digitalis activity has been reported to occur in guinea pigs. The evidential force of this study has been challenged.
Atropa belladonna *Datura stramonium* *Hyoscyamus niger* *Mandragora officinarum* *Scopalia carniolica*	All these plants contain anticholinergic, tropane alkaloids such as hyoscyamine and/or scopolamine, which can potentiate the effects of synthetic drugs with similar pharmacological activity (e.g., antidepressants, antihistaminics, antispasmodics).
Ephedra species	The tertiary alkaloid ephedrine could interfere with conventional antihypertensive therapy.
Papaver somniferum	The opium alkaloids of the oriental poppy will interact and potentiate the effects of analgesics and other CNS depressants.
Pilocarpus pennatifolius	The alkaloid pilocarpine has potent cholinergic effects which may antagonise orthodox treatment with anti-asthma and anticholinergic agents.
Rauwolfia serpentina	The reserpine-rescinnamide alkaloids of this root will interact and potentiate the effects of antihypertensives, psychotropics and CNS depressant drugs.

heavy betel nut chewing. The mechanism for this effect was suggested as antagonism of the anticholinergic agent procyclidine, by the active alkaloid ingredient of the betel, arecoline (Deahl 1989).

Taylor et al. (1992) have included betel nut chewing among the factors that dispose to asthma severity and unsatisfactory control by orthodox medicines in Asians residing in the United Kingdom. It was suggested that arecoline, or another alkaloid in betel nut, for example, guvacoline, may have a cholinergic bronchoconstrictor effect.

2. Garlic (*Allium sativum*)

Since garlic can reduce human platelet aggregation in vitro (Bordia et al. 1977), it seems difficult to exclude the possibility of untoward circumstances in patients taking oral anticoagulants. Sunter (1991) reported seeing two cases of increased international normalised ratios in patients previously stabilised on warfarin. The altered anticoagulation picture was attributed to the ingestion of garlic products, since there had been no other changes to medication or habits in either case. One patient had started taking garlic pearls, the other garlic tablets and in both cases clotting times were roughly doubled. However, an earlier review of the literature (ROCKY MOUNTAIN DRUG CONSULTATION CENTER 1986) failed to disclose any detailed and well-documented reports of an interaction between warfarin and garlic and no firm data on prolongation of prothrombin times by garlic have been published. Nonetheless, SERLIN and BRECKENRIDGE (1983) have warned that drugs, such as the agents in garlic, which cause alteration of platelet function, may potentiate warfarin's action even though the prothrombin time remains unchanged. The present situation about a possible interaction is unclear and this possibility should be explored in more detail.

3. Karela (*Momordica charantia*)

Karela is an oriental folk remedy. It is well established that oral preparations of the karela fruit have hypoglycaemic activity in a majority of non-insulin-dependent diabetic patients (LEATHERDALE et al. 1981) and that interference with conventional treatment by diet and chlorpropamide has been observed (ASLAM and STOCKLEY 1979). It has also been reported that a subcutaneously injected principle obtained from the fruit may have a hypoglycaemic effect in insulin-dependent diabetics (KHANNA et al. 1981).

4. Liquorice (*Glycyrrhiza glabra*)

Apart from its own intrinsic properties of flavouring and sweetening, liquorice has mild anti-inflammatory and mineralocorticoid properties and may cause sodium and water retention and hypokalaemia. Control of hypertension may be difficult if patients taking orthodox antihypertensive drugs also have a continued high intake of liquorice-containing products (GRIFFIN 1979). Liquorice may also have adverse effects on diabetic patients maintained on insulin; glycyrrhizin was associated, not surprisingly because of its mineralocorticoid-like effects, with hypokalaemia, and sodium retention (FUJIWARA et al. 1983).

5. *Picrorhiza kurroa*

According to BEDI et al. (1989), the rhizomes of this plant species, which are used in Ayurvedic medicine, might potentiate the photochemotherapeutic effects of methoxsalen in human patients with vitiligo.

V. Multiple or Unclarified Interactions Between Herbal Drugs and Orthodox Drugs

Medically interesting examples of multiple or unclarified interactions between herbal drugs and orthodox medications include:

1. Anthranoid Laxatives

An early German report suggests that quinidine serum levels might be reduced by the concurrent use of herbal anthranoid laxative preparations, such as Liquedepur of the Natterman company (Guckenbiehl et al. 1976). No further information is available.

2. Berberine

Berberine is an alkaloid derived from the roots and bark of the plant *Berberis aristata* (barberry bush); extracts of this plant have been used in antidiarrhoeal medication in Ayurvedic medicine in India and in the traditional medicine of China for the past 3000 years. A Burmese study on the clinical effects of berberine in acute watery diarrhoea, namely the reputed antisecretory and vibriostatic effect, showed that berberine alone did not benefit the duration of diarrhoea, frequency of stools and fluid requirements for rehydration, nor did it produce a noticeable antisecretory effect. Clinically, patients with cholera given tetracycline plus berberine were more ill, suffered longer from diarrhoea and required larger volumes of intravenous fluid than did those given tetracycline alone (Khin-Maung-U et al. 1985).

3. Ginseng (*Pamax ginseng*)

The concurrent use of ginseng and the MAOI phenelzine has been associated with adverse effects in two patients (Shader and Greenblatt 1985; Jones and Runikis 1987). However, commercial ginseng preparations are not always derived from *Pamax ginseng* and it is therefore difficult to incriminate this official source plant in the interaction.

4. Piperine

There are several studies to show that piperine, a major alkaloid of *Piper longum* and *P. nigrum*, both of which occur in Ayurvedic formulations, can enhance the bioavailability of orthodox drugs such as phenytoin, propranolol, rifampicin, sulphadiazine, tetracycline and theophylline. Among the suggested mechanisms are promotion of gastrointestinal absorption, inhibition of drug metabolism and a combination of these two (Atal et al. 1981, 1985; Bano et al. 1987, 1991; Bhat and Chandrasekhara 1987; Johri and Zutshi 1992).

5. "Shankhapusphi"

This is an Ayurvedic non-alcoholic syrup which is prepared from six herbs: *Centella asiatica*, *Convolvulus pluricaulis*, *Nardostachys jatamansi*, *Nepeta elliptica*, *Nepeta hindostana* and *Onosma bracteatum*. Dandekar et al. (1992) have reported two epileptic patients taking phenytoin who experienced an unexpected loss of seizure control and a reduction in plasma phenytoin levels when they took this herbal preparation. A follow-up study in rats to investigate this possible interaction showed that multidose co-administration (but not single dose) reduced not only the antiepileptic activity of phenytoin but also lowered plasma phenytoin levels. Shankhapusphi itself showed significant antiepileptic activity (electroshock seizure prevention) when compared with placebo.

6. Yohimbine

Yohimbine is a toxic alkaloid from *Pausinystalia yohimbe*. Preparations providing pharmacologically relevant doses of yohimbine can occasionally be encountered on the Western health food market, even though this is a toxic alkaloid which is not sufficiently safe to be freely available for uncontrolled use (De Smet and Smeets 1994). When given to healthy volunteers, 15–20 mg p.o. yohimbine is usually needed to increase blood pressure and to induce anxiety, but in patients on tricyclic antidepressants hypertension may already occur at 4 mg t.i.d. (Lacomblez et al. 1989). The toxicity of yohimbine can also be enhanced by other drugs, such as chlorpromazine, while it is attenuated by amobarbital or reserpine (Ingram 1962).

Remarkably, yohimbine may be protrayed in off-label advertising of yohimbe health food preparations as a peripheral vasodilator, which can potentiate other blood pressure-lowering agents. In reality, however, the α-adrenoreceptor antagonist properties of the alkaloid will reverse the effects of clonidine and similar antihypertensives (De Smet and Smeets 1994).

VI. Interactions Between Different Herbal Drugs

Herbal remedies are often mixtures, which raises the possibility of interactions between different herbal ingredients. Thus constituents of different botanical sources may form complexes with each other. An example of this is the alkaloid berberine, which combines with the *Glycyrrhiza* compound glycyrrhizin to form a complex with modified biopharmaceutical properties. The alleged basis of the complex is ionic interaction between glucuronic acid carboxyls and the quaternary centre of berberine (Noguchi 1978).

A further example of complex formation in Chinese herbal medicine is that the *Scutellaria* constituent, baicalin, a flavonoid glucuronide, is also complexed with berberine. Glycyrrhetinic acid is also complexed with the flavonoid glycoside rutin (Nolan and Brain 1985). However, there is no good evidence to suggest that such complexation compromises the clinical performance of these compounds (Homma et al. 1992a,b, 1993).

A good South American example of pharmacokinetic interactions between different herbal ingredients is the native custom to prepare a hallucinogenic drink called "ayahuasca" from *Banisteriopsis* vines and *Pyschotria* leaves. The former yield beta-carboline alkaloids, but these alkaloids are only hallucinogenic in high doses and it is doubtful that "ayahuasca" contains sufficient concentrations of these alkaloids to be hallucinogenic. However, the *Banisteriopsis* alkaloids are potent reversible MAOIs and selectively inhibit MAO-A. The addition of *Pyschotria* leaves provides the alkaloid dimethyltryptamine, which is hallucinogenic in low parenteral doses. The oral dimethyltryptamine would be inactivated by MAO-A but it is thought that the presence of the *Banisteriopsis* beta-carbolines in the drink prevents this untimely degradation (De Smet 1985; Ott 1994).

One of the many Oriental remedies with multiple herbal ingredients is "Sairei-to", which is actually a combination of two other multiple preparations, namely "Sho-saiko-to" and "Gorei-san".

"Sho-saiko-to" consists of seven herbs (*Bupleuri radix*, *Pinelliae tuber*, *Scutellariae radix*, *Ziziphy fructus*, *Ginseng radix*, *Glycyrrhizae radix* and *Zingiberis rhizoma*). "Gorei-san" contains five other herbs (*Alismatis rhizoma*, *Atractylodis lancea rhizoma*, *Polyporus*, *Hoelen* and *Cinnamomi cortex*) (Kimura et al. 1990; Tsumura 1991).

Although the pharmacological activity of "Sairei-to" seems to reside primarily in its "Sho-saiko-to" component, there is some evidence to suggest that "Gorei-san" is not pharmacologically inert. Studies in ethanol-treated mice suggest that "Gorei-san" may have a protective effect against certain electrolyte deficiencies and may be capable of adjusting fluid and electrolyte metabolism (Yoneyama et al. 1990). In addition, the deleterious influence of prednisolone on collagen synthesis in the mouse skin in vivo can be prevented by "Sairei-to" but not by "Sho-saiko-to" (Hanawa et al. 1987). "Sairei-to" shows renoprotective activity in rats with aminonucleoside nephrosis (Joarder et al. 1991; Ito 1993) and gentamicin-induced nephrotoxicity (Ohno et al. 1993).

H. Comment

The use of natural (herbal) and other non-orthodox medicines is a persistent aspect of present-day health care and Europeans alone are thought to spend the equivalent of US $500–600 million/year on natural remedies and food supplements. Many consumers believe that naturalness is a guarantee of harmlessness and have no qualms, when necessary, in taking their own prescribed conventional medicine as well. In the United Kingdom and Netherlands and elsewhere in Europe many immigrant races have their own traditional medicine practices, which they frequently combine with orthodox medical care. Generally too little is known about the consequences of such combinations, although the clinical reports of interactions that infrequently appear in the medical and pharmaceutical press suggest that many more interactions may be occurring that are not realised as such and are not reported in the literature.

It should be realised that the data presented in this review did not always come from studies intended and designed to evaluate adverse interactions between orthodox drugs and herbal drugs. Some data were merely obtained as a "spin-off" from pharmacological studies, in which the most useful probe happened to be a herbal constituent (sparteine, coumarin) or which were undertaken to evaluate the active compound in recreational beverages (caffeine).

It is an uncertain state and one that should be actively researched, but it is complicated at present by lack of knowledge about what many of the herbal medicines contain. It must be remembered that some herbal drugs may be contaminated or adulterated with undeclared pesticides, toxic metals, botanicals, animal substances and/or orthodox drugs which could lead to additional and unexpected adverse drug reactions and interactions (GOLDMAN and MYERSON 1991; JOSEPH et al. 1991; DE SMET 1992a; CAPOBIANCO et al. 1993). Illustrative is the recent outbreak of food poisoning associated with cucumber in Dublin (STINSON et al. 1993); this was subsequently traced to be due to the inappropriate use of "Aldicarb", an anticholinesterase pesticide, by one cucumber producer.

It must be further recalled that quality control of herbals is sometimes non-existent as indeed may be the quality of the product. However, it is impossible and unjustified to put all herbal remedies into the same box! There are reliable preparations on the market in terms of safety and quality and perhaps the best approach to improve the current uncertain situation is to promote the good ones and contrast them with the bad ones (see Table 1).

Aspects of quality have been highlighted in the "Note for Guidance on the Quality of Herbal Remedies" by the influential Committee for Proprietary Medicinal Products of the European Community, which has emphasised the need to control the purity of herbal remedies (EUROPEAN COMMUNITY 1989). They must be tested for microbiological quality and for residues of pesticides and fumigation agents, radioactivity, toxic metals and likely contaminants and adulterants. Their quality control must be improved and their content of orthodox synthetic drugs must be established. As to how this is to be achieved is a matter for national drug regulatory authorities since it is our understanding that the CPMP does not strive for a central role in the regulation of herbal drugs. Yet a general tightening up of standards is urgently needed if the alternatively medicated public is to be protected from medication hazard.

Our personal view is that the manufacture and supply of herbal medicines should be licensed and that the "license escape" category of food supplements can no longer be justified. If a non-orthodox remedy is recognised and used as a medicine by the public, and if a dose is specified, then that product should be regarded as a medicine and not a food supplement, and thus is must conform with licensing requirements. In contrast to new synthetic medicines (which should always comply with the regular requirements with respect to efficacy, safety and purity, we are prepared to accept a separate category for herbal drugs, for which the requirement of conventional proof of "efficacy" could be waived. This approach has the disadvantage that it introduces double standards. However, the present situation is that society wants to use herbal drugs anyway, and there are many herbal drugs being used that are not controlled at all. Therefore we prefer to focus on safety and purity (two requirements which should not be waived lightly) and to ensure thereby the quality and relative safety of the available products. This approach can only work, however, if it is supplemented by active herbal pharmacovigilance (De Smet 1993b), which should look not only for unknown adverse reactions but also for new adverse drug reactions (De Smet 1995).

References

Adams JF, Clarke JS, Ireland JT, Kesson CM, Watson WS (1983) Malabsorption of vitamin B_{12} and intrinsic factor secretion during biguanide therapy. Diabetologia 24:16–18

Alvares AP (1984) Environmental influences on drug biotransformations in humans. World Rev Nutr Diet 43:45–59

Anderson KE (1988) Influences of diet and nutrition on clinical pharmacokinetics. Clin Pharmacokinet 14:325–346

Anonymous (1987) Guar gum of help to diabetics. Drug Ther Bull 25:65–66

Aslam M, Stockley IH (1979) Interactions between curry ingredient (Karela) and drug (chlorpropamide). Lancet I:607

Atal CK, Zutshi U, Rao PG (1981) Scientific evidence on the role of Ayurvedic herbals on bioavailability of drugs. J Ethnopharmacol 4:229–232

Atal CK, Dubey RK, Singh J (1985) Biochemical basis of enhanced drug bioavailability by piperine: evidence that piperine is a potent inhibitor of drug metabolism. J Pharmacol Exp Ther 232:258–262

Bailey DG, Spence JD, Edgar B, Bayliff CD, Arnold JMO (1989) Ethanol enhances the hemodynamic effects of felodipine. Clin Invest Med 12:357–362

Bailey DG, Spence JD, Munoz C, Arnold JMO (1991) Interaction of citrus juices with felodipine and nifedipine. Lancet 337:268–269

Bailey DG, Arnold JM, Spence JD (1994) Grapefruit juice and drugs. How significant is the interaction? Clin Pharmacokinet 26:91–98

Bano G, Raina RK, Sharma DB (1986) Pharmacokinetics of carbamazepine in protein energy malnutrition. Pharmacology 32:232–236

Bano G, Amla V, Raina RK, Zutshi U, Chopra CL (1987) The effect of piperine on pharmacokinetics of phenytoin in healthy volunteers. Planta Med 53:568–569

Bano G, Raina RK, Zutshi U, Bdi KL, Johri RK, Sharnma SC (1991) Effect of piperine on bioavailability and pharmacokinetics of propranolol and theophylline in healthy volunteers. Eur J Clin Pharmacol 41:615–617

Barry MG, Macmathuna P, Younger K, Keeling PWN, Feely J (1991) Effect of zinc supplementation on oxidative drug metabolism in patients with hepatic cirrhosis. Br J Clin Pharmacol 31:488–491

Bedi KL, Zutshi U, Chopra CL, Amia V (1989) Picrorhiza kurroa, an Ayurvedic herb, may potentiate photochemothereapy in vitiligo. J Ethnopharmacol 27:347–352

Behall KM, Scholfield DJ, McIvor ME, Van Duyn M, Leo TA et al. (1989) Effects of guar gum on mineral balances in NIDDM adults. Diabetes Care 12:357–364

Bhat BG, Chandrasekhara N (1987) Interaction of piperine with rat liver microsomes. Toxicology 44:91–98

Bordia AK, Joshi HK, Sanadhya YK, Bhu N (1977) Effect of essential oil of garlic on serum fibrinolytic activity in patients with coronary artery disease. Atheroscelerosis 28:155–159

Bowen S, Taylor KM, Gibb IAMcL (1981) Effect of coffee and tea on blood levels and efficacy of antipsychotic drugs. Lancet I:1217–1218

Braybrooks MP, Barry BW, Abbs ET (1975) The effect of mucin on the bioavailability of tetracycline from the gastrointestinal tract in vivo, in vitro correlation. J Pharm Pharmacol 27:508–515

Brown DD, Juhl RP, Warner SL (1977) Decreased bioavailability of digoxin produced by dietary fiber and cholestyramine. Am J Cardiol 39:297

Brown GR, Greenwood JK (1987) Megavitamin toxicity. Can Pharm J 120:80–87

Callaghan TS, Hadden DR, Tomkin GH (1980) Megaloblastic anaemia due to vitamin B_{12} absorption associated with long-term metformin treatment. Br Med J 280: 1214–1215
Calne DB, Sandler M (1970) L-Dopa and Parkinsonism. Nature 226:21–24
Campbell NRC, Hasinoff BB (1991) Iron supplements: a common cause of drug interactions. Br J Clin Pharmacol 31:251–255
Capobianco DJ, Brazis PW, Fox TP (1993) Proximal-muscle weakness induced by herbs. N Engl J Med 329:1430
Carbo M, Segura J, De la Torre R, Badenas JM, Cami J (1989) Effect of quinolones on caffeine disposition. Clin Pharmacol Ther 45:234–240
Carrera G, Mitjavila S, Derache R (1973) Effect de l'acide tannique sur l'absorption de la vitamine B_{12} chez le rat. C R Acad Paris [D] 276:239–242
Chang H-M, But PPH (eds) (1987) Pharmacology and applications of Chinese materia medica. World Scientific, Singapore
Chen M-F, Shimada F, Kato H, Yano S, Kanaoka M (1990) Effect of glycyrrhizin on the pharmacokinetics of prednisolone following low dosage of prednisolone hemisuccinate. Endocrinol Jpn 37:331–341
Chen M-F, Shimada F, Kato H, Yano S, Kanaoka M (1991) Effect of oral administration of glycyrrhizin on the pharmacokinetics of prednisolone. Endocrinol Jpn 38:167–175
Chin TF, Lach JL (1975) Drug diffusion and bioavailability: tetracycline metallic chelation. Am J Hosp Pharm 32:625–629
Cotzias GC (1969) Metabolic modification of some neurological disorders. JAMA 210:1255–1262
Crewe HK, Lennard MS, Tucker GT, Woods FR, Haddock RE (1992) The effect of selective serotonoin re-uptake inhibitors on cytochrome P4502D6 (CYP2D6) activity in human liver microsomes. Br J Clin Pharmacol 34:262–265
Dandekar UP, Chandra RS, Dalvi SS, Joshi MV, Gokhale PC, Sharma MV, Shah PU, Kshirsagar NA (1992) Analysis of a clinically important interaction between phenytoin and Shankhapushpi, an Ayurvedic preparation. J Ethnopharmacol 35:285–288
D'Arcy PF (1991) Adverse reactions and interactions with herbal medicines. I. Adverse reactions. Adverse Drug React Toxicol Rev 10:189–208
D'Arcy PF (1993) Adverse reactions and interactions with herbal medicines. II. Drug interactions. Adverse Drug React Toxicol Rev 12:147–162
Davies NT (1982) Effects of phytic acid on mineral availability. In: Vahouny GV, Krtichevshky D (eds) Dietary fiber in health and disease. Plenum, New York, pp 105–116
De Freitas B, Schwartz G (1979) Effects of caffeine in chronic psychiatric patients. Am J Psychiatry 136:1337–1338
De Smet PAGM (1985) Ritual enemas and snuffs in the Americas. Foris, Dordrecht (Latin America studies no 33, Centre de Estudios y Documentacion Latinoamericanos)
De Smet PAGM (1989) Drugs used in non-orthodox medicine. In: Dukes MNG, Beeley L (eds) Side effects of drugs annual 13. Elsevier, Amsterdam, pp 442–473
De Smet PAGM (1991) Drugs used in non-orthodox medicine. In: Dukes NMG, Aronson JK (eds) Side effects of drugs annual 15. Elsevier, Amsterdam, pp 514–531
De Smet PAGM (1992a) Toxicological outlook on the quality assurance of herbal remedies. In: De Smet PAGM, Keller K, Hänsel R, Chandler RF (eds) Adverse effects of herbal drugs, vol 1. Springer, Berlin Heidelberg New York, pp 1–72
De Smet PAGM (1992b) Drugs used in non-orthodox medicine. In: Dukes MNG (ed) Side effects of drugs. Elsevier, Amsterdam, pp 1209–1232
De Smet PAGM (1993a) Legislatory outlook on the safety of herbal medicines. In: De Smet PAGM, Hänsel R, Kellere K, Chandler RF (eds) Adverse effects of herbal drugs, vol 2. Springer, Berlin Heidelberg New York, pp 1–90

De Smet PAGM (1993b) An introduction to herbal pharmacoepidemiology. J Ethnopharmacol 38:197–208
De Smet PAGM (1995) Should herbal remedies be licensed as medicines? (invited editorial). Br Med J 310:1023–1024
De Smet PAGM, Smeets OSNM (1994) Potential risks of health food products containing yohimbine extracts. Br Med J 309:958
Deahl M (1989) Betel nut-induced extrapyramidal syndrome: an unusual drug interaction. Mov Disord 4:330–333
Dent CE, Richens A, Rowe DJF, Stamp TCB (1970) Osteomalacia with long-term anticonvulsant therapy in epilepsy. Br Med J 4:69–72
Duvoisin RC, Yahr MD, Coté LD (1969) Pyridoxine reversal of L-dopa effects in Parkinsonism. Trans Am Neurol Assoc 94:81–84
Edgar B, Bailey D, Beregstrand R, Johnsson G, Regardh CG (1992) Acute effects of drinking grapefruit juice on the pharmacokinetics and dynamics of felodipine – and its potential clinical relevance. Eur J Clin Pharmacol 42:313–317
El-Shobaki FA, Saleh ZA, Saleh N (1990) The effect of some beverage extracts on intestinal iron absorption. Z Ernahrungswiss 29:264–269
European Community (1989) Quality of herbal remedies. The rules governing medicinal products in the European Community, vol 3. Guidelines on the quality, safety and efficacy of medicinal products for human use. Office for the Official Publications of the European Community, Luxembourg, pp 31–37
Evans CDH, Lacey JH (1986) Toxicity of vitamins: complications of a health movement. Br Med J 292:509–510
Fels JP, Neccairi J, Toussain P, Debry G, Luyckx A, Scheen A (1989) Effect of food intake on kinetics and bioavailability of (4-chlorophenyl) thiomethylene bisphosphonic acid (abstract). Calcif Tissue Int 44 [Suppl]:S-104
Fleisch H (1987) Experimental basis for the use of bisphosphonates in Paget's disease of bone. Clin Orthop 217:72–78
Floyd RA, Greenberg WM, Caldwell C (1977) In vitro interaction between digoxin and bran. 12th Annual ASHP Midyear Clinical Meeting, Dec 6, Atlanta
Fraser HS, Bulpitt CJ, Kahn C, Mould G, Mucklow JC, Dollery CT (1976) Factors affecting antipyrine metabolism in West African villagers. Clin Pharmacol Ther 22:369–376
Fuhr U, Klittich K, Staib AH (1993) Inhibitory effectd of grapefruit juice and its bitter principle, naringenin on CYP1A2 dependent metabolism of caffeine in man. Br J Clin Pharmacol 35:431–436
Fujiwara Y, Kikkawa R, Nakata K, Kitamura E, Takama T, Shigeta Y (1983) Hypokalaemia and sodium retention in patients with diabetes and chronic hepatitis receiving insulin and glycyrrhizin. Endocrinol Jpn 30:243–249
Fushimi R, Tachi J, Amino N, Miyai K (1989) Chinese medicine interfering with digoxin immunoassay. Lancet I:339
Fushimi R, Koh T, Tyama S, Yasuhara M, Tachi J, Kohda K, Amino N, Miyai K (1990) Digoxin-like immunoreactivity in Chinese medicine. Ther Drug Monit 12:242–245
Gallowary JH, Farmer K, Weeks GR, Marsh ID, Forrest ARW (1992) Potentially hazardous compound in herbal slimming remedy. Lancet 340:179
Gin H, Orgerie MB, Aubetin J (1989) The influence of guar gum on absorption of metformin from the gut in healthy volunteers. Horm Metab Res 21:81–83
Goldman JA, Myerson G (1991) Chinese herbal medicine; camouflaged prescription antiinflammatory drugs, corticosteroids and lead. Arthritis Rheum 34:1207
Gram LF, Debruyne C, Caillard V, Boulenger JP, Lacotte J, Moulin M, Zarifian E (1989) Substantial rise in sparteine metabolic ratio during haloperidol treatment. Br J Clin Pharmacol 27:272–275
Gram LF, Brosen K, Danish University Antidepressant Group (1993) Moclobemide treatment causes a substantial rise in the sparteine metabolic ratio. Br J Clin Pharmacol 35:649–652
Griffin JP (1979) Drug-induced disorders of mineral metabolism. In: D'Arcy PF, Griffin JP (eds) Iatrogenic diseases, 2nd edn. Oxford University Press, Oxford, pp 226–238

Griffin JP, D'Arcy PF, Speirs CJ (1988) A manual of adverse drug interactions, 4th edn. Wright, (Butterworths) London
Guckenbiehl W, Gilfrich HJ, Just H (1976) Einfluß von Laxantien und Metoclopramid auf die Chinidin-Plasmakonzentration während Langzseittherapie bei Patienten mit Herzrythmusstörungen. Med Welt 27:1273
Hanawa T, Hirama N, Kosoto H, Ja Hyun S, Ohwada S, Hasegawa R, Haranaka R, Nakagawa S (1987) Effects of Saiko-zai on collagen metabolism in mice administered with glucocorticoid. Nohon Univ J Med 29:197–205
Hansten PD, Horn JR, Koda-Kimble MA, Young LY (eds) (1993) Anticonvulsant drug interactions – folic acid (Folvite). Drug Interact Updates Q 13:355–356
Harder S, Fuhr U, Staib AH, Wolff T (1989) Ciprofloxacin-caffeine: a drug interaction established using in vivo and in vitro investigations. Am J Med 87:89S–91S
Hasegawa T, Yamaki K, Nadai M, Muraoka I, Wang L, Takagi K, Nabeshima T (1994) Lack of effect of Chinese medicines on bioavailability of ofloxacin in healthy volunteers. Int J Clin Pharmacol Ther 32:57–61
Heald GE, Poller L (1974) Anticoagulants and treatment for chilblains. Br Med J 2:455
Healy DP, Polk RE, Kanawati L, Rock DT, Mooney ML (1989) Interaction between oral ciprofloxacin and caffeine in normal volunteers. Antimicrob Agents Chemother 33:474–478
Helsing E (1992) Vitamins. In: Dukes MNG (ed) Side effect of drugs, 11 edn. Elsevier, Amsterdam, pp 959–976
Hesseling PB, Klopper JF, van Heerden PD (1979) Die effek van Rooibostee op ysterabsorpsie. S Afr Med J 55:631–632
Hirsch SR (1979) Precipitation of antipsychotic drug in interactions with coffee or tea. Lancet II:1130–1131
Hogan RP (1983) Hemorrhagic diathesis caused by drinking a herbal tea. JAMA 249:2679–2680
Homma M, Oka K, Ikeshima K, Takahashi N, Yamamoto S, Niitsuma T, Itoh H (1992a) Different effects of three Kampo-remedies which compromise of saiko on prednisolone pharmacokinetics. 5th World Conference on Clinical Pharmacology and Therapeutics, July 26–31, Yokohama
Homma M, Oka K, Yamada T, Niitsuma T, Itoh H, Takahashi N (1992b) A strategy for discovering biologically active compoounds with high probability in traditional Chinese herb remedies: an application of Saiboku-to in bronchial asthma. Anal Biochem 203:179–187
Homma M, Oka K, Niitsuma T, Itoh H (1993) Pharmacokinetic evaluation of traditional Chinese herbal remedies. Lancet 341:1595
Homma M, Oka K, Niitsuma T, Itoh H (1994) A novel 11 beta-hydroxysteroid dehydrogenase inhibitor contained in Saiboku-To, a herbal remedy for steroid-dependent bronchial asthma. J Pharm Pharmacol 46:305–309
Hoppe HA (1975) Angiospermen, 8 edn. Gruyter, Berlin, pp 91–92, 435, 519–520, 700–701 (Drogenkunde, vol 1)
Huupponen R, Karhuvaara S, Seppälä P (1985) Effect of guar gum on glipizide absorption in man. Eur J Clin Pharmacol 28:717–719
Ingram CG (1962) Some pharamcologic actions of yohimbine and chlorpromazine in man. Clin Pharmacol Ther 3:345–352
Ito K (1993) Effects of Sairei-to on nephrotic syndrome. J Tokyo Womens Med Coll 63:464–468
Joarder ZH, Ogawa T, Yorioka N, Yamakido M (1991) Studies on the effectiveness of Sairei-to on puromycin aminonucleoside nephrosis in rats. Hiroshima J Med Sci 40:127–135
Johri RK, Zutshi U (1992) An Ayurvedic formulation "Trikatu" and its constituents. J Ethnopharmacol 37:85–91
Jones BD, Runikis AM (1987) Interaction with ginseng. J Clin Psychopharmacol 7:201–202
Jori A, Bianchetti A, Prestini PE (1969) Effect of essential oils on drug metabolism. Biochem Pharmacol 18:2081–2085

Joseph AM, Biggs T, Garr M, Singh J, Lederle FA (1991) Stealth steroids. N Engl J Med 324:62
Karunanithy R, Sumita KP (1991) Undeclared drugs in Chinese antirheumatoid medicine. Int J Pharm Pract 1:117–119
Kato H, Nakanishi K, Kanoaka M (1984) The enzyme immunoassay of glycyrrhizin and glycyrrhetinic acid and its clinical application. Minophagen Med Rev 15 [Suppl]:3–11
Kempin SJ (1983) Warfarin resistance caused by broccoli. N Engl J Med 308:1229–1230
Khanna P, Jain SC, Panagariya A, Dixit VP (1981) Hypoglycemic activity of polypeptide-P from plant source. J Nat Prod 44:648–655
Khin-Maung-U, Myo-Khin, Nyunt-Nyunt-Wai, Aye-Kyaw, Tin-U (1985) Clinical trial of berberine in acute watery diarrhoea. Br Med J 291:1601–1605
Kimura K, Nanba S, Tojo A, Matsuoka H, Sugimoto T (1990) Effects of Sairei-to on the relapse of steroid-dependent nephrotic syndrome. Am J Clin Med 18:45–50
Koren G, Boichis H, Keren G (1982) Effect of tea on the absorption of pharmacologic doses of an oral iron preparation. Isr J Med Sci 18:547
Kulhanek F, Linde OK (1981) Coffee and tea influence pharmacokinetics of antipsychotic drugs. Lancet II:359–360
Kulhanek F, Linde OK, Meisenberg G (1979) Precipitation of antipsychotic drugs in interaction with coffee and tea. Lancet II:1130
Kunin CM, Finland M (1961) Clinical pharmacology of the tetracycline antibiotics. Clin Pharmacol Ther 2:51–69
Kwan T, Paiusco AD, Kohl L (1992) Digitalis toxicity caused by toad venom. Chest 102:949–950
Lacomblez L, Bensimon G, Isnard F, Diquet B, Lecrubier Y, Puech AJ (1989) Effect of yohimbine on blood pressure in patients with depression and orthostatic hypotension induced by clomipramine. Clin Pharmacol Ther 45:241–251
La Piana Simonsen L (1989) Top 200 drugs of 1988. Pharmacy Times 40
Lasswell WL Jr, Weber SS, Wilkins JM (1984) In vitro interaction of neuroleptics and tricyclic antidepressants with coffee, tea and gallotanic acid. J Pharm Sci 73:1056–1058
Leatherdale BA, Panesar RK, Sing G, Atkins TW, Bailey CJ, Bignell AH (1981) Improvement in glucose tolerance due to Momordica charantia (karela). Br Med J 282:1823–1824
Lever PG, Hague JR (1964) Observations on phenothiazine concentrates and diluting agents. Am J Psychiatry 120:1000–1002
Lewis WH, Elvin-Lewis MPF (1977) Medical botany. Plants affecting man's health. Wiley, New York, p 192
Lin C-S, Lin M-C, Chen K-S, Ho C-C, Tsai S-R, Ho C-S, Shieh W-H (1989) A digoxin-like immunoreactive substance and atrioventricular block induced by a Chinese medicine "Kyushin". Jpn Circ J 53:1077–1080
Loeper J, Descatoire V, Letteron P, Moulis C, Degott C, Dansette P, Fau D, Pessayre D (1994) Hepatotoxicity of germander in mice. Gastroenterology 106:464–472
Mahr G, Soergel F, Granneman GR, Kinzig M, Muth P, Patterson K, Fuhr U, Nickel P, Stephan U (1992) Effects of temafloxacin and ciprofloxacin on the pharmacokinetics of caffeine. Clin Pharmacokin 22 [Suppl 1]:90–97
McIvor ME, Cummings CC, Mendeloff AI (1985) Long-term ingestion of guar gum is not toxic in patients with noninsulin-dependent diabetes mellitus. Am J Clin Nutr 41:891–894
Medical Letter (1971) Vitamin C – were the trials well conducted and are large doses safe? Med Lett Drugs Ther 13:48
Merkel U, Sigusch H, Hoffmann A (1994) Grapefruit juice inhibits 7-hydroxylation of coumarin in healthy volunteers. Eur J Clin Pharmacol 46:175–177
Monder C, Stewart PM, Lakshmi V, Valentino R, Burt D, Edward CRW (1989) Licorice inhibits corticosteroid 11-dehydrogenase of rat kidney and liver: in vitro and in vivo studies. Endocrinology 125:1046–1053

Nakada T, Kato H, Yano S, Kanaoka M (1986) Pharmacokinetics of glycyrrhizin and glycyrrhetinic acid in healthy humans with long-term treatment of glycyrrhizin preparations (1). J Med Pharm Soc WAKAN-YAKU 3:278–279

Neugebauer G, Akpan W, Abshagen U (1983) Interaktion von Guar mit Glibenclamide und Bezafibrat. Beitr Infusionsther Klin Ernahrung 12:40–47

Neuvonen PJ (1976) Interactions with the absorption of tetracyclines. Drugs 11:45–54

Noguchi M (1978) Studies on the pharmaceutical quality evaluation of crude drug preparations used in orient medicine "Kampoo". II. Precipitation reaction of berberine and glycyrrhizin in aqueous solution. Chem Pharm Bull (Tokyo) 26:2624–2629

Nolan JE, Brain KR (1985) Component interaction in Chinese herbal medicines. Acta Agronom 34 [Suppl]:113

Ohno I, Shibasaki T, Nakano H, Matsuda H, Matsumoto H, Misawa T, Ishimoto F, Sakai O (1993) Effects of Sairei-to on gentamicin nephrotoxicity in rats. Arch Toxicol 67:145–147

Ott J (1994) Ayahuasca analogues. Pangaean entheogens. Natural Products, Kennewick

Panesar NS (1992) Bufalin and unidentified substance(s) in traditional Chinese medicine cross-react in commercial digoxin assay. Clin Chem 38:2155–2156

Patriarca PA, Kendal AP, Stricof RL, Weber JA, Meissner MK, Dateno B, (1983) Influenza vaccination and warfarin or theophylline toxicity in nursing-home residents. N Engl J Med 308:1601–1602

Perlman BB (1990) Interaction between lithium salt and ispaghula husk. Lancet 335:416

Pierpaoli PG (1972) Drug therapy and diet. Drug Intell Clin Pharm 6:89–99

Reissell P, Manninen V (1982) Effect of administration of activated charcoal and fibre on absorption, excretion and steady state blood levels of digoxin and digitoxin. Evidence for intestinal secretion of the glycosides. Acta Med Scand 668 [Suppl]:88–90

Richter WO, Jacob BG, Schwandi P (1991) Interaction between fibre and lovastatin. Lancet 338:706

Robinson MF, Thomson CD, Huemmer PK (1985) Effect of a megadose of ascorbic acid, a meal and orange juice on the absorption of selenium as sodium selenite. N Z Med J 98:627

Rocky Mountain Drug Consultation Center (1986) Warfarin-garlic interaction. Micromedex, Denver (Drugdex, vol 61)

Roe DA, Kalkwarf H, Stevens J (1988) Effect of fiber supplements on the apparent absorption of pharmacological doses of riboflavin. J Am Diet Assoc 88:211–213

Rossander L (1987) Effect of dietary fiber on iron absorption in man. Scand J Gastroenterol 129 [Suppl]:68–72

Schellens JH, Ghabrial H, van der Wart HH, Bakker EN, Wilkinson GR, Breimer DD (1991) Differential effects of quinidine on the disposition of nifedipine, sparteine, and mephenytoin in humans. Clin Pharmacol Ther 50:520–528

Seawright AA, Steele DP, Menrath RE (1972) Seasonal variation in hepatic microsomal oxidation metabolism in vitro and susceptibility to carbon tetrachloride in a flock of sheep. Aust Vet J 48:488–494

Serlin MJ, Breckenridge AM (1983) Drug interactions with warfarin. Drugs 25:610–620

Shader RI, Greenblatt DJ (1985) Phenelzine and the dream machine – ramblings and reflections. J Clin Psychopharmacol 5:65

Shimada F, Chen MF, Kato H, Yano S, Kanaoka M (1989) Pharmacokinetics of glycyrrhizin and glycyrrhetinic acid in healthy humans with long-term treatment of glycyrrhizin preparations (II). J Med Pharm Soc WAKAN-YAKU 6:402–403

Siegel RK (1976) Herbal intoxication – psychoactive effects from herbal cigarettes, tea, and capsules. JAMA 236:473–476

Sokol RJ, Johnson KE, Karrer KE, Narkewicz MR, Smith D, Kam I (1991) Improvement of cyclosporing absorption in children after liver transplantation by means of water-soluble vitamin E. Lancet 338:212–214

Soons PA, Vogels BAPM, Roosemalen CM, Schoemaker HC, Uchida E, Edgar B, Lundahl J, Cohen AF, Breimer DD (1991) Grapefruit juice and cimetidine inhibit stereoselective metabolism of nitrendipine in humans. Clin Pharmacol Ther 50:394–403
Stewart DE (1992) High-fiber diet and serum tricyclic antidepressant levels. J Clin Psychopharmacol 12:438–440
Stinson JC, O'Gharabhain F, Adebayo G, Chambers PL, Feely J (1993) Pesticide-contaminated cucumber. Lancet 341:64
Stockley IH (1984) Warfarin resistance caused by over-the-counter drugs, food supplements and vegetables. Pharm Int 5:165–167
Stockley IH (1991) Drug interactions. A source book of adverse interactions, their mechanisms, clinical importance and management, 2nd edn. Blackwell, Oxford
Sunter WH (1991) Warfarin and garlic. Pharm J 246:722
Tamura Y, Nishikawa T, Yamada K, Yamamoto M, Kumagai A (1979) Effects of glycyrrhetinic acid and its derivatives on Δ^4-5α- and 5β-reductase in rat liver. Arzneimittelforschung 26:647–649
Taylor RFH, Al-Jarad N, John LME, Conroy DM, Barnes NC (1992) Betel-nut chewing and asthma. Lancet 339:1134–1136
Teuscher E, Lindequist U (1988) Biogene Gifte. Biologie-Chemie-Pharmakologie. Academie, Berlin, pp 228–229
Todd PA, Benfield P, Goa KL (1990) Guar gum. A review of its pharmacological properties and use as a dietary adjunct in hypercholesteraemia. Drugs 39:917–928
Tsumura A (1991) Kampo – how the Japanese updated traditional herbal medicine. Japan Publications, Tokyo
Udall JA, Krock LB (1968) A modified method of anticoagulant therapy. Curr Ther Res 10:207–211
Uusitupa M, Siitonen O, Savolainen K, Silvasti M, Penttila I, Parviainen M (1989) Metabolic and nutritional effects of long-term use of guar gum in the treatment of noninsulin-dependent diabetes of poor metabolic control. Am J Clin Nutr 49:345–351
Vesell ES, Shively CA, Passananti GT (1979) Failure of cola nut chewing to alter antipyrine disposition in normal male subjects from a small town in South Central Pennsylvania. Clin Pharmacol Ther 26:287–293
Walker FB (1984) Myocardial infarction after diet-induced warfarin resistance. Arch Intern Med 144:2089–2090
Watson AJM, Pegg M, Green JRB (1984) Enteral feeds may antagonise warfarin. Br Med J 288:557
Wesley-Hadzija B (1971) A note on the influence of diet in West Africa on urinary pH and excretion of amphetamine in man. J Pharm Pharmacol 23:366–368
Yoneyama Y, Usukura Y, Arai H, Hasegawa R, Haranaka R, Nakagawa S (1990) Effects of Sairei-to on glutathione metabolism in alcohol-treated mice. Nihon Univ J Med 32:379–387

Subject Index

Springer-Verlag and the Environment

We at Springer-Verlag firmly believe that an international science publisher has a special obligation to the environment, and our corporate policies consistently reflect this conviction.

We also expect our business partners – paper mills, printers, packaging manufacturers, etc. – to commit themselves to using environmentally friendly materials and production processes.

The paper in this book is made from low- or no-chlorine pulp and is acid free, in conformance with international standards for paper permanency.

MIX
Papier aus verantwortungsvollen Quellen
Paper from responsible sources
FSC® C105338

If you have any concerns about our products,
you can contact us on
ProductSafety@springernature.com

In case Publisher is established outside the EU,
the EU authorized representative is:
Springer Nature Customer Service Center GmbH
Europaplatz 3, 69115 Heidelberg, Germany

Printed by Libri Plureos GmbH
in Hamburg, Germany